High-Efficiency and High-Performance Power Electronics for Power Grids and Electrical Drives

High-Efficiency and High-Performance Power Electronics for Power Grids and Electrical Drives

Editor

Massimiliano Luna

MDPI • Basel • Beijing • Wuhan • Barcelona • Belgrade • Manchester • Tokyo • Cluj • Tianjin

Editor
Massimiliano Luna
CNR-INM
Palermo, Italy

Editorial Office
MDPI
St. Alban-Anlage 66
4052 Basel, Switzerland

This is a reprint of articles from the Special Issue published online in the open access journal *Energies* (ISSN 1996-1073) (available at: https://www.mdpi.com/journal/energies/special_issues/high-efficiency_and_high-performance_power_electronics_for_power_grids_and_electrical_drives).

For citation purposes, cite each article independently as indicated on the article page online and as indicated below:

LastName, A.A.; LastName, B.B.; LastName, C.C. Article Title. *Journal Name* **Year**, *Volume Number*, Page Range.

ISBN 978-3-0365-8080-7 (Hbk)
ISBN 978-3-0365-8081-4 (PDF)

Contents

About the Editor

Massimiliano Luna

Massimiliano Luna, Ph.D., is a permanent research scientist at the National Research Council of Italy (CNR). He obtained his Ph.D. in Energy Science (2009) and his M.Sc. degree in Electrical Engineering (2003) from the University of Palermo, Italy. His research has been mostly focused on Power Electronics and electrical drives, static and dynamic modeling of renewable sources, FPGA- and DSP-based signal processing, and control and management of smart microgrids. He participated in ten research projects and is currently leading the following research projects for CNR: "Mission Innovation – MISSION (Smart, multi-vector and integrated microgrids to accelerate the energy transition) - LA1.7" and "NAUSICA (Efficient ships using innovative and low carbon technological solutions) - OR3". Furthermore, he holds two patents and has been the supervisor of the related technology transfer project. In 2021, he was visiting researcher at Universidad Politécnica de Madrid (UPM) – Centro de Electrónica Industrial (CEI) as a winner of the Short-Term Mobility competition held by CNR. Dr. Luna co-authored more than 60 scientific papers in international journals and conferences, a book chapter, and two software applications. One of his co-authored papers won the Third Prize Paper Award 2020, recognized by the IEEE IAS Industrial Drives Committee; another was shortlisted for the Best Paper Award at the 2017 IEEE International Symposium on Industrial Electronics. In addition, Dr. Luna serves as a reviewer for several journals and conferences and is a member of two Technical Committees of the IEEE and IMACS societies. Finally, he was called as an expert peer reviewer for the evaluation of research project funding on behalf of the Italian Ministry of University and Research.

Preface to "High-Efficiency and High-Performance Power Electronics for Power Grids and Electrical Drives"

Power electronics have enabled the transition from rotating to static power conversion and radically transformed the way we condition electrical energy in stationary and non-stationary applications. As a result, power electronic converters are pervasive in today's society. They are available in many variants, each with pros and cons, and their rated power keeps increasing significantly. Many successful applications have already been deployed. However, the urge for a sustainable future requires further performance increase in terms of efficiency, power density, power quality, cost, robustness to faults, as well as stable operation and high dynamic performance in several operating conditions. These results can be obtained by leveraging different approaches such as advanced converter topologies, innovative control techniques, and wide bandgap power devices.

Based on the above premises, this Special Issue collects papers addressing techniques to improve the efficiency and performance of the application of power electronics in power grids and electrical drives. Such papers can be grouped into three categories:

(a) Power converters used to interface Renewable Energy Sources (RESs) and Energy Storage Systems (ESSs) with microgrids or to recharge ESSs in electric vehicles.

(b) Performance increase in variable speed electrical drives and direct online induction motors.

(c) Application of electrical loss minimization techniques to microgrids based on RESs.

The articles published in this Special Issue contribute to expanding the scientific knowledge in the field of power electronics and offer valuable insights into the recent developments for researchers, scholars, and practitioners working in such a field.

The Guest Editor is grateful to the MDPI publisher for the invitation to hold such a position for this Special Issue of *Energies*. He would also like to thank all the authors and reviewers for their commitment that led to the success of this Special Issue. Furthermore, he is very grateful to the editorial staff of Energies, especially to Ms. Vickie Zhang for her valuable cooperation and support.

Massimiliano Luna
Editor

Editorial

High-Efficiency and High-Performance Power Electronics for Power Grids and Electrical Drives

Massimiliano Luna

Istituto di Ingegneria del Mare (INM), Consiglio Nazionale delle Ricerche (CNR), Via Ugo La Malfa 153, 90146 Palermo, Italy; massimiliano.luna@cnr.it

Citation: Luna, M. High-Efficiency and High-Performance Power Electronics for Power Grids and Electrical Drives. *Energies* 2022, 15, 5844. https://doi.org/10.3390/en15165844

Received: 1 August 2022
Accepted: 10 August 2022
Published: 11 August 2022

Publisher's Note: MDPI stays neutral with regard to jurisdictional claims in published maps and institutional affiliations.

1. Introduction

Since the invention of the light bulb by T. A. Edison, the use of electricity has kept growing during the recent century. In fact, electrical energy is the most versatile among the various forms of energy. It can be easily generated, transmitted across long distances, distributed as AC or DC, and converted into other forms with high efficiency to be used on demand or stored for later use. In particular, power electronics is the enabling technology that allowed for the transition from rotating to static power conversion, thus radically transforming how we condition electrical energy in stationary and non-stationary applications. Currently, power electronic converters are available in many variants, each with pros and cons. They generally exhibit good efficiency, dynamic performance, and reliability, especially when they supply passive loads. The most recent power electronic converters are very sophisticated and cover different application domains: from grid-connected inverters for renewable energy sources (RESs) to optimal power flow controllers for microgrids under the supervision of an energy management system (EMS); from high-dynamics variable speed electrical drives (VSDs) for industrial applications to smart and efficient drives for e-mobility; and from wireless charging of electric vehicles to bidirectional converters for the integration of energy storage systems (ESSs) into AC or DC microgrids.

Although such applications have been successfully deployed, the urge for a sustainable future requires further performance increase in terms of efficiency, power density, power quality, cost, robustness to faults, as well as stable operation and high dynamic performance in several operating conditions. These results can be obtained by leveraging different approaches. For example, wide bandgap power devices such as silicon carbide (SiC) or gallium nitride (GaN) allow for faster dynamics, reduced power loss, and increased power density. In addition, advanced converter topologies such as multilevel/multiphase converters or partial power converters can also offer several benefits. The first can improve power quality and can reach higher power levels through modular design; the others allow for processing only a fraction of load power, thus achieving increased power density and, again, higher power levels.

Moreover, it is also possible to take advantage of innovative control techniques. For example, electrical loss minimization (ELM) techniques can reduce losses for a given load power, whereas maximum torque per ampere (MTPA) techniques for electrical drives can reduce the current needed to obtain a given load torque level. As a result, both techniques lead to increased efficiency; in addition, MTPA improves the dynamic performance.

Based on these premises, this Special Issue was launched to gather technical and scientific contributions on various techniques aimed at improving the efficiency and performance in power electronics applications for power grids and electrical drives. The Guest Editor of this Special Issue was pleased to observe a positive response from the scientific community, which contributed ten high-quality articles. Such a response confirms the importance of the subject and the need for investigating new solutions in this field. The articles published in this Special Issue contribute to expanding the scientific knowledge in the field of power electronics and will be briefly reviewed in the following section.

2. Overview of the Contributions in This Special Issue

This Special Issue includes ten contributions on various techniques aimed at improving the efficiency and performance in power electronics applications for power grids and electrical drives [1–10]. Such articles were authored by international research teams from several countries and covered different topics. For the reader's convenience, they were grouped into three categories:

(a) Power converters used to interface RESs and ESSs with microgrids [1,3,5,7,8,10] or to recharge ESSs in electric vehicles [9];

(b) Performance increase in VSDs and direct online induction motors [4,6];

(c) Application of electrical loss minimization techniques to microgrids based on RESs [2].

In the following, after a short description of each contribution, some perspectives for future developments are briefly given.

As for category (a), the first three contributions deal with multilevel converters. The study presented in [1] by Wang et al. is focused on modular multilevel converters (MMC). Although such converters are well-suited for high-power and medium-voltage applications, they exhibit a quite complex topology based on several submodules (SMs). Therefore, their performance can be affected by SM failures, leading to unequal power distribution and undesirable internal dynamics. Wang et al. [1] studied the impact of SM failures on the MMC internal dynamics performance and proposed two implementations of auxiliary controllers that can regulate such dynamics. The performance of the presented techniques was assessed using time-domain simulations and through experiments. In general, the authors found that appropriate control countermeasures can minimize the adverse impact of SM failures in terms of control range, power quality on the AC and DC sides, and distribution of voltage stress across the SM capacitors and semiconductor switches. Furthermore, they presented the trade-offs of the two proposed control methods. The first control scheme is better than the second from the AC side viewpoint but pollutes the DC current with a fundamental frequency ripple. In contrast, the second scheme is better than the other from the DC side viewpoint; nevertheless, it increases the risk of DC injection into the AC grid and reduces the AC control range. Future development could be devoted to devising control strategies that can improve the performance on both the DC and AC sides without trade-offs.

In [3], Di Benedetto et al. designed two multilevel converters that exhibited high efficiency and power density, as required for microgrid applications, as well as good performance in terms of quality of voltage and current waveforms using only silicon power semiconductors. In particular, the paper focused on a three-phase five-level E-type multilevel–multicell rectifier (3Φ5L E-Type MMR) and a three-phase five-level E-type multilevel–multicell inverter (3Φ5L E-Type MMI). The E-type topology was chosen because it allows for implementing an interleaved configuration using an intercell transformer, thus reducing the volume of the output filter. First, the proposed hardware design and control strategy were validated by simulations. Then, a prototypal 3Φ5L E-Type MMR plus 3Φ5L E-Type MMI was built using silicon MOSFETs or IGBTs switching at 20 kHz, and its performance was assessed experimentally. The results obtained showed excellent performance. The THDv and THDi were 0.88% and 1.95%, respectively. Furthermore, the converter exhibited a peak efficiency of nearly 99% using only silicon power semiconductors, a power density of 8.4 kW/dm^3, and specific power of 3.24 kW/kg. In comparison, a similar SiC-based converter proposed by another author exhibited the following figures of merit: 99.2%, 1.4 kW/dm^3, and 2.5 kW/kg; furthermore, another SiC-based converter proposed in the technical literature exhibited 99.26% efficiency and a power density of 4 kW/dm^3. This comparison shows that remarkable results can be obtained using an appropriate topology, even without using wide-bandgap devices. On the one hand, this concept could also be applied to other converter topologies. On the other hand, the performance of the converter proposed in [3] could have been even higher if SiC devices had been used.

Another contribution about multilevel converters was given by Foti et al. [5]. The authors presented a multi-input N-level power converter topology for grid-connected applications, which can perform independent management of N-1 power sources and an ESS without requiring additional DC–DC power converters. The proposed topology encompasses a three-phase neutral point clamped multilevel inverter (MLI), a three-phase transformer in an open-end winding configuration, and a conventional two-level inverter (TLI). The MLI can interface several DC sources such as photovoltaic (PV) modules and wind energy conversion systems with rectified output, providing active power to the grid. It operates at a low switching frequency (<1 kHz), thus featuring low switching power losses. The TLI is operated at a higher switching frequency but with a lower DC bus voltage, and its DC port can be connected to a bulk capacitor or an ESS. The TLI works as an active filter to compensate for the low-order harmonics generated by the MLI and for imbalances among the source voltages, provided that they do not exceed the DC bus voltage. The proposed topology reduces the complexity and cost of a multi-input power converter. Moreover, it exhibits reduced losses and lower switch count with the same power quality level. Both features contribute to increasing the efficiency and the power density of the converter. The validity of such a concept was confirmed first theoretically and then by simulations and experimental tests. Even when the N-1 input voltages were imbalanced, the THD was as low as 1.5%. Furthermore, the peak efficiency of the converter reached a maximum value of 95%. In future work, it could be interesting to apply the proposed concept to other application domains, such as electric vehicles and the aerospace industry.

Other contributions that fall into category (a) are [7,8] by Luna et al. These papers were focused on the use of bidirectional DC–DC converters to integrate an ESS into a DC microgrid, ensuring stable operation and good dynamic performance. The authors showed that this integration deserves careful design because the ESS converter is required to work in different scenarios depending on the microgrid configuration. They chose the Split-Pi converter as a case study because it presents several merits at the only cost of non-isolated operation; however, the proposed approach has general validity. In [7], the authors analyzed the five possible operating scenarios, showing how to model the equivalent load of the ESS converter, and devised the state-space model of the ESS converter supplying such a load. Each scenario requires a different control scheme with reference to the number of control loops and the controller design procedure to obtain high performance. Proportional-integral-derivative (PID) controllers were employed, and the related design criteria were given. Then, in [8], the authors validated the theoretical analysis presented in [7] by performing several simulations and experimental tests. Future development of this work could be devoted to designing unconventional control systems for ESS converters suitable for operating in more than one microgrid scenario or, possibly, in all the scenarios.

The main objective of the work by Aouichak et al. was to present and experimentally validate a bidirectional DC–AC converter for ESSs, connected to the AC grid and operated under the supervision of a smart home electricity management system (HEMS) to reduce the cost of electricity [10]. The proposed converter can operate in both grid-connected and off-grid modes. It is based on two cascaded bidirectional stages and can generate a full sinusoidal wave from a sinusoidal half wave thanks to the high performance of wide bandgap devices. First, a modified half-bridge DC–DC stage regulates the DC voltage and establishes the positive half of the AC waveform; such a stage is based on two SiC power MOSFETs controlled at a high frequency (150 kHz). Then, an H-bridge DC–AC stage periodically reverses such a waveform; this stage is composed of four MOSFETs on a silicon substrate, controlled at line frequency (50 Hz). The converter was validated experimentally in inverter mode and in rectifier mode with power factor correction (PFC). The use of SiC switches combined with appropriate control strategies allowed for an increase in the compactness of the converter while ensuring good performance, especially in terms of efficiency, which exceeded 95% over the entire power range from 100 W to 1.5 kW. The main advantages of the proposed converter are twofold: (1) it is based on a reasonably complex topology and, thus, can be recommended as an alternative solution for HEMS applications; (2) in stand-alone operations, it does not require a bulky output filter, so it is more compact than a traditional H-bridge inverter. Future development could be focused on analyzing the electromagnetic compatibility aspects related to the high switching frequency to ensure a safe connection to the AC grid.

The last contribution falling into category (a) is the one by Madzharov et al. and discusses the application of the energy dosing technique to improve the performance of a charging station for electric vehicles based on a resonant DC–DC converter [9]. The proposed converter comprises a half-bridge resonant inverter with energy dosing (RI with ED) without reverse diodes, a high-frequency matching transformer, and an output rectifier with a capacitive filter. The switching devices operate with zero current switching (ZCS) and zero voltage switching (ZVS). The main advantage of the energy dosing schemes is that the power does not depend on the load resistance value but is a function of the operating frequency, the resonant capacitance, and the supply voltage. Therefore, the power transferred to the vehicle can be kept constant despite variations in the magnetic coupling coefficient during stops and driving. The characteristics of the RI with ED were compared with two other competing schemes often used in charging stations as high-frequency sources. The proposed converter presents a higher intrinsic regulation capability because the frequency variation required to maintain the nominal power against load variation is lower. A further extension of this work could include optimizing the component size to achieve different goals, such as minimum losses, maximum efficiency, and minimum weight and volume.

The two contributions that fall into category (b) focused on performance increases in VSDs and direct online induction motors (DOL-IMs). In [4], Di Benedetto et al. focused on implementing a high-performance power conversion system to reduce overvoltage in VSDs supplied by SiC-based inverters using long power cables. In particular, a three-phase two-level inverter with snubber circuits based on capacitors and diodes was investigated, designed, and tested to mitigate the overvoltage without sacrificing the conversion efficiency. Furthermore, an additional circuit was used to recover the energy from the snubbers avoiding increased losses. The proposed analysis was validated through experimental tests performed on a prototypal converter. The experimental results showed that, in the absence of the snubbers, the voltage at the motor terminals could reach twice the DC bus voltage when the cable length was equal to 15 m or 30 m: using a 400 V DC bus, the peak voltage at the motor side was 845 V, and the dV/dt was 9.95 V/ns. Instead, the proposed converter reduced the dV/dt to 1.47 V/ns and limited the overvoltage to 562 V. The power density was much higher than that for classical solutions such as the three-phase two-level inverter complemented by a bulky passive filter. The only drawback was a slight reduction in the conversion efficiency (0.976% against 0.982%). The proposed technique is promising, but further work should be devoted to finding a good compromise between the efficiency, dV/dt, and overvoltage of the entire electrical drive system.

In [6], Tornello et al. presented a technique to improve the performance of induction motors by exploiting an auxiliary winding set supplied by a partial power inverter. In fact, three-phase DOL-IMs are the most dominant solution in the industrial sector, providing a variety of constant speed and variable load applications where dynamic requirements are not critical. However, they exhibit a low power factor at partial loads, which is mitigated by adding suitable capacitors; moreover, a significant in-rush current is generated at start-up, leading to voltage dips, speed losses, torque pulsations, and possible activation of protection devices. In such applications, the additional cost implied by a VSD with a full-size inverter is not justified. However, the authors of [6] showed that it is possible to compromise cost and performance by using an auxiliary winding and a partial power inverter. The auxiliary winding features fewer turns than the main one and thus has lower voltage and power ratings. If suitably controlled by an inverter sized for a fraction of the motor's rated power and supplied through a floating capacitor, it allows for the following advantages to be obtained: (1) the machine power factor and, thus the efficiency, can be increased; (2) torque oscillations produced by the mechanical load or by distorted grid voltages can be mitigated; and (3) grid current peaks at motor start-up can be mitigated as well. The proposed technique was assessed by simulating a machine with a turn ratio of five. The results obtained showed that the power handled by the auxiliary winding and the inverter in the worst-case scenario was roughly one-fifth of the machine's rated power. Given that the inrush current mitigation depends on the capacitance of the floating capacitor, future extensions of this study could be focused on finding alternative approaches, such as performing the start-up at a higher DC-bus voltage to increase the energy stored in the floating capacitor.

Lastly, regarding category (c), Kumar et al. applied an ELM technique to a microgrid with power converters interfacing RESs such as solar photovoltaics (SPV) and wind turbine generators (WTG), as well as battery storage systems (BSSs) [2]. The authors' goal was to minimize the detrimental effect of active and reactive power losses in the microgrid to exploit the full potential of green energy generation and use. First, the optimal locations for accommodating the distributed generators (DGs) were obtained by considering two specific indices. Then, an optimization problem was formulated to perform the optimal sizing of the DGs for achieving the ELM. The problem was solved using a constriction factor-based particle swarm optimization (CF-PSO) technique, which has overtaken many algorithms, including the genetic algorithms (GA). Finally, a reliability analysis (RA) of the microgrid was performed by evaluating five indices and using the optimal location and sizing of the RESs and BSSs. The outcomes of the study showed an enhancement in the electrical loss minimization and an improvement in the bus voltage profile compared with a system without DGs. In addition, the RA was repeated considering the uncertainties in the reliability data of the SPV and WTG interfaced by power converters, including the failure rate and the time to repair. It was shown that the microgrid's reliability significantly improved by considering the reliability data of the optimally integrated DGs. Some perspectives for future developments include the extension of the analysis to include aspects such as CO_2 emissions, economics, system protection and reconfiguration, and system security.

3. Conclusions

This Special Issue is composed of ten papers that presented various techniques aimed at improving the efficiency and performance of power electronics applications in power grids and electrical drives. The contributions provided offer valuable insights into the recent developments in such a field. The Guest Editor briefly summarized each contribution and highlighted some perspectives for future developments. It is hoped that the proposed techniques will soon be implemented in the energy industry and further improved.

Funding: This research received no external funding.

Acknowledgments: The Guest Editor is grateful to the MDPI Publisher for the invitation to hold such a position for this Special Issue of *Energies*. He also thanks all of the authors and reviewers for their commitment, which led to the success of this Special Issue. Furthermore, he is very grateful to the editorial staff of *Energies* for the valuable cooperation and support.

Conflicts of Interest: The author declares no conflict of interest.

References

1. Wang, S.; Alsokhiry, F.; Adam, G. Impact of Submodule Faults on the Performance of Modular Multilevel Converters. *Energies* **2020**, *13*, 4089. [CrossRef]
2. Kumar, S.; Sarita, K.; Vardhan, A.; Elavarasan, R.; Saket, R.; Das, N. Reliability Assessment of Wind-Solar PV Integrated Distribution System Using Electrical Loss Minimization Technique. *Energies* **2020**, *13*, 5631. [CrossRef]
3. Di Benedetto, M.; Lidozzi, A.; Solero, L.; Crescimbini, F.; Grbović, P. High-Performance 3-Phase 5-Level E-Type Multilevel–Multicell Converters for Microgrids. *Energies* **2021**, *14*, 843. [CrossRef]
4. Di Benedetto, M.; Bigarelli, L.; Lidozzi, A.; Solero, L. Efficiency Comparison of 2-Level SiC Inverter and Soft Switching-Snubber SiC Inverter for Electric Motor Drives. *Energies* **2021**, *14*, 1690. [CrossRef]
5. Foti, S.; Testa, A.; De Caro, S.; Tornello, L.; Scelba, G.; Cacciato, M. Multi-Level Multi-Input Converter for Hybrid Renewable Energy Generators. *Energies* **2021**, *14*, 1764. [CrossRef]

6. Tornello, L.; Foti, S.; Cacciato, M.; Testa, A.; Scelba, G.; De Caro, S.; Scarcella, G.; Rizzo, S. Performance Improvement of Grid-Connected Induction Motors through an Auxiliary Winding Set. *Energies* **2021**, *14*, 2178. [CrossRef]
7. Luna, M.; Sferlazza, A.; Accetta, A.; Di Piazza, M.; La Tona, G.; Pucci, M. Modeling and Performance Assessment of the Split-Pi Used as a Storage Converter in All the Possible DC Microgrid Scenarios. Part I: Theoretical Analysis. *Energies* **2021**, *14*, 4902. [CrossRef]
8. Luna, M.; Sferlazza, A.; Accetta, A.; Di Piazza, M.; La Tona, G.; Pucci, M. Modeling and Performance Assessment of the Split-Pi Used as a Storage Converter in All the Possible DC Microgrid Scenarios. Part II: Simulation and Experimental Results. *Energies* **2021**, *14*, 5616. [CrossRef]
9. Madzharov, N.; Hinov, N. High-Performance Power Converter for Charging Electric Vehicles. *Energies* **2021**, *14*, 8569. [CrossRef]
10. Aouichak, I.; Jacques, S.; Bissey, S.; Reymond, C.; Besson, T.; Le Bunetel, J. A Bidirectional Grid-Connected DC-AC Converter for Autonomous and Intelligent Electricity Storage in the Residential Sector. *Energies* **2022**, *15*, 1194. [CrossRef]

Article

Impact of Submodule Faults on the Performance of Modular Multilevel Converters

Shuren Wang [1], Fahad Saeed Alsokhiry [2] and Grain Philip Adam [1,*]

[1] Department of Electronic and Electrical Engineering, University of Strathclyde, Glasgow G1 1RD, UK; shuren.wang@strath.ac.uk
[2] Renewable Energy Research Group and Department of Electrical and Computer Engineering, Faculty of Engineering, King Abdulaziz University, Jeddah 21589, Saudi Arabia; falsokhiry@kau.edu.sa
* Correspondence: grain.adam@strath.ac.uk

Received: 21 May 2020; Accepted: 24 July 2020; Published: 6 August 2020

Abstract: Modular multilevel converter (MMC) is well suited for high-power and medium-voltage applications. However, its performance is adversely affected by asymmetry that might be introduced by the failure of a limited number of submodules (SMs) or even by severe deviations in the values of SM capacitors and arm inductors, particularly when the number of SMs per arm is relatively low. Although a safe-failed operation is easily achieved through the incorporation of redundant SMs, the SMs' faults make MMC arms present unequal impedances, which leads to undesirable internal dynamics because of unequal power distribution between the arms. The severity of these undesirable dynamics varies with the implementation of auxiliary controllers that regulate the MMC internal dynamics. This paper studied the impact of SMs failure on the MMC internal dynamics performance, considering two implementations of internal dynamics control, including a direct control method for suppressing the fundamental component that may arise in the dc-link current. Performances of the presented and widely-appreciated conventional methods for regulating MMC internal dynamics were assessed under normal and SM fault conditions, using detailed time-domain simulations and considering both active and reactive power applications. The effectiveness of control methods is also verified by the experiment. Related trade-offs of the control methods are presented, whereas it is found that the adverse impact of SMs failure on MMC ac and dc side performances could be minimized with appropriate control countermeasures.

Keywords: ac/dc converter for medium and high-voltage applications; modular multilevel converter; submodule fault; fault-tolerant control and operation

---

1. Introduction

To date, modular multilevel converter (MMC) is a preferred technology for high-power medium/high-voltage distribution and transmission applications due to ease of scalability, modularity, reduced power losses, and high-quality ac and dc side waveforms [1]. However, the use of submodules (SMs) with floating distributed capacitors in MMC results in complex internal dynamics, which necessitates the adoption of a complex multi-layer control system compared to that of conventional voltage source converters.

Typically, the common-mode current of balanced MMC consists of dc and harmonic components, and its dc component acts as a link between the dc power and the power flowing through SMs, which constitute the MMC arms. The harmonic components of the common-mode current circulate between the arms, which are widely referred to as circulating currents representing undesirable parasitic components that add losses and increase SM capacitor voltage ripples [1]. In contrast, the differential-mode currents of an internally balanced MMC are the fundamental output currents that flow in the ac side and responsible for power transfer between the MMC arm and ac-side output circuit. Normally, the SM capacitor voltages affect the synthesis of ac and dc side voltages of the MMC, i.e., the differential- and common-mode voltages, respectively. To reduce such couplings, several control methods exist in the open literature for suppression of the circulating current and regulation of the capacitor voltages independent of dc-link voltage [2].

SM, arm, and phase-leg voltages/energies represent three important elements or layers that must be controlled to minimize MMC internal dynamic interactions during normal and abnormal conditions. To manage the voltage differences between SMs within each arm, SM voltage balancing algorithms based on either centralized sorting or distributed control are employed [3,4]. Generally, MMCs with de-coupled internal-external dynamics respond faster to active power and dc voltage set-points; exhibit reduced output voltage distortion during major transients and overall impacts of cross-modulation on ac and dc side waveforms [5,6]. Wide-ranging research efforts have been invested into high-level controllers (capacitor voltage/energy sum controllers in arm and phase-leg levels) [7,8], in which predominantly ideal (identical) passive components are usually assumed. Although a large number of SMs may reduce the adverse effects of capacitance tolerances, passive component tolerances remain a prevalence issue, particularly, for MMCs, with a relatively low number of SMs per arm as anticipated in medium-voltage (MV) applications. Therefore, it is imperative to account for uncertainties due to passive component tolerances during MMC design and maintenance stages [9–12]. In this line, the adverse effects of asymmetrical cell capacitances on the ac output voltage of MMC that employs three-level flying capacitor SMs have been identified when the energy-based balancing approach was used in the high-level controllers that regulate the MMC internal dynamics [10]. The work in [11] has revealed the inducement of fundamental frequency ripple in the dc-link current of the MMC with asymmetrical arm inductances and proposed a voltage-based active control method to suppress the induced fundamental frequency ripple from the dc-link current.

On the other hand, a fault that happened in MMC SMs may cause operational issues ranging from distortion of ac and dc side voltages and currents to total disruption of power transfer. Methods for SMs fault detection and identification are proposed in [13–15], while increased MMC resiliency to SM faults through the concept of redundant SMs are discussed in [16] and [17]. With regard to the SM fault tolerance capability of the MMC, several SM-fault control methods have been discussed [18–21]. Two approaches are proposed, and the respective features are presented in [18]. The fundamental common-mode current ripple during SM faulty has been claimed in [19], and a dq-frame-based control method is proposed. However, this method depends on phase lock-loop for the ripple. Further research in [20,21] has proposed proportional resonant (PR) based control methods to suppress the fundamental ripple, but the control effects on the overall operation are not analyzed. Besides, in most previous works, the predominant assumption is that the number of faulty SMs is less than the number of redundant SMs or within a hot redundancy configuration, while the extreme condition, in which the number of faulty SMs is larger than the redundant SMs, is seldom discussed. In these cases, SM capacitor voltages would be increased, within a safe margin, for continuous operation.

Based on the preliminary findings in [22], this paper considered the adverse effects of MMC asymmetry caused by SM fault on both dc and ac sides when mainstream balancing control methods are employed. The differential-mode voltage balancing control method reduces the fundamental component in the common-mode and dc-link currents of the MMC with significant passive component tolerances. However, its effectiveness is limited since it is not direct control but via minor ripple injection, and its control objective is to nullify the differential-mode capacitor voltage sums. Considering the random nature of the passive component tolerance distribution within one phase-leg of an MMC-based power conversion system, the proportional-resonant (PR)-based controller that operates at the fundamental frequency was studied to suppress any fundamental circulating component that may arise in the three-phase common-mode currents. Comparatively, control effects on both ac and dc side performances were investigated in this paper. This paper is organized as follows. Section 2 provides a brief review of MMC operation fundamentals, and Section 3 discusses the issues that arise due to component tolerance and SM fault. Section 4 introduces the internal control methods against internal imbalance. Section 5 verifies the effectiveness and assesses the performances of control methods. Finally, Section 6 summarizes this paper.

2. MMC Basic Operation

Figure 1 shows a three-phase half-bridge MMC, with V_{dc} and I_{dc} representing dc bus voltage and current, respectively. Each phase-leg consists of upper and lower arms, and each arm comprises a reactor with nominal inductance $\overline{L_{ARM}}$ and N series-connected SMs. Each SM consists of a capacitor with nominal capacitance $\overline{C_{SM}}$ and an IGBT-based half-bridge circuit. The term circulating current represents the ac component of the common-mode current, i_{cm}. The i_{cm} is mainly caused by cross-modulation of the upper and lower arms or, simply, the interaction of the arm voltages, currents, and switching functions. Strategies of the inner-arm SM voltage balancing and the second-order circulating current suppression have been widely discussed [2–5]. Figure 1 also shows a generic MMC connected into the ac grid through an interfacing transformer.

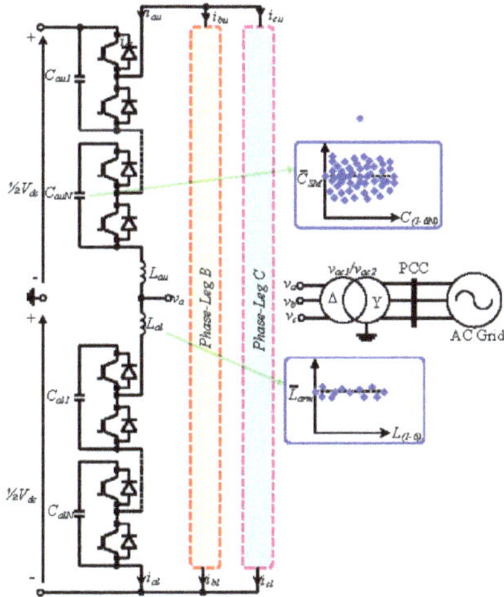

Figure 1. Modular multilevel converter (MMC) topology configuration.

Under the ideal condition, each MMC arm has an equivalent capacitance C_{ARM}, which can be calculated as:

$$C_{ARM} = \frac{\overline{C_{SM}}}{N} \tag{1}$$

As the arm equivalent capacitance is alternatively charging and discharging, the MMC internal power can be dynamically balanced as it transfers power between ac and dc sides.

Figure 2 visualizes the MMC power paths, which can be divided into ac and dc loops. The dc loop depicted in Figure 2 represents the conduction path through dc output to MMC phase-leg, in which the common-mode current of each phase follows, which consists of ac and dc components ($i_{cm} = I_d + i_h$, where I_d and i_h are dc and harmonic components of i_{cm}). In a three-phase MMC, the dc components of the common-mode currents of the phase legs add to the dc-link currents that flow in the positive and negative dc poles at the upper/positive and lower/negative dc nodes. In this way, the MMCs exchange power with the dc side. The ac components of the common-mode currents of the MMC phase legs are predominantly second-order harmonics and add to zero at upper and lower dc-link nodes; they represent parasitic currents, which increase the MMC SM capacitance requirement for a given voltage ripple and semiconductor losses. In contrast, the depicted ac loop represents the conduction path through which each MMC phase-leg exchanges active power with ac side and fundamental frequency ac current flows. The output phase current represents the differential-mode current; for example, for a phase A, $i_a = i_{au} - i_{al}$, where i_a, i_{au}, and i_{al} are output and upper and lower arm currents. In a balanced three-phase MMC, the fundamental frequency components of the currents add to zero at positive and negative dc nodes; while in an internally unbalanced MMC with asymmetrical ac and dc loops in the phase-legs, the asymmetries introduce zero and negative sequence fundamental frequency components in the common-mode currents of the phase-legs, which may leak into dc side and appear as fundamental frequency ripples in the dc current.

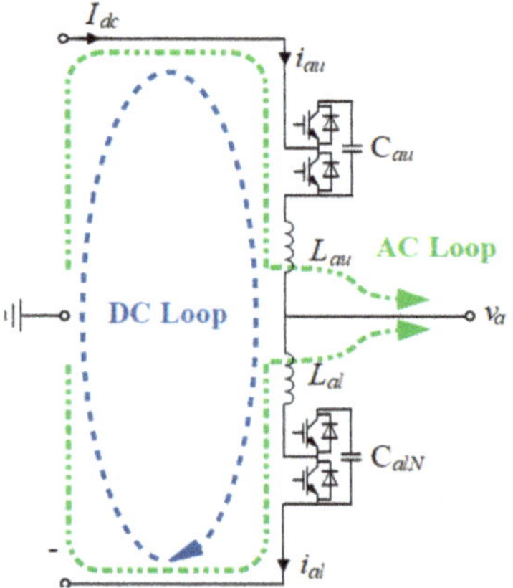

Figure 2. Diagram of MMC internal dynamics (phase A).

3. Internally Unbalanced MMC

The manufacturing process introduces significant inaccuracies and deviations in the values of passive components from the nominal. Therefore, instead of ideal and identical parameters, the practical MMC passive parameters can be assumed as a random distribution shown in Figure 1, with the inherent tolerances included. Commonly, circuit parameter tolerances are neglected to facilitate modeling, analysis, and controller design of complex energy conversion systems, such as the MMC. Nonetheless, it is essential to take countermeasures to neutralize any potential adverse implications of circuit parameter tolerances on MMC performance [10], for the following reasons:

1. Because of a large number of passive components, parameter uncertainties due to manufacturing tolerances are inevitable and vary with several factors, including lifetimes.

2. Although the SM capacitor voltage balancing controller or algorithm distributes the total dc voltage across each arm equally amongst the SM capacitors, the switching devices of each SM only withstand an SM capacitor voltage. However, asymmetry due to parameter tolerances or SM faults may cause unequal power contribution and voltage distribution between the MMC arms. Thus, the current and voltage stresses in the switching devices and heat distribution may differ between arms.

3. Besides, capacitance and inductance differences between arms may actuate unbalanced fundamental frequency common-mode currents, which tend to leak into the dc side and appear as an undesirable current ripple in the dc current [23]. This is mainly because of the inherent power balancing mechanism of the MMC operation: in order to keep ac-dc power balance, different capacitor charge/discharge rate is induced, leading to fundamental frequency current difference between upper and lower arms.

The imbalance between upper and lower arms of one phase-leg, created by SM faults on the MMC internal dynamics, can be explained with the aid of the ac and dc loop described earlier. It is well-established from the literature that the incorporation of redundant SMs can protect active and passive components of SMs from damage due to excessive over-voltage. However, with the hot redundancy approach, which is extensively studied, the post-fault voltage stresses on the MMC switching device and SM capacitor may differ from the pre-fault condition [18]. When an SM fails, it will be bypassed, and MMC continues to operate as normal, following a brief period of transients. For the simplicity of the analysis, this paper assumed that all the N SMs in the MMC arms are in use, and N_F SMs become faulty. Therefore, the equivalent arm capacitance is:

$$C'_{ARM} = \frac{\overline{C_{SM}}}{N - N_F} \tag{2}$$

Hence, 10% of the total number of SMs in a particular arm have failed and bypassed, leading to an 11% increase in the arm equivalent capacitance; hence, this scenario resembles an extreme case of passive component tolerances. Therefore, the consequences of SM faults are similar to that caused by passive components' tolerances, in which SM fault appears as a vertical asymmetry between the upper and lower arms of the same phase-leg and horizontal asymmetries relative to the healthy phase-legs. The work in [23] has demonstrated that the horizontal asymmetries between the phase-legs have negligible effects on dc-loop dynamics during steady-state, even though the phase-legs equivalent capacitances store different energies. In contrast, the vertical asymmetry has several operational implications, specifically, contamination of common-mode current by circulating current at the fundamental frequency, which increases semiconductor losses and contributes to the inducement of fundamental frequency ripple on the dc side, and reduced exploitable modulation index range, which generates the differential-mode voltage in the ac output. Throughout this paper, the sum and difference of arm capacitor voltage sums are referred to as the common- and differential-mode capacitor voltage sums of each phase-leg, respectively. For example, the common- and differential-mode capacitor voltage sums for phase-leg A are $\sum V_{c,a} = \sum_k^N V_{c,au} + \sum_k^N V_{c,al}$ and $\Delta V_{c,a} = \sum_k^N V_{c,au} - \sum_k^N V_{c,al}$, respectively, where $\sum_k^N V_{c,au}$ and $\sum_k^N V_{c,al}$ are phase-leg A upper and lower arm capacitor voltage sums.

4. MMC Internal Dynamics Controllers

Figure 3 shows a generic depiction of the MMC control structure, which consists of internal dynamic and external system-level controllers, and modulator that generates the gating signals for the SMs. The internal dynamic controllers of the MMC operate at three levels, namely, SM, arm, and phase-leg. These controllers ensure adequate distribution of voltage stresses across the followings: SMs semiconductor switches and capacitors; upper and lower arms, to avoid narrowing of modulation index control range; phase-legs, to enable SM capacitor voltage regulation independent of the dc-link voltage and to prevent the development of ac and dc circuit currents between the phase-legs. In other words, the latter reduces interactions between MMC internal and external dynamics. The ac output (external) controllers include the following: positive and negative sequence separation stage; outer system-level controllers that regulate active power or dc voltage and reactive powers or ac voltage; inner positive and negative currents controllers that generate the principle modulation functions for the arms.

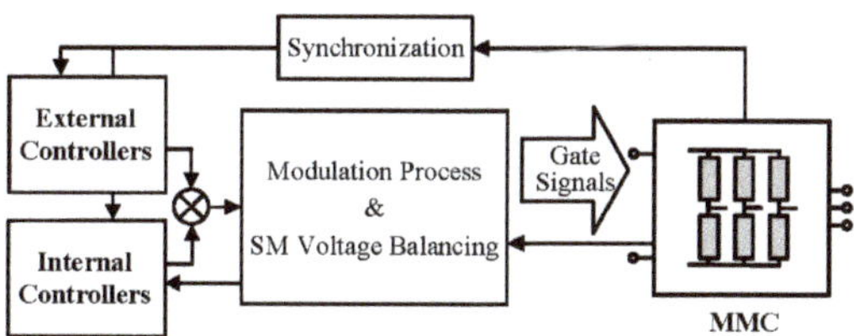

Figure 3. Diagram of MMC control structure.

Although theoretically, each MMC arm exchanges zero average power during one period, tolerances of the passive components lead to different levels of background energy stored in the SM capacitors, and presentation of different impedances by the MMC arms. Thus, the instantaneous power of the passive components deviates from their nominal values at the fundamental frequency. As voltage (or energy) of each arm is usually controlled by corresponding inter-arm balancing controllers, (also known as differential-mode capacitor voltage sum controller or vertical controllers), the power imbalance due to asymmetries of conduction paths will be compensated largely by injection fundamental frequency circulating currents into the common-mode loops. The asymmetries of energy levels and conduction path impendence lead to unequal contributions from the vertical balancing controllers of three phase-legs; hence, unequal fundamental currents in the common-mode loops. Recall that the vertical balancing controllers prioritize nullification of vertical voltage/energy asymmetry between the upper and lower arms over unintended consequences of injection odd-order harmonics into the common-mode currents [23]. Without a dedicated controller, a vertical asymmetry of any kind results in unbalanced fundamental common-mode currents in three phase-legs and thereby the development of fundamental frequency oscillations in the dc-link and increased semiconductor losses.

Two internal controllers to be assessed are:

4.1. Scheme-A

Conventionally, the common- and differential-mode capacitor voltage sums are controlled equally across all three-phase legs using the internal dynamic controllers shown in Figure 4. The regulation of average SM capacitor voltage (common-mode capacitor voltage sum) is achieved through the manipulation of common-mode current in conjunction with the common-mode capacitor voltage/energy sum controller (where $\sum_{j=1}^{N} V_{cj}$ and $\sum_{j=1}^{N} E_{cj}$ represent the common-mode capacitor voltage and energy sums per phase-leg, respectively). To eliminate the steady-state dc mean value error and suppress the 2ω circulating current, a PIR (proportional integral and resonant) controller with a resonant frequency at 2ω is adopted. This controller decouples SM capacitor voltage from the input dc-link voltage, hence the synthesis of the ac voltage from the dc-link voltage. A differential-mode capacitor voltage (or energy) sum control regulates arm active power using fundamental common-mode current injection, with another resonant frequency ω component. Since the performances of energy-based internal dynamic control methods deteriorate as the stored energies in the arms vary rapidly with SM capacitance tolerances [22], the voltage-based balancing method shown in Figure 4 was adopted in this paper. From this point, it is referred to as control scheme A. From the ac side point of view, this method is able to maximize the use of full modulation index control range for the synthesis of the arm and ac output voltages, as determined by the minimum arm capacitor voltage sum. However, the by-product of this method is slightly higher fundamental current ripples in the dc loops and in the dc-link current.

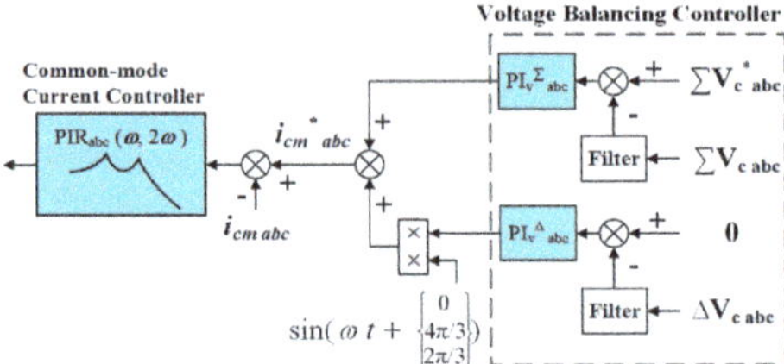

Figure 4. Diagram of the common- and differential-mode capacitor voltage sum controller (Scheme-A) [23].

4.2. Scheme-B

Since SM faults in one or multiple arms of the MMC lead to vertical and horizontal capacitance asymmetries, with the former cases, fundamental frequency ripples to appear in the common-mode currents, and a direct fundamental current ripple suppressing method (referred to as Scheme-B) displayed in Figure 5 is proposed. Instead of targeting differential-mode capacitor voltage sums, the Scheme-B suppresses the fundamental frequency ripple in the common-mode current directly. This method uses an additional fundamental frequency PR controller to cancel the fundamental components in common-mode currents, thereby suppressing the dc-link current ripple, which is predominantly of fundamental frequency in the internally asymmetric MMC. However, the potential disadvantage of the proposed direct suppression of fundamental circulating current method is that the modulation-index control range might be reduced. In this method, the MMC arm with a smaller capacitor voltage sum would limit the ac voltage synthesis. Therefore, a dedicated design margin is needed to enable full exploitation of the modulation index control range in practical MMC under all operating conditions. Figure 6 illustrates potential impact of unequal or imbalance upper and lower capacitor voltage sums in the same phase-leg on modulation index control range (synthesis of the maximum arm or output ac voltage), assuming the SM capacitor of voltage sum of each arm is regulated at least at V_{dc0}, where V_{dc0} represents the nominal dc-link voltage. Notice that under the ideal case of internally balanced MMC, when the upper and lower arm voltages are limited by the nominal input dc-link voltage V_{dc0}, the maximum peak output phase ac voltage is $\frac{1}{2}V_{dc0}$ for sinusoidal references and can rise to 0.577 V_{dc0} ($V_{dc0}/\sqrt{3}$) with 3rd harmonic injection. Notice that when upper and lower arms of a phase-leg suffer from unequal SM capacitor voltage sums, the arm with the least capacitor voltage sum defines the maximum safe limits for the synthesis of the output ac voltage. Should the arm with the largest capacitor voltage sum successfully synthesizes the requested voltage set by its modulation function, and the arm with the least capacitor voltage sum fails to synthesize the target voltage, the common-mode voltage that the phase-leg presents at the dc side may contain fundamental component besides the dc voltage. These may further exacerbate fundamental frequency ripples in the dc current. Besides, the imbalance in the SMs capacitor voltage sums, particularly, between the upper and lower arms, may appear as an imbalance in the differential-mode voltages of the phase-legs, which resemble three-phase output voltages. In Figure 6, $v_{arm,u0}$ and $v_{arm,l0}$ denote the upper and lower arm voltages in the ideal case, corresponding to balanced upper and lower SMs capacitor voltage sums; $v_{arm,u}$ and $v_{arm,l}$ stand for the upper and lower arm voltages under unequal upper and lower SMs capacitor voltage sums, in which the upper arm has the least SMs capacitor voltage sum and limits the maximum attainable ac voltage.

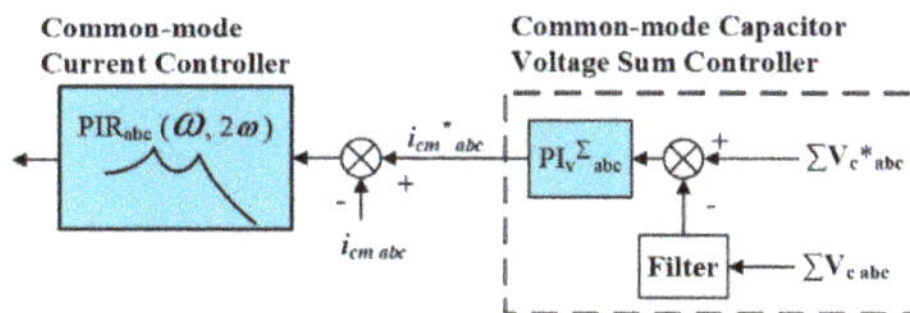

Figure 5. Diagram of internal control structure against inter-arm parametric imbalance (Scheme-B) [23].

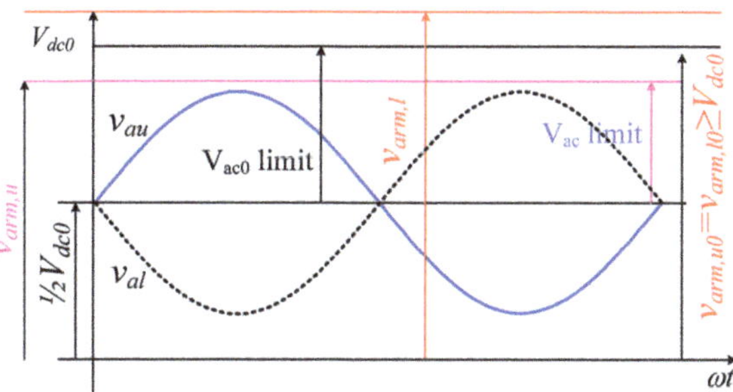

Figure 6. Arbitrary arm voltage of the MMC that suffers from unequal capacitor voltage sums in the upper and lower arms of the same phase-leg.

In summary, since trade-offs exist, it is important to conclusively establish and quantitatively evaluate the performance of the above methods, under SM faults or tolerances conditions. The two trade-off cases are:

1. Prioritization of balanced capacitor voltage sums of the arms over the suppression of fundamental frequency current into the dc loops of three phase-legs;
2. Prioritization of suppressing fundamental frequency currents in the dc loop over active balancing of the capacitor voltage sums of the upper and lower arms.

5. Simulations

This section presents detailed MATLAB-based simulations that assess the MMC performance during SM faults for two different implementations of internal dynamic controllers described earlier, i.e., control schemes A and B, considering two distinct operating points. The simulated MMC models include the following external controllers: active and reactive powers; positive and negative sequence currents. Two control methods for internal dynamics considered in this section are:

1. Scheme-A consists of common- and differential-mode capacitor voltage sum balancing controllers, which include a circulating current suppression controller.
2. Scheme-B consists of common-mode capacitor voltage sum control in conjunction with the proposed direct fundamental circulating current suppression. Circulating current suppression controller is also included.

Sorting-based SM level voltage balancing method is adopted, and modulators based on level-shifted pulse width modulation and phase disposition carriers are used to generate the gating signals for the SMs.

5.1. Impact of SM Faults on MMC Operation for Different Control Schemes

This subsection presents time-domain simulations that illustrate the behaviors of the 20-cell MMC in Figure 7 under SM faults. The parameters of the simulated MMC are listed in Table 1. In this study, two faulted SMs represented 10% of the total SMs per arm, and the SM fault was applied at the upper arm of phase-leg A at 0.4 s, and the two faulty SMs were bypassed.

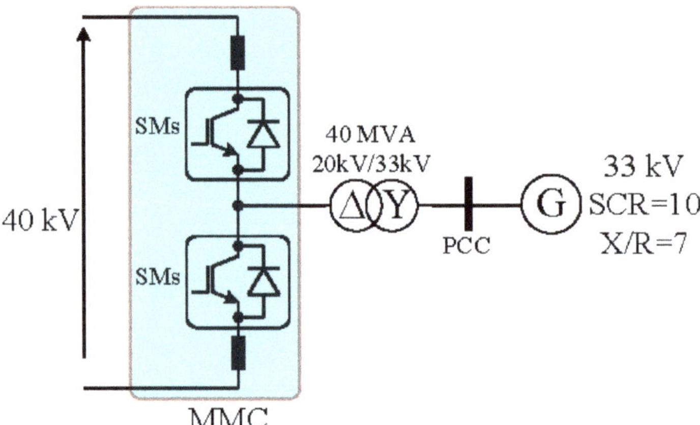

Figure 7. Simplified single line diagram of the 40 kV, 40 MVA, and 20-cell MMC connected to ac grid.

Table 1. Simulation Parameters.

System Parameters		Value
DC voltage	V_{dc}	40 kV
Rated power	S	40 MVA
AC grid line to line voltage	v_{ac2}	33 kV
AC grid frequency	F	50 Hz
Transformer ratio	v_{ac1}/v_{ac2}	20 kV/33 kV
Transformer leakage-inductance	L_T	0.2 pu
Numbers of SMs per arm	N	20
Expected arm inductance value	L	0.18 pu
Expected SM capacitance value	C	6.7 mF
DC cable length	D_{cable}	10 km
DC cable resistance per km	R_{cable}	10 mΩ/km
DC cable inductance per km	L_{cable}	1.4 mH/km
DC cable capacitance per km	C_{cable}	0.1 μF/km

Figures 8 and 9 show simulation waveforms for the control schemes A and B, respectively. These results are obtained when the MMC injects a 1.0 pu (40 MW) active power into the ac grid and regulates reactive power at zero. In contrast, Figures 10 and 11 display simulation waveforms when the MMC injects 0.6 pu (24 MVAr) capacitive reactive power and zero active power into the ac grid, with control schemes A and B, respectively. The 0.6 pu (24 MVAr) represents the maximum capacitive reactive power, which the simulated MMC can inject into the ac grid.

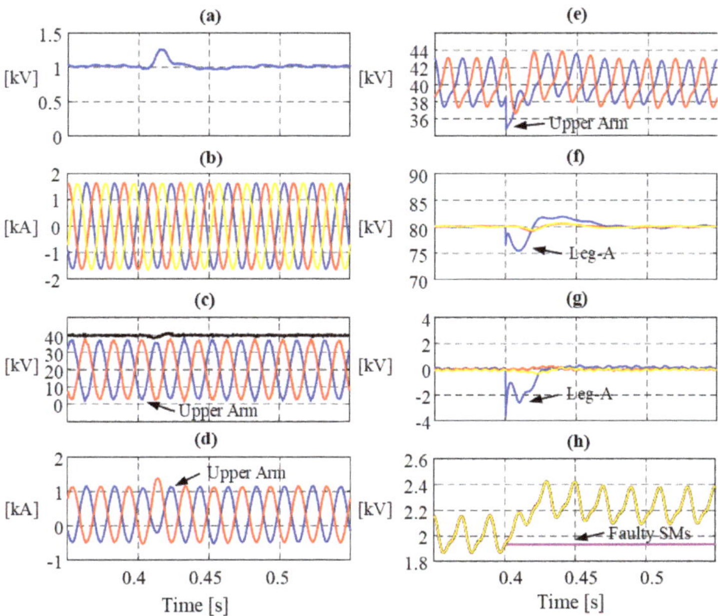

Figure 8. Simulation waveforms of the MMC under submodule (SM) fault and control scheme-A operates in active power mode: (**a**) dc current, (**b**) ac current, (**c**) leg A upper and lower arm voltages superimposed on its common-mode voltage, (**d**) leg A upper and lower arm currents, (**e**) leg A upper and lower capacitor voltage sums, (**f**) common-mode capacitor voltage sums of three phase-legs, (**g**) differential-mode capacitor voltage sums of three phase-legs, and (**h**) leg A upper arm SM voltages.

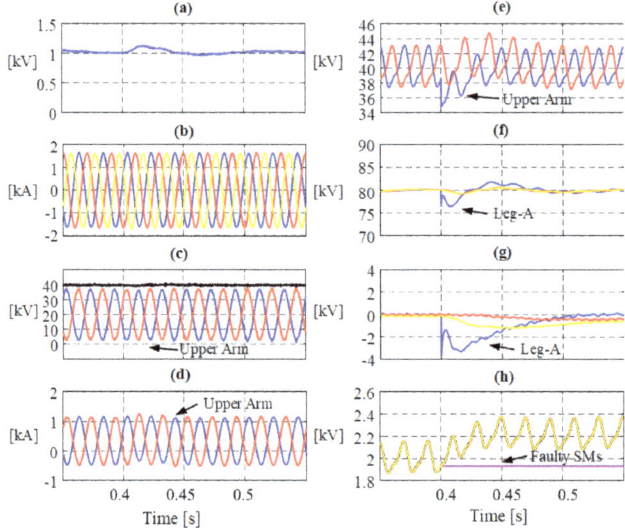

Figure 9. Simulation waveforms of the MMC under SM fault and control scheme-B operates in active power mode: (**a**) dc current, (**b**) ac current, (**c**) leg A upper and lower arm voltages superimposed on its common-mode voltage, (**d**) leg A upper and lower arm currents, (**e**) leg A upper and lower capacitor voltage sums, (**f**) common-mode capacitor voltage sums of three phase-legs, (**g**) differential-mode capacitor voltage sums of three phase-legs, and (**h**) leg A upper arm SM voltages.

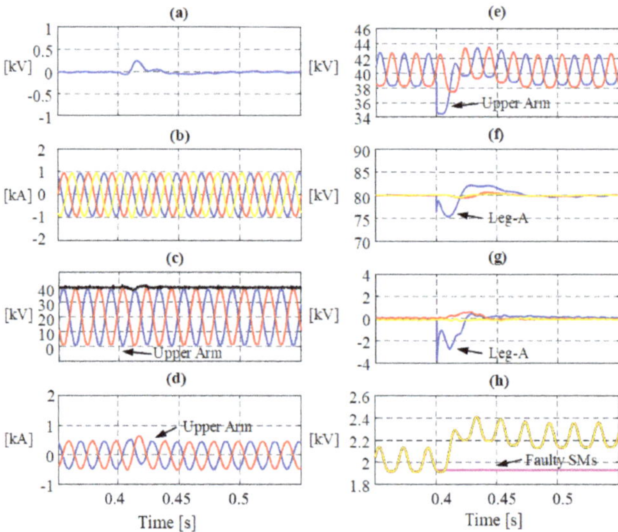

Figure 10. Simulation waveforms of the MMC under SM fault and control scheme-A operates in reactive power mode: (**a**) dc current, (**b**) ac current, (**c**) leg A upper and lower arm voltages superimposed on its common-mode voltage, (**d**) leg A upper and lower arm currents, (**e**) leg A upper and lower capacitor voltage sums, (**f**) common-mode capacitor voltage sums of three phase-legs, (**g**) differential-mode capacitor voltage sums of three phase-legs, and (**h**) leg A upper arm SM voltages.

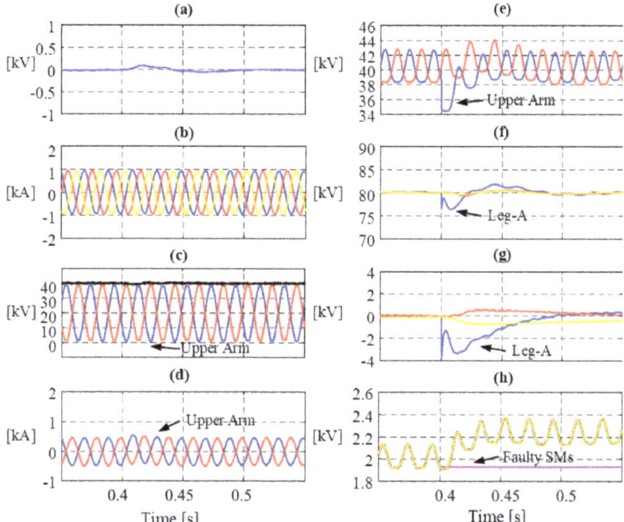

Figure 11. Simulation waveforms of the MMC under SM fault and control scheme-B operates in reactive power mode: (**a**) dc current, (**b**) ac current, (**c**) leg A upper and lower arm voltages superimposed on its common-mode voltage, (**d**) leg A upper and lower arm currents, (**e**) leg A upper and lower capacitor voltage sums, (**f**) common-mode capacitor voltage sums of three phase-legs, (**g**) differential-mode capacitor voltage sums of three phase-legs, and (**h**) leg A upper arm SM voltages.

The observations drawn from results of Scheme-A presented in Figure 8 can be summarized as follows:

1. When the SM fault occurs in the upper arm of leg-A at 0.4 s, the upper arm capacitor voltage sum decreases briefly and quickly recovers by sinking extra dc current, largely, due to the actions of differential-mode capacitor voltage sum or arm voltage balancing controllers of Scheme-A. During the transient period, the upper and lower arm voltages of the faulty phase exhibit brief disturbances, while the three-phase ac output currents remain unaffected, see Figure 8a–e.

2. Following the SM fault, the common-mode capacitor voltage sums of three legs return to their pre-fault set-points, with the deviations emerged in both common- and differential-mode capacitor voltage sums of the three phase-legs (including the faulty phase-leg A) at 0.4 s that are quickly eliminated, see Figure 8f,g. These results demonstrate the importance of inter-arm vertical controllers implemented in Scheme-A.

3. The plots in Figure 8h show that the healthy SM capacitor voltages of the faulty arm (upper arm of phase-leg A) increase due to a reduced number of active SMs that could contribute to the target capacitor voltage sums enforced by collective actions of common- and differential-mode controllers of Scheme-A. After bypass of the faulty SMs, their capacitor voltages do not exhibit any fluctuations associated with fundamental and remnant of 2nd harmonic currents as anticipated.

Simulation waveforms of the 20-cell MMC in Figure 7, with the Scheme-B, are presented in Figure 9. The observations drawn from the results in Figure 9 are as follows:

1. MMC with the use of Scheme-B exhibits slightly different behavior from that with the control scheme A. The dc-link current displayed in Figure 9a exhibits smaller increase, and SM capacitor voltage sum of the faulty leg (an upper arm of phase-leg A) exhibits slightly small under-shoot and over-shoot compared to those with Scheme-A. The smaller increase in the dc current with control scheme B can be attributed to the absence of vertical controllers, which actively inject

additional active powers and fundamental currents into the common-mode loops in order to enforce equalization of the arm capacitor voltage sums.

2. The Scheme-B makes the MMC less sensitive to SM faults and remains capable of synthesizing the output ac voltages that the ac grid imposes at its ac terminals, with three-phase ac output currents remain unaffected, see Figure 9a,e.

3. The common-mode capacitor voltage sums of the three phase-legs remain tightly controlled and less affected throughout as the control scheme B only enforces equal dc voltages across the MMC phase-legs, with the SM level capacitor voltage controller ensuring the imposed dc voltage across the phase-legs being equally distributed across the SM capacitors. These are achieved without paying any attention to the symmetry of upper and lower arm capacitor voltage sums, see Figure 9f,g.

4. The healthy SMs have slightly lower capacitor voltages with Scheme-B than that of Scheme-A due to lower dc voltage across the faulty arm in post-fault condition, as demonstrated by the clear drift the differential-mode voltage displayed in Figure 9g. As in the previous case, the capacitor voltages of the faulty SMs become flat under the post-fault condition.

In addition, the results of fast Fourier transform (FFT) analysis presented in the appendix for the MMC in Appendix A, when it is controlled using the control schemes A and B, respectively, show that the amplitudes of the fundamental frequency component in dc-link current are 11.3 A and 4.1 A, respectively. These results confirm the effectiveness of the Scheme-B in dealing with fundamental frequency ripples in the dc loop (common-mode and dc-link currents) compared to that of the Scheme-A.

The observations, drawn from Figures 10 and 11 when control schemes A and B are used to control the 20-cell MMC in Figure 7, are as follows:

Figures 10a and 11a show that the MMC dc current for the two control schemes is zero, as anticipated during pure reactive power exchange. In line with previous cases, these results show that the dc current of Scheme-A exhibits a brief period of larger temporary over-current due to actions of arm balancing controllers, which force the dc voltages across all arms to be equal.

Figures 10b–d and 11b–d display the three-phase output currents and phase-leg A upper and lower arm voltages superimposed on the common-mode capacitor voltage sum of phase A. Notice that reactive power output reaches the limit as the arm voltage synthesized nearly touches the limits in both cases.

Although the common-mode capacitor voltage sums of the schemes A and B exhibit similar behaviors, the capacitor voltage sums of the upper and lower arms of the faulty phase-leg exhibit slightly different behaviors, see Figure 10e,f and Figure 11e,f. Besides, the differential-mode capacitor voltage sums in Figures 10g and 11g present different convergence patterns, as Scheme-A controls the differential-mode capacitor voltage sum actively.

5.2. Parametric Studies of Asymmetric MMC Operation

To further generalize the findings of the above detailed quantitative studies on asymmetries introduced by the MMC SM faults, additional parametric studies are conducted on a 50-cell MMC using control schemes A and B. The test system parameters are listed in Table 2. To reproduce more representative MMC asymmetries that may arise from SM faults or severe deviations in the values of passive parameters from their nominal values, the SM capacitance (C_u and C_l) and arm inductance (L_u and L_l) of the upper and lower arms are varied within $\pm10\%$ of their nominal values, C_0 and L_0 (i.e., $0.9C_0 \leq C_{u,l} \leq 1.1C_0$ and $0.9L_0 \leq L_{u,l} \leq 1.1L_0$). In the parametric studies, the 50-cell MMC in Figure 12 injects 100 MW into the ac grid at a unity power factor.

Table 2. Simulation Parameters.

System parameters		Value
DC voltage	V_{dc}	100 kV
Rated power	S	100 MVA
AC grid line to line voltage	v_{ac2}	66 kV
AC grid frequency	F	50 Hz
Transformer ratio	v_{ac1}/v_{ac2}	50 kV/66 kV
Transformer leakage-inductance	L_T	0.2 pu
Numbers of SMs per arm	N	50
Expected arm inductance value	L	0.18 pu
Expected SM capacitance value	C	6.7 mF

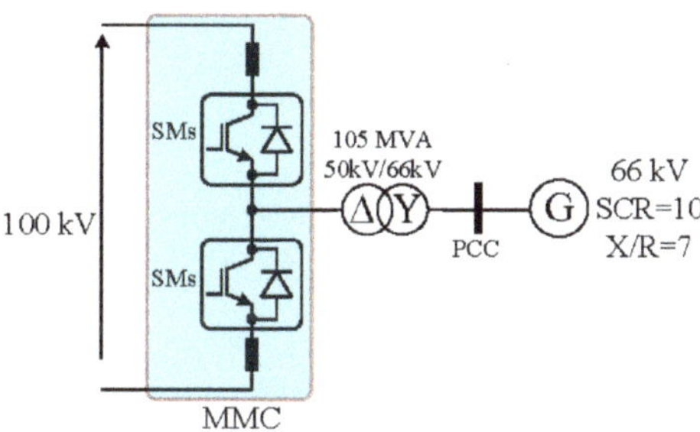

Figure 12. Test system employed in parametric studies with a 50-cell MMC, in which 5 faulty SMs represent 10% of the total number of SMs per arm.

Figures 13 and 14 present side-by-side comparisons of the normalized fundamental frequency ripple in the dc current by its average and maximum achievable modulation range versus average asymmetries in the SM capacitances and inductances between the upper and lower arms of the same phase-leg, for control schemes A and B. The observations drawn from these parametric studies are:

1. Figure 13a,b show that the highest fundamental frequency ripple in the dc current is observed when the upper and lower arms of the same phase-leg simultaneously present a higher mismatch in both SM capacitance and arm inductance, i.e., at extrema of $(T_C$ and $T_L) = (-10\%$ and $-10\%)$ and $(+10\%$ and $+10\%)$. In contrast, the lowest fundamental current ripples are observed when T_C and T_L are (0 and 0), (+10% and −10%), and (−10% and +10%), entailing the polarities of SM capacitance, and arm inductance tolerances affect the magnitude of the fundamental current ripple on dc-side current. Notice that these observations are applied to both control schemes.

2. Quantitatively, comparative parametric studies shown in Figure 13a,b confirm that the Scheme-B, which prioritizes suppression of fundamental frequency ripple in the common-mode currents, exhibits a lower residual ripple in the dc current compared to Scheme-A that prioritizes strict management of voltage stress within the converter over dc side waveform quality.

3. Figure 14a,b display the variations of maximum attainable modulation index in faulty leg-A, with control schemes A and B. The active arm balancing control of scheme A helps to preserve the maximum achievable modulation index range during severe SM capacitance and arm inductance asymmetries. In contrast, with Scheme-B, the MMC modulation index range experiences small reduction as the levels of passive parameter mismatch increase. The worst-case reduction observed in the modulation index range is about 1.5%, which corresponds to a small reduction in MMC ac voltage synthesis and reactive power generation capabilities.

4. The Scheme-A ensures a dc offset free in ac output voltage independent of the severity of MMC internal asymmetries at the expense of compromised dc current quality. While Scheme-B increases the risk of dc injection into the ac grid, particularly, when the arms with minimum capacitor

voltage sums are no longer sufficient to synthesize the required ac voltage to exchange the desired active and reactive powers. These observations are in line with the discussions above.

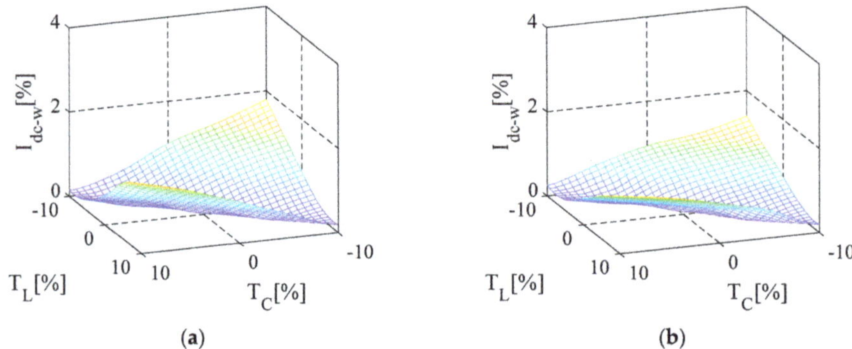

Figure 13. Variation of fundamental frequency dc-link current ripple magnitude (pu) with SM capacitance tolerances T_C and arm inductor tolerances T_L: (**a**) Scheme-A and (**b**) Scheme-B.

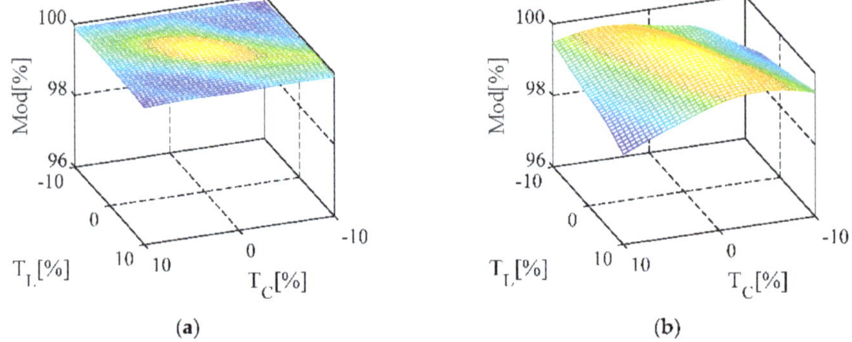

Figure 14. Variation of fundamental frequency dc-link current ripple magnitude (pu) with SM capacitance tolerances T_C and arm inductor tolerances T_L: (**a**) Scheme-A, and (**b**) Scheme-B.

It is worth emphasizing that the 20-cell MMC in previous studies is a good representation of the scenario with low SMs numbers per arm. In such a scenario, the loss of 10% of total SMs per arm in absolute term, namely, two SMs, is of practical significance, particularly from the following point of views: (1) Increased duties on remaining and healthy SMs (frequency of insertion and bypass of SM capacitors). (2) Increased sampling errors in the synthesis of the dc, arm, and output ac voltages. (3) Average switching frequency per device. (4) Increased likelihood of occurrence in a practical system. On the other hand, the 50-cell MMC is a relatively good representation for upper MV applications, in which the likelihood of loss of 10% of the total number of SMs per arm is much lower but remains possible with less impact on the quality of the ac and dc side waveforms and duties on SM capacitors.

6. Experimental Verifications

Since SM faults make the MMCs present unequal equivalent capacitances per arm and per phase-leg bases, this section emulates the steady-state impacts of SM faults through the deliberate use of unequal cell capacitances and arm inductances on experimental test rig of a single-phase grid-connected MMC with three cells per arm, see Figure 15. The MMC test rig is equipped with the following controllers: 1) active and reactive power controller, which sets P = 2kW and Q = 0, 2) inner current controller, circulating current suppression controller, 3) horizontal/common-mode capacitor voltage sum, 4) level-shifted pulse width modulation with 1 kHz carrier frequency and sorting-based capacitor voltage balancing.

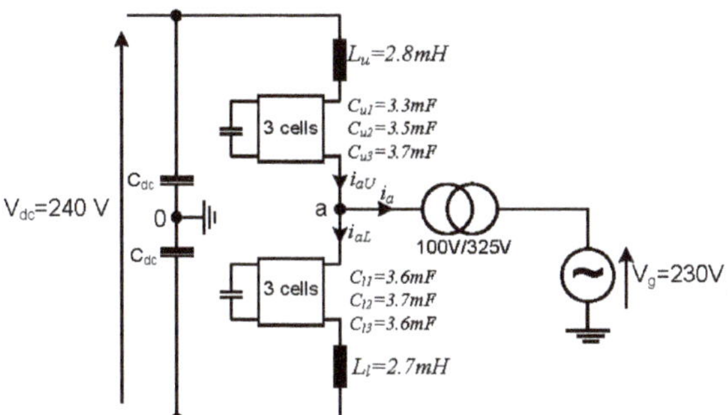

Figure 15. Experimental test of a 3-cell HB-MMC.

Figures 16 and 17 display experimental results when control schemes A and B are employed. Figure 16a,b and Figure 17a,b present upper and lower arm capacitor voltage sums and upper and lower arm currents for the control schemes A and B, respectively. Although both methods appear to work well, it is worth noticing that the Scheme-B exhibits slightly lower ripples in the common-mode current. In line with simulations, the FFT shows that the Scheme-B exhibits slightly lower contamination of the common-mode current by the fundamental frequency component compared to that of the control method A. Despite the few SM numbers of the experimental test rig, the presented experimental results point toward the same conclusions as simulation cases discussed earlier.

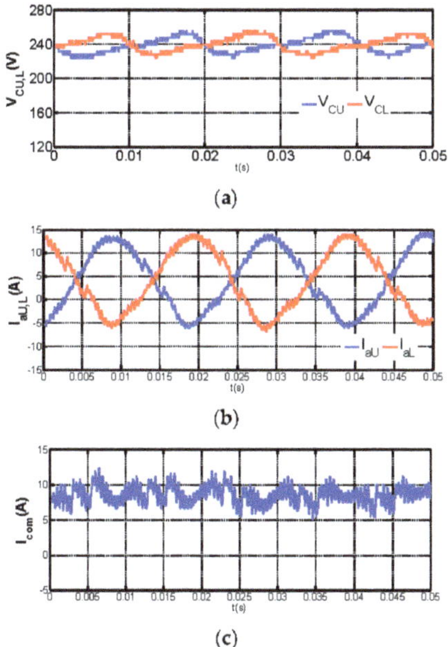

Figure 16. Experimental waveforms of the 3-cell MMC that employs internal Scheme-A: (**a**) upper and lower capacitor voltage sums, (**b**) upper and lower arm currents, and (**c**) common-mode current with the measured fundamental frequency content of 6.6%.

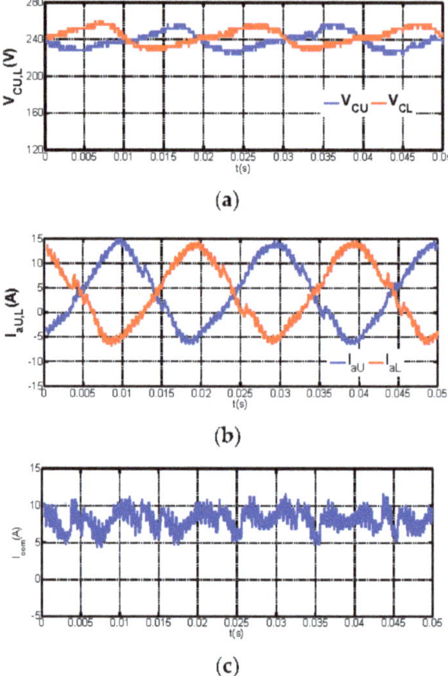

Figure 17. Experimental waveforms of the 3-cell MMC that employs internal Scheme-B: (**a**) upper and lower capacitor voltage sums, (**b**) upper and lower arm currents, and (**c**) common-mode current with the measured fundamental frequency content of 5.4%.

7. Conclusions

This paper has investigated the impacts of MMC internal asymmetries due to SM faults or severe deviations of the SM capacitance or arm inductance values from their nominal values, particularly, on the following aspects: control range, power quality in ac and dc sides, and distribution of voltage stresses across the SM capacitors and semiconductor switches. Moreover, the paper has presented a direct method for fundamental frequency ripple suppression from the dc link, and its performance is compared to a well-performing conventional controller for managing MMC internal dynamics. The investigation has found that the control for actively balancing arm capacitor voltage sums (Scheme-A) is better than the direct fundamental ripple suppression method (Scheme-B) from the ac side viewpoint, but it pollutes dc current with a fundamental frequency ripple. In contrast, the Scheme-B is better than its counterpart A, from the dc side viewpoint; nevertheless, it increases the risk of dc injection into the ac grid and reduction of ac control range.

Author Contributions: Conceptualization: G.P.A. and S.W.; Methodology: S.W. and F.S.A.; Software: S.W. and F.S.A.; Validation: G.P.A. and S.W.; Formal Analysis: F.S.A. and S.W.; Investigation: S.W. and G.P.A.; Resources: F.S.A.; Data Curation: S.W.; Writing—Original Draft Preparation: S.W. and F.S.A.; Writing—Review and Editing: G.P.A.; Visualization: S.W. and F.S.A.; Supervision: G.P.A.; Project Administration: G.P.A and F.S.A.; Funding Acquisition: F.S.A. and G.P.A. All authors have read and agreed to the published version of the manuscript.

Funding: This study was funded by King Abdulaziz University, Jeddah, Saudi Arabia, and King Abdulah City for Atomic and Renewable Energy, Riyadh, Saudi Arabia, under grant no. (KCR-KFL-14-20). Therefore, the authors gratefully acknowledge their technical and financial support.

Conflicts of Interest: The authors declare no conflict of interest.

Appendix A

Zoomed waveforms of the dc currents displayed in Figures 8a and 9a and their FFT analyses are shown in Figures A1 and A2, respectively, to further consolidate the previous claims.

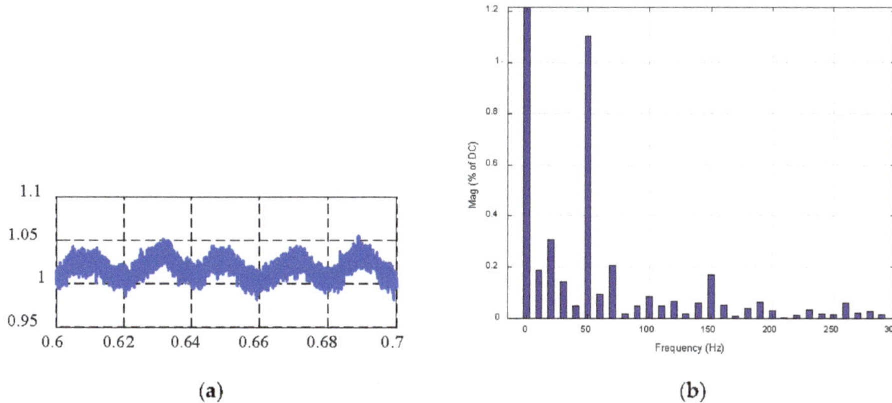

(a)   (b)

Figure A1. Additional simulation waveforms. (**a**) Zoomed version of dc-link current displayed in Figure 8a, (**b**) FFT spectrum.

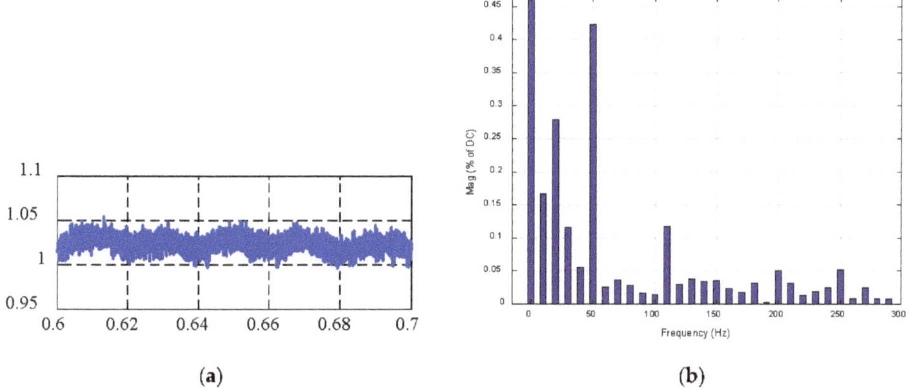

(a)   (b)

Figure A2. Additional simulation waveforms. (**a**) Zoomed version of dc-link current displayed in Figure 9a, (**b**) FFT spectrum.

References

1. Perez, M.A.; Bernet, S.; Rodriguez, J.; Kouro, S.; Lizana, R. Circuit topologies, modeling, control schemes, and applications of modular multilevel converters. *IEEE Trans. Power Electron.* **2015**, *30*, 4–17. [CrossRef]
2. Wang, J.; Liang, J.; Wang, C.; Dong, X. Circulating current suppression for MMC-HVDC under unbalanced grid conditions. *IEEE Trans. Ind. Appl.* **2017**, *53*, 3250–3259. [CrossRef]
3. Xiaoqian, L.; Qiang, S.; Jianguo, L.; Wenhua, L. Capacitor Voltage Balancing Control based on CPS-PWM of Modular Multilevel Converter. In Proceedings of the Energy Conversion Congress and Exposition (ECCE00), Phoenix, AZ, USA, 17–22 September 2011; pp. 4029–4034.
4. Siemaszko, D. Fast sorting method for balancing capacitor voltages in modular multilevel converters. *IEEE Trans. Power Electron.* **2015**, *30*, 463–470. [CrossRef]
5. Yang, W.; Song, Q.; Liu, W. Decoupled Control of Modular Multilevel Converter Based on Intermediate. *IEEE Trans. Ind. Electron.* **2016**, *63*, 4695–4706. [CrossRef]

6.   Adam, G.P.; Ahmed, K.H.; Finney, S.J.; Williams, B.W. AC fault ride-through capability of a VSC-HVDC transmission systems. In Proceedings of the 2010 IEEE Energy Conversion Congress and Exposition, Atlanta, GA, USA, 12–16 September 2010; pp. 3739–3745.

7.   Sasongko, F.; Sekiguchi, K.; Oguma, K.; Hagiwara, M.; Akagi, H. Theory and Experiment on an Optimal Carrier Frequency of a Modular Multilevel Cascade Converter with Phase-Shifted PWM. *IEEE Trans. Power Electron.* **2016**, *31*, 3456–3471. [CrossRef]

8.   Fan, S.; Zhang, K.; Xiong, J.; Xue, Y. An Improved Control System for Modular Multilevel Converters with New Modulation Strategy and Voltage Balancing Control. *IEEE Trans. Power Electron.* **2014**, *30*, 358371.

9.   Mehrabadi, N.R.; Burgos, R.; Boroyevich, D.; Roy, C. Modeling and design of the modular multilevel converter with parametric and model-form uncertainty quantification. In Proceedings of the 2017 IEEE Energy Conversion Congress and Exposition (ECCE), Cincinnati, OH, USA, 1–5 October 2017; pp. 1513–1520.

10.   Dekka, A.; Wu, B.; Lizana, R.; Zargari, N.R. A comparison of voltage balancing versus energy balancing approach for modular multilevel converters. In Proceedings of the IECON Proceedings (Industrial Electronics Conference), Florence, Italy, 24–27 October 2016; pp. 3117–3122.

11.   Zeng, R.; Xu, L.; Yao, L.; Finney, S.J. Analysis and Control of Modular Multilevel Converters under Unbalanced Conditions. *IEEE Trans. Ind. Electron.* **2016**, *63*, 71–81. [CrossRef]

12.   Zhang, F.; Zhao, C.; Xu, J. New control strategy of decoupling the AC/DC voltage offset for modular multilevel converter. *IET Gener. Transm. Distrib. Spec.* **2016**, *10*, 1382–1392. [CrossRef]

13.   Ghazanfari, A.; Mohamed, Y.A.R.I. A Resilient Framework for Fault-Tolerant Operation of Modular Multilevel Converters. *IEEE Trans. Ind. Electron.* **2016**, *63*, 2669–2678. [CrossRef]

14.   Deng, F.; Chen, Z.; Khan, M.R.; Zhu, R. Fault detection and localization method for modular multilevel converters. *IEEE Trans. Power Electron.* **2015**, *30*, 2721–2732. [CrossRef]

15.   Zhou, W.; Sheng, J.; Luo, H.; Li, W.; He, X. Detection and Localization of Submodule Open-Circuit Failures for Modular Multilevel Converters with Single Ring Theorem. *IEEE Trans. Power Electron.* **2019**, *34*, 3729–3739. [CrossRef]

16.   Konstantinou, G.; Pou, J.; Ceballos, S.; Agelidis, V.G. Active redundant submodule configuration in modular multilevel converters. *IEEE Trans. Power Deliv.* **2013**, *28*, 2333–2341. [CrossRef]

17.   Konstantinou, G.S.; Ciobotaru, M.; Agelidis, M. Effect of redundant sub-module utilization on modular multilevel converters. In Proceedings of the 2012 IEEE Int. Conf. on Ind. Technol. ICIT 2012, Proc., Athens, Greece, 19–21 March 2012; pp. 815–820.

18.   Ahmed, N.; Ängquist, L.; Antonopoulos, A.; Harnefors, L.; Norrga, S.; Nee, H.P. Performance of the modular multilevel converter with redundant submodules. In Proceedings of the IECON 2015 - 41st Annual Conference of the IEEE Industrial Electronics Society, Yokohama, Japan, 9–12 November 2015; pp. 003922–003927.

19.   Deng, F.; Tian, Y.; Zhu, R.; Chen, Z. Fault-Tolerant Approach for Modular Multilevel Converters under Submodule Faults. *IEEE Trans. Ind. Electron.* **2016**, *63*, 7253–7263. [CrossRef]

20.   Li, B.; Xu, Z.; Ding, J.; Xu, D. Fault-Tolerant Control of Medium-Voltage Modular Multilevel Converters with Minimum Performance Degradation under Submodule Failures. *IEEE Access* **2018**, *6*, 11772–11781. [CrossRef]

21.   Wang, J.; Tang, Y. A Fault-Tolerant Operation Method for Medium Voltage Modular Multilevel Converters with Phase-Shifted Carrier Modulation. *IEEE Trans. Power Electron.* **2019**, *34*, 9459–9470. [CrossRef]

22.   Wang, S.; Adam, G.; Massoud, A.; Holliday, D.; Williams, B. Strategies for Decoupling Internal and External Dynamics Resulting From Inter-Arm Passive Component Tolerances in HVDC-MMC. In Proceedings of the 2019 IEEE Energy Conversion Congress and Exposition (ECCE), Baltimore, MD, USA, 29 September–3 October 2019; pp. 209–213.

23.   Wang, S.; Adam, G.P.; Massoud, A.M.; Holliday, D.; Williams, B.W. Analysis and Assessment of Modular Multilevel Converter Internal Control Schemes. *IEEE J. Emerg. Sel. Top. Power Electron.* **2020**, *8*, 697–719. [CrossRef]

Article

Reliability Assessment of Wind-Solar PV Integrated Distribution System Using Electrical Loss Minimization Technique

Sachin Kumar [1], Kumari Sarita [1], Akanksha Singh S Vardhan [2], Rajvikram Madurai Elavarasan [3,*], R. K. Saket [1] and Narottam Das [4,5,*]

[1] Department of Electrical Engineering, Indian Institute of Technology (BHU), Varanasi 221005, India; sachinkumar.rs.eee18@itbhu.ac.in (S.K.); kumarisarita.rs.eee19@itbhu.ac.in (K.S.); rksaket.eee@iitbhu.ac.in (R.K.S.)
[2] Department of Electrical Engineering, Shri G.S. Institute of Technology and Science, Indore 452003, Madhya Pradesh, India; aakankshasingh843@gmail.com
[3] Electrical and Automotive Parts Manufacturing Unit, AA Industries, Chennai 600123, Tamilnadu, India
[4] School of Engineering and Technology, Central Queensland University, Melbourne, VIC 3000, Australia
[5] Centre for Intelligent Systems, School of Engineering and Technology, Central Queensland University, Brisbane, QLD 4000, Australia
* Correspondence: rajvikram787@gmail.com (R.M.E.); n.das@cqu.edu.au (N.D.)

Received: 31 August 2020; Accepted: 22 October 2020; Published: 28 October 2020

Abstract: This article presents the Reliability Assessment (RA) of renewable energy interfaced Electrical Distribution System (EDS) considering the electrical loss minimization (ELM). ELM aims at minimizing the detrimental effect of real power and reactive power losses in the EDS. Some techniques, including integration of Renewable Energy Source (RES), network reconfiguration, and expansion planning, have been suggested in the literature for achieving ELM. The optimal RES integration (also referred to as Distributed Generation (DG)) is one of the globally accepted techniques to achieve minimization of electrical losses. Therefore, first, the locations to accommodate these DGs are obtained by implementing two indexes, namely Index-1 for single DG and Index-2 for multiple DGs. Second, a Constriction Factor-based Particle Swarm Optimization (CF-PSO) technique is applied to obtain an optimal sizing(s) of the DGs for achieving the ELM. Third, the RA of the EDS is performed using the optimal location(s) and sizing(s) of the RESs (i.e., Solar photovoltaic (SPV) and Wind Turbine Generator (WTG)). Moreover, a Battery Storage System (BSS) is also incorporated optimally with the RESs to further achieve the ELM and to improve the system's reliability. The result analysis is performed by considering the power output rating of WTG-GE's V162-5.6MW (IECS), SPV-Sunpower's SPR-P5-545-UPP, and BSS-Freqcon's BESS-3000 (i.e., Battery Energy Storage System 3000), which are provided by the corresponding manufacturers. According to the outcomes of the study, the results are found to be coherent with those obtained using other techniques that are available in the literature. These results are considered for the RA of the EDS. RA is further analyzed considering the uncertainties in reliability data of WTG and SPV, including the failure rate and the repair time. The RA of optimally placed DGs is performed by considering the electrical loss minimization. It is inferred that the reliability of the EDS improves by contemplating suitable reliability data of optimally integrated DGs.

Keywords: battery storage system; distributed generation; electrical loss minimization; particle swarm optimization; reliability analysis; solar photovoltaic, wind turbine generator

---

1. Introduction

1.1. Literature Survey

The incorporation of distributed power sources in the EDS can support the increasing load demands. To satisfy the increasing load demands, an extremely thorough comparison and assessment has been performed on exploring the potentiality of the distributed energy sources, including the wind and solar energies [1,2]. The EDS is the combination of electrical loads and Distributed Generation (DG) (especially, WTG and SPV) integrated with electrical storage system (namely BSS) [3]. In this regard, several comprehensive studies are available on sustainable energy production and management. The state policies, renewable energy plants, and the development of renewable energy sources including solar energy, wind energy, small hydroelectric energy, biomass energy, tidal energy, and geothermal energy have been analyzed for different states of India [4–6]. As the energy crisis deepens day by day, the EDS accompanied by RESs is a better solution and also it acts as a complement to the central power grid system [7]. Therefore, a study on renewable energy potential has been performed for five countries, namely China, Iceland, India, Sweden, and the United States of America. The strengths, weaknesses, opportunities, and threats related to green energy generation and use have been reviewed for these countries using a SWOT analysis [8,9]. Sometimes an arrangement including the battery storage is needed due to high penetration of renewable resources into an electrical network to fulfill the load demands [10,11]. Thus, integration of BSS in the distribution system prevails as a better solution for obtaining a steady power output, especially from WTG and SPV owing to the uncertainties involved in the energy harvesting sources such as wind speed [1], solar irradiation, and ambient temperature [12]. Therefore, it is a requisite to determine the siting(s) and sizing(s) of WTG, SPV, and BSS. The system reliability, stability, and power quality are thus, improved substantially. Therefore, to obtain improvements in system's attributes, the multi-objective optimization methods for problems including operational cost, siting and sizing of distributed energy resources (DERs), Carbon Dioxide (CO_2) emission, total power loss, voltage deviation, demand-side management, charging-discharging of BSS, total harmonic distortion, and system reliability are also established [13–17].

Improvement in network reliability is observed when congestion management algorithms are implemented to identify the transmission line congestion [18]. The DGs and BSSs are scheduled optimally to alleviate this transmission line congestion. A two-step optimization approach is used for solving the congestion problem [19]. In this optimization method, the optimum location and the size of the SPV array are observed, and then the BSS size with location is determined for accomplishing further reduction in electrical losses and voltage deviation. The authors in [20] have proposed an index to obtain the optimal siting of DGs in EDS. This index is implemented to resolve the multiple problems, including total ELM, ENS, and voltage deviation. Another objective, namely loss of yearly energy minimization, is observed through the integration of DERs-based DGs and network reconfiguration [21]. Furthermore, a novel two-stage stochastic programming is proposed, and the uncertainty considerations together with the load variation are also studied especially for wind energy and solar power generation [22]. In this method, the total cost is reduced by incorporating BSS into the EDS and by considering the demand response programs in planning. Simultaneously, the enhancement of the power system reliability is achieved due to the obtained optimal BSS size and location. The reliability improvement and reduction in network losses are also observed using compound co-optimization strategic plan [23]. The reliability indices such as expected ENS, EIR, LOLE, and LOLP, are defined in the co-optimization strategy. Furthermore, a moth-flame optimization [24], Olympic games ranking process [25], firefly algorithm [26], lightning search algorithm [27], crow search optimization [28], and an improved variant PSO [29] techniques have been implemented and discussed in the literature so as to obtain the optimal site, optimal size, optimal parameters of DGs, and ELM.

The RA provides a better evaluation of any power system's performance [30]. For assessing the power system's reliability, some indices have been mentioned in the available literature. These indices are categorized into load-based and system-based indices. The indices values decrease if the ageing of the sub-components is considered [31]. Reliability is considered to be a primary requirement in the designing phase of EDS. Thus, the optimal siting and optimal sizing of RESs for further RA have also been considered. To fulfill the RA in EDS [15], has introduced a restoration strategy for ENS calculation. The optimization of reliability indices has been considered in [20,23], and the improvement in the system's reliability is observed.

1.2. Motivation

The complexity of the electrical power system relies in its ability to tackle the unexpected load variations. These load variations are managed by the electrical distribution system (EDS) for continuous electrical power supply to the loads. This system is categorized into radial and loop structured distribution systems. The use of radial DS (RDS) is preferred in this study as it is simple, cheap, and mostly applicable to sparsely distributed loads. Furthermore, to meet the rapid growth in demand as well as to promote sustainability, integration of renewable energy source (RES) in the distribution system is desideratum. The RESs, including Wind Turbine Generator (WTG) and Solar photovoltaic (SPV), provide minimal electrical losses, improves system bus voltages, possesses less operation costs, and more significantly, it emits less CO_2 emissions. Thus, the optimization of these parameters would ultimately improves the reliability of the EDS. Reliability in an electrical power system is defined as the ability to provide adequate, stable, and reliable power for a particular distribution system. Therefore, the study on Reliability Assessment (RA) and performance analysis of RDS considering the optimal RES siting and sizing is crucial. It is also observed that the integration of renewable sources in EDS has considerable impacts on systems' operation and planning strategy [32]. Some issues co-related to RES integration are explained as follows.

- Electrical loss minimization using system reconfiguration [13].
- Electrical loss minimization [33,34],
- Reduction in investment during system capacity enhancement [35],
- Improvement in bus voltage [36],
- Mitigation of greenhouse gases [37],
- Improvement in voltage stability [38],
- Improvement in system's reliability by considering the reliability indices [39–41],
- Enhancement in system security [42],
- Facilitate system restoration [43],
- Reduction in harmonic distortion [44],
- Optimal load management strategy [45].

The mentioned issues have been analyzed individually that is without considering the effect on another issue(s). The concept of analyzing the effect of the mentioned issues motivates the authors for carrying out this research work. In the view of strong motivation, this work is concentrated on the reliability assessment of IEEE 33 bus distribution system by considering the effect of electrical loss minimization through optimal siting and sizing of distributed generation system.

1.3. Contribution

The available literature focus on optimal siting and sizing of DGs obtaining a better power system performance. Also, the DG integration of SPV, WTG, and BSS in the distribution system has not been discussed as of authors knowledge. The system RA is discussed individually without analyzing the impact of optimal DG integration. Thus, the two indexes have been implemented for obtaining the optimal location(s) of DG(s) in the selected EDS. Then, a Constriction Factor (CF)-based PSO technique for optimal sizing(s) of DG(s) is presented. Simultaneously, all the parameters considered in this study (as given in Table 1) are not discussed in the existing literature and thus, this work provides completeness in contribution. According to the outcomes of the study, the results are found coherent with those obtained using other techniques that are available in the literature. However, the results can be claimed to be even better since the proposed approach considers all the parameters given in Table 1, including the ELM and the reliability aspects. The comparative analysis for a reduction in electrical power loss and voltage deviation is performed by integrating the WTG, SPV, and BSS in IEEE 33 bus EDS. The results obtained are compared with the results in the literature, as described in Table 2. After obtaining the reduced power loss and improved bus voltages, the reliability assessment is accomplished. The performance analysis is done for the selected distribution system and then the system's RA is carried out with and without the integration of DGs.

Table 1. Published work.

No. of Bus	Parameters Considered							Reference
	Size	Location	Voltage	Loss	Reliability	Power Factor	DG	
34, 69		✓	✓	✓		✓	PV/WTG	[33]
33, 69, 119	✓	✓	✓	✓			PV/WTG	[34]
12		✓	✓	✓			DGen	[38]
33, 69	✓	✓	✓	✓			PV	[44]
13	✓	✓	✓	✓			PV/BSS	[19]
33, 118	✓	✓		✓			PV/WTG	[21]
38	✓	✓					PV/WTG	[27]
33, 69	✓	✓	✓	✓			PV/ESS	[28]
38, 69	✓	✓	✓				GT/WTG	[46]
69, 118	✓	✓	✓	✓			DGen/Cap	[47]
33	✓	✓	✓	✓		✓	PV/WTG	[48]
33	✓	✓	✓	✓	✓	✓	SPV/WTG/BSS	Present Work

Table 2. Work related to IEEE 33 bus with multiple DGs.

Method	# of DG	DG Position	Total DG Size (MW)	Loss (MW)	Reference
MOGA	1(SPV)	8	1.6333	0.113	[34]
	2(SPV)	14, 30	0.8337, 0.99851	0.08435	
	1(WTG)	8	1.85	0.08556	
	2(WTG)	14, 30	1.1, 0.75	0.04791	
GA	3(SPV)	14, 24, 28	0.6947, 1.1844, 1.4628	0.0756	[44]
ABC	3(SPV)	9, 24, 32	1.1372, 1.0674, 0.8031	0.0752	
PSO	3(SPV)	9, 24, 30	1.0625, 1.0447, 0.9518	0.0744	
BBO	3(SPV)	14, 24, 30	0.7539, 1.0994, 1.0714	0.0715	
CSO	5(BSS)	1, 4, 11, 12, 18	0.15, 0.4117, 0.6705, 0.1, 8.9055	0.02379	[28]
DMA	1(SPV)	6	2	0.0908	[48]
		18	1	0.1175	
	1(WTG)	33	1.65	0.1068	

Though the research on DG siting and sizing is abundant in the literature, the RA in the EDS still prevails as an emerging area of research. Thus, the work mainly focuses on the reliability assessment of an IEEE 33 bus system is carried out in the view of loss minimization technique of WTG, SPV, and BSS integration. For the accomplishment of the mentioned tasks, the research work is performed according to the framed flowchart as shown in Figure 1. Furthermore, the following contributions are highlighted in this work.

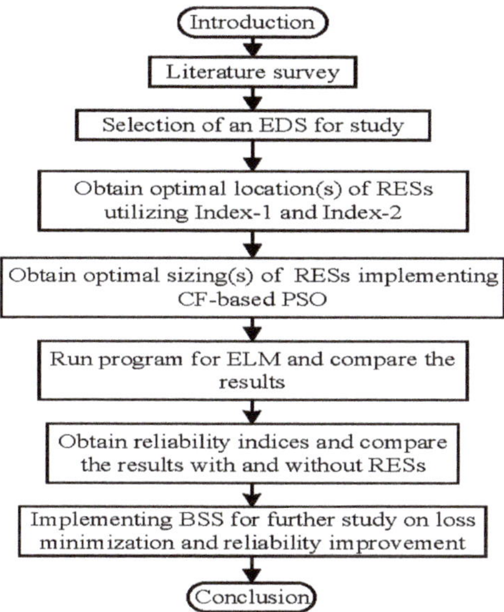

Figure 1. Workflow of the research work.

- The optimal locations for SPV, WTG, and BSS is obtained by considering Index-1 and Index-2.
- Optimal sizes of SPV, WTG, and BSS are derived by employing CF-PSO technique.
- Distribution system performance, including minimization of I^2R loss and minimization of deviation in bus voltages, is analyzed with and without DGs.
- Comparison of CF-PSO technique with other nature-inspired optimization techniques for achieving a sound reliability assessment.
- A brief study is done on the inclusion of uncertainties in WTG and SPV reliability data such as failure rate (λ_p) and time to repair (RT).
- Reliability assessment is done by evaluating the five indices namely EENS, AENS, SAIDI, SAIFI, and ASAI for Case 1, Case 2, and Case 3; where Case 1 is for integrating WTG only, Case 2 is for WTG+SPV, and Case 3 is for WTG+SPV+BSS (adding BSS optimally).

The work is further continued to accomplish the mentioned contributions as follows. The problem statement, objective function, and methodologies implemented, including Index-1, Index-2, and CF-based PSO techniques are delineated in Section 2. The basics of EDS reliability assessment is formulated and explained in Section 3. In Section 4, an overview of SPV, WTG, and BSS is explained using mathematical expressions. Furthermore, the results of several cases and scenarios are elucidated with supporting graphical depictions and tables in Section 5. Finally, the conclusions with future scopes are drawn in Section 6. The Appendix A of some important data and figure is provided at the end of the work.

1.4. Parameters Considered for the Study

The parameters considered for the study are described as follows.

- **DG siting and DG sizing**: The determination of bus voltages and the flow of powers is done by an Optimal Power Flow (OPF) method. The optimal siting of DGs is required for ELM during this power flow results. The performance of the power network is affected by an inappropriate location of DG. The IEEE 1547 standards for integration and operation of DG into EDSs are presented in [49].

- **Power loss**: The occurrence of Active Power Loss (APL) is greater than Reactive Power Loss (RPL) in EDS. Hence, distribution companies should reduce these losses and this can be accomplished by means of reconfiguration of feeder, capacitor allocation, high voltage distribution system, grading of the conductor, DG placement, and many other methods.
- **Bus voltage**: It is expected to maintain bus voltages nearly 1 pu with an angle of 0^0. The power loss occurring in EDS during OPF creates a voltage drop at each bus of the system. Therefore, the DG integration technique is implemented for voltage profile (VP) improvement.
- **DG type**: The three DGs have been considered, which are categorized as WTG, SPV, and BSS. The classification of the several DG technologies is based on the generation of active power 'P' and reactive power 'Q', as illustrated in Figure 2.
- **Reliability**: The reliability indices considered for the distribution system reliability are as follows.

  - Expected Energy Not Supplied (EENS); MWh per year
  - Average Energy Not Supplied (AENS); MWh per customer per year
  - System Average Interruption Duration Index (SAIDI); hour per customer per year
  - System Average Interruption Frequency Index (SAIFI); failure per customer per year
  - Average System Availability Index (ASAI); pu

The calculations of these five indices are performed using Equations (A1)–(A7) of the Appendix A.2. The reliability indices considered in this work can be used in obtaining other indices, including IEAR, CAIDI and ASUI, as illustrated in Equations (42), (45) and (46c).

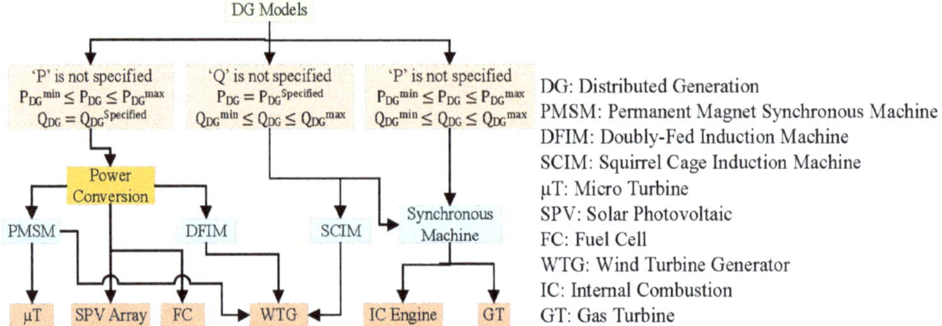

Figure 2. Models of several DG technologies.

2. Problem Formulation, Objective Function (OF), and Methodology

The bus voltage and system reliability are the most affecting factors for the EDS when losses are considered. It becomes necessary to improve these two factors by implementing distributed generation (DG) into the EDS. DG siting is one of the favored techniques used in the EDS for the improvement of the system's reliability and bus Voltage Profile (VP), including ELM. The DG location, DG size, DG power factor (pf), DG penetration, and DG type are required for the effective implementation of DGs in the EDS. Simultaneously, it is required to study the mathematical expressions, and modeling of related parameters and DGs integrated into the system. An overview of parameters considered for the ELM and mathematical modeling are elaborated in the upcoming subsections.

2.1. Optimal Location

Obtaining the optimal location of DGs is crucial part of EDS. To identify the optimal locations, two indexes are used. Index-1 is implemented only for placing the single DG and Index-2 is incorporated for placing more than one DGs in the EDS. Power loss is minimized by using Index-1 for placing 1DG (viz. Case 1). However, Index-2 provides minimum power loss for multiple DGs (viz. Case 2 and Case 3). These two indicators are represented by Equations (17) and (18), respectively [50,51]. It can be observed from the equation of Index-1 that the large index value depicts the weakest node of the system because the complex power injected at bus *i* is large. It implies that the single DG can be placed at this particular bus. On the other hand, Index-2 shows the voltage stability, which concludes that the reduced values of this index give the weakest bus of the EDS. Table 3 shows the values of both the indexes with corresponding five buses to arrange DG optimally in a given EDS. Therefore, DGs can be placed hierarchically at these buses.

Table 3. Values of indexes with corresponding bus number [17,20].

Index-1		Index-2	
Value	Bus No.	Value	Bus No.
1.350×10^{-3}	6	41.52×10^{-3}	30
0.928×10^{-3}	29	16.44×10^{-3}	13
0.867×10^{-3}	30	16.43×10^{-3}	24
0.864×10^{-3}	5	7.36×10^{-3}	31
0.735×10^{-3}	28	6.49×10^{-3}	20

In Figure 3, V_a and V_b are the magnitudes of the voltages at buses *a* and *b*, respectively. δ_a and δ_b are the phase angles of the voltages at buses a and b, respectively. Z_l and Y_l are the impedance and admittance of l-line, respectively. R_l and X_l are the resistance and reactance of a l-line. I_l is the current in the l-line. The electrical power loss in the l-line is given by (1).

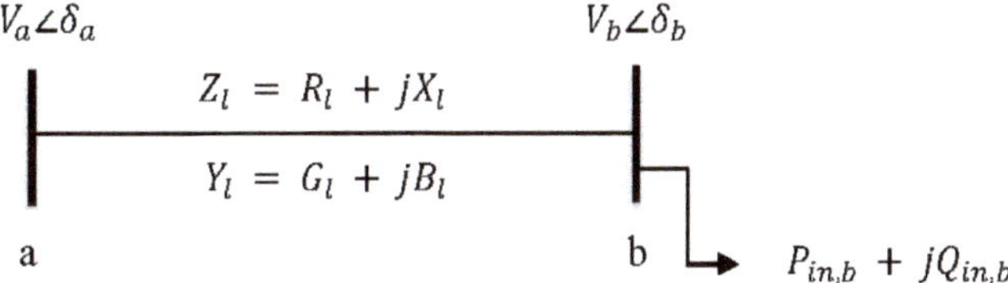

Figure 3. General 2-bus system to formulate the line loss and load factor.

$$S_{LOSS_l} = (V_a - V_b) \times I_l^* \tag{1}$$

$$I_l = (V_a - V_b) \times Y_l \tag{2}$$

Then the bus voltage matrix is formed by using Equation (3) where $[Z_{bus}]$ is the network impedance matrix, $[I_{bus}]$ is the bus injection matrix, and n_{bus} is the number of total buses in EDS.

$$[V_{bus}]_{n_{bus} \times 1} = [Z_{bus}]_{n_{bus} \times n_{bus}} [I_{bus}]_{n_{bus} \times 1} \tag{3}$$

By expanding Equation (3), the node voltages can be obtained by using Equations (4) and (5).

$$V_a = \sum_{i=1}^{n_{bus}} Z_{ai} \times I_i \tag{4}$$

$$V_b = \sum_{i=1}^{n_{bus}} Z_{bi} \times I_i \tag{5}$$

where i is $1,2,\ldots,n_{bus}$. Z_{ai}, Z_{bi} and I_i are the element of impedance matrix that signify the a_{th} row and i_{th} column, b_{th} row and i_{th} column, and current injection at bus-i, respectively.

$$\text{Current Injection, } I_i = \frac{(P_{in,i} + jQ_{in,i})^*}{V_i^*} \tag{6}$$

where $P_{in,i}$ and $jQ_{in,i}$ are active power and reactive power injected at bus-i, respectively. V_i^* is the voltage at bus-i. Now, put (2) and (4) to (6) in (1) then the electrical loss of the line-l is derived as (7).

$$S_{LOSS,l} = [V_a - V_b]\left(\frac{\sum^{n_{bus}}(Z_{ai} - Z_{bi})Y_l}{V_i^*}\right)^* [P_{in,i} + jQ_{in,i}] \tag{7}$$

For an electrical system with n_l number of branch/lines, the line loss is given by (8).

$$[B_{LOSS_l}] = \sum_{i=1}^{n_{bus}} \frac{(V_a - V_b)(Z_{ai} - Z_{bi})^* Y_l^*}{V_i} S_{in,i} \tag{8}$$

where B_{LOSS_l} is line loss, $S_{in,i}$ is apparent power injected at bus-i, l is 1 to n_l.

$$[B_{LOSS_l}] = \sum_{i=1}^{n_{bus}} [LF_{li}][S_{in,i}] \tag{9}$$

$$LF_{li} = \sum_{i=1}^{n_{bus}} \frac{(V_a - V_b)(Z_{ai} - Z_{bi})^* Y_l^*}{V_i} \tag{10}$$

$$LF_{li} = \begin{cases} \text{non-zero,} & \text{if l-line is in the path of bus-i} \\ 0, & \text{Else} \end{cases} \tag{11}$$

where load factor (LF) is given by $\frac{(V_a - V_b)(Z_{ai} - Z_{bi})^* Y_l^*}{V_i}$. LF_{li} is load factor of the lth line due to the ith bus injection (it is non-zero if lth line is in the path of ith bus else zero) as described in (11). For example, a 6-bus distribution is taken for explanation as shown in Figure 4 and a general branch loss formula is derived as (12).

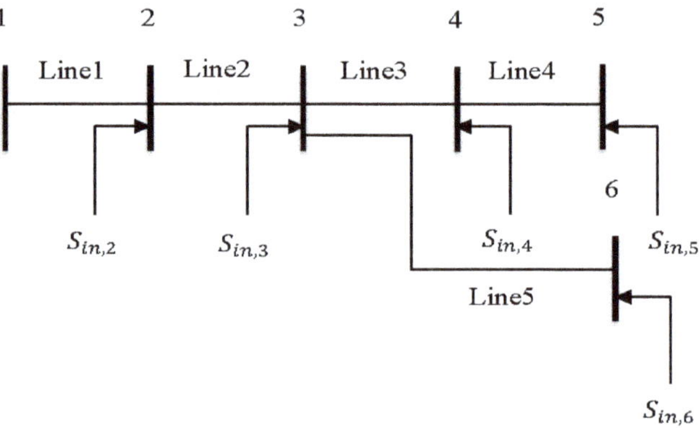

Figure 4. 6-bus EDS for Index-1 calculation.

$$
\begin{bmatrix} B_{LOSS_1} \\ B_{LOSS_2} \\ B_{LOSS_3} \\ B_{LOSS_4} \\ B_{LOSS_5} \end{bmatrix}_{5\times1}
=
\begin{bmatrix}
LF_{11} & LF_{12} & LF_{13} & LF_{14} & LF_{15} & LF_{16} \\
LF_{21} & LF_{22} & LF_{23} & LF_{24} & LF_{25} & LF_{26} \\
LF_{31} & LF_{32} & LF_{33} & LF_{34} & LF_{35} & LF_{36} \\
LF_{41} & LF_{42} & LF_{43} & LF_{44} & LF_{45} & LF_{46} \\
LF_{51} & LF_{52} & LF_{53} & LF_{54} & LF_{55} & LF_{56}
\end{bmatrix}_{5\times6}
\times
\begin{bmatrix} S_{in_1} \\ S_{in_2} \\ S_{in_3} \\ S_{in_4} \\ S_{in_5} \\ S_{in_6} \end{bmatrix}_{6\times1}
\tag{12}
$$

$$
[B_{LOSS}M]_{5\times1} = [LFM]_{5\times6} \times [S_{in}M]_{6\times1}
\tag{13}
$$

where $[B_{LOSS}M]$ is branch/line loss matrix, $[LFM]$ is the load factor matrix, $[S_{in}M]$ is complex power injection matrix. Equation (13) is reduced according to the Figure 4 where power injection at all the buses except the source bus are available.

$$
\begin{bmatrix} B_{LOSS_1} \\ B_{LOSS_2} \\ B_{LOSS_3} \\ B_{LOSS_4} \\ B_{LOSS_5} \end{bmatrix}_{5\times1}
=
\begin{bmatrix}
0 & LF_{12} & LF_{13} & LF_{14} & LF_{15} & LF_{16} \\
0 & 0 & LF_{23} & LF_{24} & LF_{25} & LF_{26} \\
0 & 0 & 0 & LF_{34} & LF_{35} & 0 \\
0 & 0 & 0 & 0 & LF_{45} & 0 \\
0 & 0 & 0 & 0 & 0 & LF_{56}
\end{bmatrix}_{5\times6}
\times
\begin{bmatrix} S_{in_1} \\ S_{in_2} \\ S_{in_3} \\ S_{in_4} \\ S_{in_5} \\ S_{in_6} \end{bmatrix}_{6\times1}
\tag{14}
$$

$$
A_i = \sum_{l=1}^{n_l} LF_{li}
\tag{15}
$$

The calculation of effective power injections is done as given in (16).

$$
\left.
\begin{aligned}
S_{eff,6} &= S_{in,6} \\
S_{eff,5} &= S_{in,5} \\
S_{eff,4} &= S_{in,4} + S_{eff,5} \\
S_{eff,3} &= S_{in,3} + S_{eff,4} + S_{eff,6} \\
S_{eff,2} &= S_{in,2} + S_{eff,3} \\
S_{eff,1} &= S_{in,1} + S_{eff,2}
\end{aligned}
\right\}
\tag{16}
$$

Equation (17) shows the final equation for the calculation of Index-1. The index is implemented for attaining the optimal position of one renewable energy source as DG. In Equation (15), $|A_i|$ is fully dependent on LF values of all the branches (or lines) connected between bus and the source bus (main station). The closeness of the i_{th} bus from the source bus can be observed in the LF_{li} as guided in Equation (10). If the i_{th} bus is not near, the number of lines between i_{th} bus and source bus is being plenty and the corresponding Z_{ai}, Z_{bi}, and Y_l parameters will account in the electrical loss component. Furthermore, if the node voltage is high, the value of LF_{li} will be small and vice-versa as observed in the derived equation. The equation of Index-1 is also accounted for the effective complex power supplied by the i_{th} bus. The Index-1 will be high only when both the terms are high in (17). Thus, the value of Index-1 represents its main contribution in the total electrical loss and hence, in finding the optimal siting of one RES.

$$
(Index - 1)_i = |A_i| \times |S_{in_{eff_i}}|
\tag{17}
$$

$S_{in_{eff_i}}$ is the effective injection of complex power, which is the sum of injected powers from other buses connected to *ith* bus as shown in Figure 4.

$$
(Index - 2)_{i+1} = |V_i|^4 - 4(P_{i+1}X_j - Q_{i+1}R_j)^2 - 4(P_{i+1}R_j - Q_{i+1}X_j)|V_i|^2
\tag{18}
$$

where j is branch number, V_i is sending bus voltage, P_{i+1} and Q_{i+1} are the Active Power (AP) and Reactive Power (RP) at the receiving end bus, respectively. R_j and X_j are the resistance and reactance between sending and receiving end bus, respectively.

2.2. Power Balance

The AP and RP balance expressions are shown in Equations (19) and (20).

$$P_{net_i} = P_{dg_i} - P_{dem_i} - V_i \sum_{j=1}^{N_{bus}} V_j Y_{i,j} cos(\delta_i - \delta_j - \theta_i + \theta_j) \tag{19}$$

$$Q_{net_i} = Q_{dg_i} - Q_{dem_i} - V_i \sum_{j=1}^{N_{bus}} V_j Y_{i,j} sin(\delta_i - \delta_j - \theta_i + \theta_j) \tag{20}$$

where $P_{net_i} = 0$ and $Q_{net_i} = 0$ are the net AP and RP at i-bus, respectively. P_{dg_i} and Q_{dg_i} represent DG AP and RP at i-bus, respectively. Active and reactive load demands are mentioned by P_{dem_i} and Q_{dem_i}, respectively. V_j is the bus voltage at j-bus, $Y_{i,j}$ is the branch admittance between i, j-buses, δ_i and δ_j represent the phase angles of i-bus and j-bus voltages, respectively. $(\theta_i - \theta_j)$ are the impedance angle of branch connected between i and j-buses.

2.3. Objective Function (OF)

In this research the OF is considered to be APL minimization in the EDS. The reliability indices are then evaluated by fixing the optimal location and size of the DGs. The OF of the problem is given in Equation (21).

2.3.1. Active Power Loss (APL)

The minimization of APL occurred in EDS is the OF considered. The primary aim of the OPF technique is to minimize the system APL as given as Equation (21).

$$min AP_{loss} = \sum_{i=1}^{N_{bus}} \sum_{j=1}^{N_{bus}} C_{1_{ij}}(P_{real_i} P_{real_j} + Q_{real_i} Q_{real_j}) + C_{2_{ij}}(Q_{real_i} P_{real_j} - P_{real_i} Q_{real_j}) \tag{21}$$

where P_{real_i}, P_{real_j}, Q_{real_i}, Q_{real_j} are the AP and RP at i and j-buses, respectively. N_{bus} = number of buses or nodes, $C_{1_{ij}}$ and $C_{2_{ij}}$ are defined as follows.

$$C_{1_{ij}} = \frac{R_{ij}}{V_i V_j} cos(\delta_i - \delta_j) \tag{22a}$$

$$C_{2_{ij}} = \frac{R_{ij}}{V_i V_j} sin(\delta_i - \delta_j) \tag{22b}$$

where V_i, δ_i and V_j, δ_j are the voltages and corresponding angles at ith and jth buses, respectively, R_{ij} = resistance of a branch between i and j-buses.

2.3.2. Reactive Power Loss (RPL)

The availability of RP ensures the AP transmission from source to load. Voltage stability margin or bus voltages are also dependent on this RP support. The RPL is obtained at different pf of DG using Equation (23).

$$RP_{loss} = \sum_{i=1}^{N_{bus}} Q_{gen_i} - \sum_{i=1}^{N_{bus}} Q_{dem_i} \tag{23}$$

where Q_{gen_i} and Q_{dem_i} are the RP generation and demand at the *ith* bus (including the slack bus), respectively. Q_{dem_i} = RP demand at the *ith* bus.

2.3.3. Reliability Indices

The indices are assessed for divergent DG reliability data by fixing the site and size of DGs. Furthermore, the reliability improvement of distribution network has been accomplished by integrating one DG and multiple DGs in the EDS. Several reliability indices exist to observe the system's reliability such as EENS, AENS, SAIDI, SAIFI, and ASAI which are also used in this study to analyze the reliability improvement. The indices of the network reliability are dependent function of failure rate (λ_p) and repair time (RT) as given in Equation (24) [20,23,52].

$$\text{Reliability Indices} = f(\lambda_p, RT) \tag{24}$$

2.4. Constraints

The OF minimization is a primary task to obtain the optimal results. OF minimization is subjected to design the constraints so that the requirements of the EDS must be satisfied with DG operation. Thus, the constraints are discussed in the succeeding subsection [53,54].

2.4.1. Equality Constraints

These constraints follow the Kirchhoff's current rule as the algebraic sum of powers in and powers out should be equal in an EDS [54,55]. Two of these constraints are described as follows.

$$\sum_{i=1}^{N_{bus}} AP_{gen_i} = \sum_{i=1}^{N_{bus}} AP_{dem_i} + AP_{loss} \tag{25}$$

where AP_{gen_i} = AP generated by the generation units at *ith* bus, AP_{dem_i} = AP demand at *ith* bus.

$$\sum_{i=1}^{N_{bus}} RP_{gen_i} = \sum_{i=1}^{N_{bus}} RP_{dem_i} + RP_{loss} \tag{26}$$

where RP_{gen_i} = RP generated by the generation units at *ith* bus, RP_{dem_i} = RP demand at *ith* bus.

2.4.2. Inequality Constraints

These constraints are associated with the limits applied to the system parameters for the operation of EDS. Some of these constraints are described as follows.

A.   Power flow

To maintain the line capacity within limits, these constraints ensure the apparent power to be within limits at the ends of a line [53,54].

$$AP_{a_{ij}} \leq AP_{a_{ij}}^{max} \tag{27}$$

where $AP_{a_{ij}}^{max}$ = the highest permissible apparent powers (AP_a) for lines i to j, $AP_{a_{ij}}$ = the actual AP_a transmitted from i to j.

B.   DG capacity

These limits ensure the non-reversal of power flow. The power from the substation is provided to the EDS must be greater than the DG power. Also, the DG has the minimum and maximum power generation boundaries [56].

$$\sum_{i=1}^{n_{DG}} AP_{DG_i} \leq 0.85 \times \sum_{i=1}^{n_{bus}} AP_{dem_i} + AP_{loss} \tag{28}$$

$$\sum_{i=1}^{n_{DG}} RP_{DG_i} \leq 0.85 \times \sum_{i=1}^{n_{bus}} RP_{dem_i} + RP_{loss} \tag{29}$$

$$AP_{DG_p}^{min} \leq AP_{DG_p} \leq AP_{DG_p}^{max} \tag{30}$$

$$RP_{DG_p}^{min} \leq RP_{DG_p} \leq RP_{DG_p}^{max} \tag{31}$$

where $p = 1, 2, \ldots, n_{DG}$, $AP_{DG_p}^{min}$ (set to zero) and $AP_{DG_p}^{max}$ (from Equation (28)) are the lower and upper AP outputs of DG unit p, respectively. $RP_{DG_p}^{min}$ (set to zero) and $RP_{DG_p}^{max}$ (from Equation (29)) are the lower and upper RP outputs of DG unit p, respectively. n_{DG} = number of DGs present in the distribution network.

C.   Bus voltage

The voltages at buses present in the EDS must be limited within minimum and maximum limits [57,58].

$$|V_{i_{minimum}}| \leq |V_i| \leq |V_{i_{maximum}}| \tag{32}$$

where $|V_{i_{minimum}}|$ and $|V_{i_{maximum}}|$ = the lower and upper boundaries of the bus voltage $|V_i|$ which are set to 95% and 105%, respectively.

D.   Branch current

It refers as thermal capacity of the EDS lines. The current in the distribution lines must be within limits and should exceed the maximum current as given in Equation (33) [20].

$$I_i \leq I_i^{max} \tag{33}$$

2.5. Constriction Factor-Based PSO (CF-PSO) Technique

PSO is a novel progression computational technique which is in the frame since 1995. The use of this method is seen in RP dispatch [59], generation scheduling [60], renewable source integrated power system [61], and cost analysis [62]. In basic PSO method, the candidate solution is improved iteratively under any given constraint. The PSO algorithm is shown in Figure 5. Due to the reduction in computational time and requirement of less memory, PSO has overtaken many algorithms including the Genetic algorithm (GA) as PSO is mutation free. It searches the optimized value globally with the help of several particles present in a swarm based on specific constraints. As all particles have its local and global best values because of its own and global positions. This method updates the particle position and velocity as described in Equations (34) and (35).

$$V_n^{p+1} = W' \times V_n^p + C_1' \times R_1' \times (Personal_{BEST_i} - X_n^p) + C_2' \times R_2' \times (Global_{BEST} - X_n^p) \tag{34}$$

$$X_n^{p+1} = X_n^p + C_f \times V_n^{p+1} \tag{35}$$

where V_n^{p+1} = n^{th} particle velocity at $(p+1)^{th}$ iteration, W' = particle inertial weight, V_n^p = n^{th} particle velocity at p^{th} iteration, C_1', C_2' = constants (0, 2.5), R_1', R_2' = numbers generated randomly (0, 1), $Personal_{BEST_i}$ = the nth particle's best position considering its own property, $Global_{BEST}$ = the nth particle's best position considering the whole population, X_n^{p+1}, X_n^p = n^{th} particle position at $(p+1)^{th}$ and p^{th} iterations, respectively. C_f = constriction factor (CF) assures efficient convergence [63,64].

Due to faster convergence to the global point, the basic PSO faces the difficulty of premature convergence. The particles have started oscillating around the optimal point without providing any type of restriction to the highest velocity of the particles available in swarm. Therefore, the optimal global solution is rare to obtain. The use of properly defined CF is briefly described for advance convergence of the PSO [65]. This can also be applied for the DG siting and DG size in the EDS. It reduces the computation time and requires little memory. Although this technique suffers from partial optimization, by altering its parameter during problem solving will produce an improved result [66–68]. To obtain the improved result, a constriction factor (CF) is used and thus, the method is known as CF-based PSO technique. The parameters set for the CF-PSO are as follows. The values of initial weight, final weight, C_1, C_2, R_1, R_2, and CF are considered to be 9×10^{-1}, 4×10^{-1}, 201×10^{-2}, 201×10^{-2}, 0 to 1, 0 to 1, and 729×10^{-3}, respectively. A flowchart is provided in Figure 5 to obtain the DG location, DG sizing and system reliability of the DGs in 33 bus EDS.

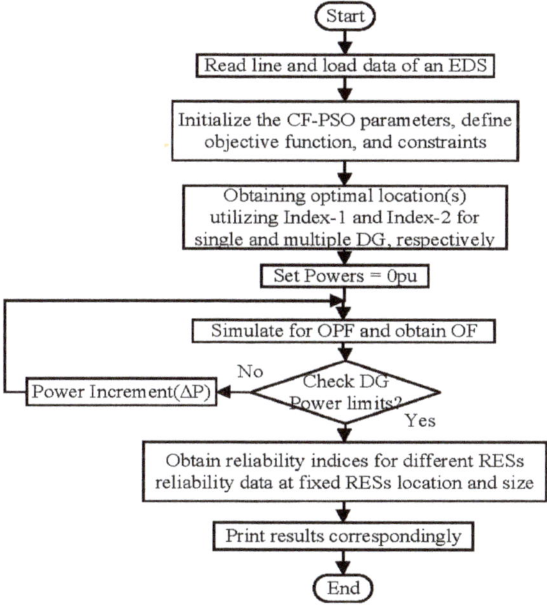

Figure 5. Algorithm implemented for the research work.

3. Reliability Assessment of Distribution System

The RA of an EDS is as important as contrasted to other components and parts of the EDS. The IEEE guide for EDS reliability is given by standard number 1366-2012 [69]. According to given standard, reliability of an EDS can be analyzed using some reliability indices. The reliability indices considered for EDS reliability assessment include EENS, AENS, SAIDI, SAIFI, and ASAI. These indices are mainly classified in two categories which are elaborated in Equations (39)–(46c).

3.1. Reliability Parameters at Load Point 'p'

The reliability indices are the function of reliability parameters mentioned in Equations (36)–(38). The reliability parameters have been calculated at load point 'p' as follows.

$$\text{Failure rate (average)}; \lambda_p = \sum_{k \in n} num_k \times F_k \quad \text{failure per year} \tag{36}$$

$$\text{Outage duration (annual)}; U_p = \sum_{k=n} F_k D_{pk} \quad \text{hour per year} \tag{37}$$

$$\text{Outage duration (average)}; D_p = \frac{U_p}{\lambda_p} \quad \text{hour} \tag{38}$$

where F_k = failure rate (average) of the *kth* element, n = number of elements in the EDS, num_k = number of *kth* elements in the EDS, D_{pk} = duration of failure at p_{th} load point due to k_{th} failed element. The calculation equations of λ_p and U_p are given in Appendix A.2.

3.2. System-Based Indices

These indices are further categorized in load-oriented indices and customer oriented indices as given in Equations (39)–(40) and Equations (41)–(46c), respectively.

3.2.1. Load-Oriented Indices

Load-Oriented Indices have been calculated at load point 'p', as mentioned in Equations (39)–(40).

$$EENS_p = P_p U_p \quad \text{megawatt hour per year} \tag{39}$$

$$AENS_p = \frac{\sum_{p=1}^{n_p} EENS_p}{\sum_{p=1}^{n_p} N_p} \quad \text{megawatt hour per customer per year} \tag{40}$$

where P_p = demand/load (average) of the *pth* load point, $EENS_p$ = expected ENS at *Pth* load or customer point. where n_p = total load points, N_p = number of customers at *Pth* load point.

3.2.2. Customer Oriented Indices

These indices have allowed to enhance the EDS's reliability related to the improvement of customer or load services. The two of the indices namely ECOST and IEAR are related to the cost reliability and thus, termed as reliability worth of the system.

$$ECOST_p(= LOEE_p) = P_p \sum_{k=n} f(D_{pk}) F_k \quad \text{k\$ per year} \tag{41}$$

$$IEAR_p = \frac{ECOST_p}{EENS_p} \quad \text{\$ per kilowatt hour} \tag{42}$$

where $ECOST_p$ = expected interrupted cost at *Pth* load point, $IEAR_p$ = interrupted energy assessment rate at *Pth* load point, $LOEE_p$ = loss of expected energy, $f(D_{pk})$ = system composite customer damage function (\$ per kilowatt) as provided in Table A3 of Appendix A.

$$SAIFI = \frac{\sum_{p=1}^{n_p} \lambda_p N_p}{\sum_{p=1}^{n_p} N_p} \quad \text{failure per customer per year} \tag{43}$$

$$SAIDI = \frac{\sum_{p=1}^{n_p} U_p N_p}{\sum_{p=1}^{n_p} N_p} \quad \text{hour per customer per year} \tag{44}$$

$$\text{CAIDI}\left(= \frac{SAIDI}{SAIFI}\right) = \frac{\sum_{p=1}^{n_p} U_p N_p}{\sum_{p=1}^{n_p} \lambda_p N_p} \quad \text{hour per customer per interruption} \tag{45}$$

where CAIDI = Customer Average Interruption Duration Index.

$$ASAI = \frac{8760 \sum_{p=1}^{n_p} N_p - \sum_{p=1}^{n_p} U_p N_p}{8760 \sum_{p=1}^{n_p} N_p} \quad \text{per unit} \tag{46a}$$

Also, *ASAI* can be derived as follows.

$$ASAI = 1 - \frac{SAIDI}{8760} \tag{46b}$$

$$ASUI = 1 - ASAI \quad \text{per unit} \tag{46c}$$

where *ASUI* = Average Service Unavailability Index.

4. Modeling of WTG, SPV, and BSS

The reliability assessment of the IEEE 33 bus EDS is accomplished, considering the optimal siting(s) and sizing(s) of SPV, WTG, and BSS. In this regard, a brief modeling and specifications of these RESs are illustrated in Sections 4.1–4.3.

4.1. Wind Turbine Generator

The V162-5.6MW(IECS based on IEC IIB), a WTG, manufactured by General Electric Company is considered for its output power rating. The specifications of the WTG considered in this study are provided in Table 4. The mechanical power of WTG (P_{mech}) is a function of generator rotor speed and wind speed as formulated in Equations (47)–(51) [70].

$$P_{mech}(v_{wind}, \omega_{rotor}) = \frac{1}{2} \times \rho \times v_{wind}^3 \times C_p(\lambda, \theta) \tag{47}$$

$$\lambda = \frac{\omega_{rotor} \times GR \times R_{rotor}}{v_{wind}} \tag{48}$$

$$C_p(v_{wind}, \omega_{rotor}, \theta) = C_1(C_2 \frac{1}{\alpha} - C_3\theta - C_4\theta^x - C_5) \times exp(\frac{-C_6}{\alpha}) \tag{49}$$

$$\frac{1}{\alpha} = \frac{1}{(\lambda + 0.08\theta)} - \frac{0.035}{1 + \theta^3} \tag{50}$$

where ρ = air density, A_s = area swept by the turbine rotor blades, $Vwind$ = speed of the wind, C_p is the non-linear function of the tip speed ratio (λ) and pitch angle (θ), ω_{rotor} = generator rotor speed, GR = gear ratio, R_{rotor} = rotor radius at the turbine blades, C_1 to C_6 and x are constants and computed in [70].

$$P_{WTG} = \begin{cases} 0; & 0 \leq V \leq V_{cin} \text{ or } V \geq V_{cout} \\ P_{WTG,rated} \times \left(\frac{V - V_{cin}}{V_{rated} - V_{cin}}\right); & V_{cin} \leq V \leq V_{rated} \\ P_{WTG,rated}; & V_{rated} \leq V \leq V_{cout} \end{cases} \tag{51}$$

where P_{WTG} = output WTG power, $P_{WTG,rated}$ = rated output WTG power, V_{rated} = rated wind speed.

Table 4. Wind Turbine (V162-5.6 MW) specifications.

Parameter	Rating (Unit)
Rated output power	5.6 MW
Cut-in Speed (V_{cin})	3 m/s
Cut-out Speed (V_{cout})	25 m/s
Temperature	$-20\ ^\circ$C to 45 $^\circ$C
Diameter	162 m
Swept Area	20612 m^2
Frequency	50/60 Hz
Hub Height	119 m, 125 m, 148 m, 149 m, and 166 m

4.2. Solar Photovoltaic

The SPR-P5-545-UPP, a Solar PV Module, manufactured by Sunpower Company is considered for its output power rating. The specifications of the SPV module considered in this study are provided in Table 5. The SPV module is developed by implementing several cells. The power output for the SPV module can be derived as described by Equations (52)–(65) [71–73].

$$P_{SPV(AC)}(t) = P_{SPV(out)}(t) \times \eta_{inverter} \tag{52}$$

$$P_{SPV(out)}(t) = FF_A(t) \times I_{short}(t) \times V_{open}(t) \tag{53}$$

$$I_{short}(t) = \frac{SR}{1000}[I_{short,STC} + C_I(T_{SM}(t) - 25)] \tag{54}$$

$$V_{open}(t) = V_{open,STC} + C_V[T_{SM}(t) - 25] \tag{55}$$

$$FF_A(t) = FF_i(t) \times [1 - R_s(t)] \tag{56}$$

$$FF_i(t) = \frac{V_{open,0}(t) - ln[V_{open,0}(t) + 0.72]}{V_{open,0}(t) + 1} \tag{57}$$

$$V_{open,0}(t) = V_{open}(t) \times \frac{Q}{N \times K[T_{SM}(t) + 273.15]} \tag{58}$$

$$R_s(t) = R_s \times \frac{I_{short}(t)}{V_{open}(t)} \tag{59}$$

$$T_{SM}(t) = T_{amb}(t) + SR\frac{(T_{nom} - 20)}{0.8} \tag{60}$$

where t = time instant, $P_{SPV(AC)}(t)$ = AC output power, $P_{SPV(out)}(t)$ = maximum output power, $\eta_{inverter}$ = inverter efficiency, $FF_A(t)$ = fill factor actual, $I_{short}(t)$ = short circuit current under operating conditions, $V_{open}(t)$ = open circuit voltage under operating conditions, SR = solar radiation intensity (W/m^2), STC = standard test conditions, C_I = temperature coefficient for current (A/$^\circ$C), $T_{SM}(t)$ = solar module temperature ($^\circ$C), C_V = temperature coefficient for voltage (V/$^\circ$C), $FF_i(t)$ = fill factor ideal, $R_s(t)$ = normalized series resistance of solar module, $V_{open,0}(t)$ = normalized open circuit voltage, Q = an electron charge, $N(\approx 1)$ = ideality factor, K = Boltzmann's constant, R_s = series resistance of SPV module, $T_{amb}(t)$ = ambient temperature of SPV module, T_{nom} = nominal operating cell temperature.

The R_s is being evaluated as follows [74].

$$R_s = R_{s,STC} = r_{s,STC} \times \frac{V_{open,STC}}{I_{short,STC}} \tag{61}$$

$$R_{s,STC} = 1 - \frac{FF_{A,STC}}{FF_{i,STC}} \tag{62}$$

$$FF_{A,STC} = \frac{V_{mpp,STC} \times I_{mpp,STC}}{V_{open,STC}} \times I_{short,STC} \tag{63}$$

$$FF_{i,STC} = \frac{V_{open,0,STC} - ln[V_{open,0,STC} + 0.72]}{V_{open,0,STC}} + 1 \tag{64}$$

$$V_{open,0,STC} = V_{open,STC} \times \frac{Q}{N \times K[T_{SM,STC}}(t) + 273.15] \tag{65}$$

where $r_{s,STC}$ = normalized series resistance under STC, and all other parameters in Equations (61)–(65) are evaluated at STC.

Table 5. Bifacial Solar Panel (SPR-P5-545-UPP) specifications.

Parameter	Rating (Unit)	Parameter	Rating (Unit)
nominal power	545 W	Maximum Series Fuse	25 A
Tolerance of Power	±3/0%	Temperature	−40–85 °C
Efficiency	21.1%	Power Temperature Coefficient	−0.34%/°C
Rated voltage	46.1 V	Voltage Temperature Coefficient	−0.28%/°C
Rated current	11.84 A	Current Temperature Coefficient	0.06%/°C
Open circuit voltage	55.8 V	Weight	31.5 kg
Short circuit current	12.62 A	Solar Cells	Mono-crystalline
Maximum System Voltage (IEC)	1500 V	L × B × H mm^3	2362 × 1092 × 35

4.3. Battery Storage System

The BESS 3000, a Lithium-Ion Battery System, manufactured by Freqcon Company is considered for its output power rating. The specifications of the BSS considered in this study are provided in Table 6. The flowchart of BSS dispatch modeling is illustrated in Figure 6. Mathematical modeling is further defined. The BSS dispatch strategy starts functioning by monitoring the peak load hours. If the peak load is greater/less than the capacity of WTG and SPV, the BSS discharges/charges to support the distribution system; otherwise the BSS operates as a neutral device. Also, the BSS SOC will decide the charging or neutral operation during the off-peak hours. The battery model is derived with the help of Equations (66)–(68) assuming that no battery is self-discharging [75].

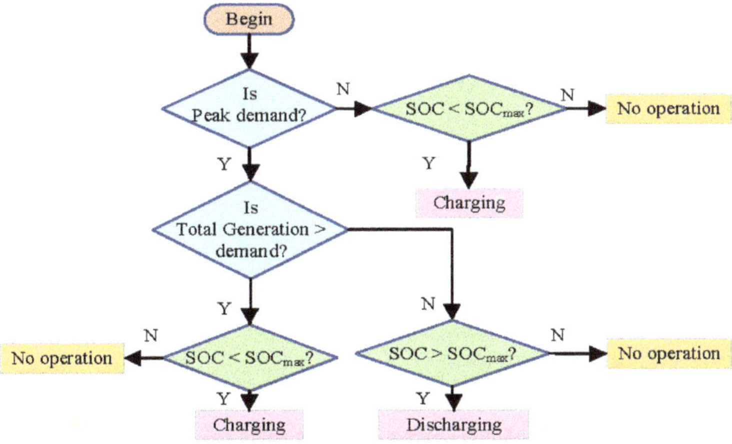

Figure 6. General BSS dispatch strategy opted for battery operation.

$$P_{Battery(DC)}(t) = \frac{E_{Battery}(t) - E_{Battery}(t - \Delta t)}{\Delta t} \tag{66}$$

$$P_{Battery(AC)}(t) = \begin{cases} \dfrac{P_{Battery(DC)}(t)}{\eta_{Battery}}; P_{Battery(DC)}(t) > 0 \\ P_{Battery(DC)}(t) \times \eta_{Battery} \times PF_{Inverter}; Otherwise \end{cases} \tag{67}$$

$$SOC(t) = SOC(t - \Delta t) + \eta_{Charging} \times \frac{P_{Battery(DC)}(t)}{C_{Battery} \times V} \times \Delta t \tag{68}$$

where $P_{Battery(DC)}(t)$ = DC charging/discharging power of the battery in Δt interval (W), $E_{Battery}(t)$ = energy of battery (Wh), $P_{Battery(AC)}(t)$ = AC power discharged/charged state of battery, $\eta_{Battery}$ = efficiency of the battery, $PF_{Inverter}$ = inverter power factor, $SOC(t)$ = state of charge of the battery, $\eta_{Charging}$ = charging efficiency of the battery, $C_{Battery}$ = battery capacity (Ah), V = nominal voltage of the battery (V).

Table 6. Battery Storage System (BESS 3000) specifications.

Parameter	Rating
Rated output power	3000 kW
Storage Capacity	1000 kWh
Rated output current	2795 A
Rated output AC voltage	620 V
Power factor	0.95 Cap . . . 0.95 Ind
Total harmonic distortion	<3%
Efficiency	>98%
Type	Lithium-ion
IGBT Switching Frequency (Converter)	2–4 kHz

5. Results and Discussion

The DG location, DG size, and EDS reliability are obtained and analyzed. The 33 bus EDS (Figure A1 of Appendix A.1) is considered in this study. The branch and load data for this EDS are adopted from [76]. It contains 33 buses and 32-branches with a total of 3.715 MW and 2.3 MVAr AP and RP loads, respectively. This EDS operates at 12.66 kV, 100 MVA base values. The APL and RPL for without DG case are obtained as 0.211009 MW and 0.143056 MVAr, respectively. The objective function (OF) minimization is done by implementing the CF-based PSO, as explained in Section 2. The following steps are followed to obtain the results.

Step 1: Optimal siting(s) and sizing(s) of WTG, SPV, and BSS are evaluated considering electrical loss minimization (ELM). The technical ratings of WTG, SPV, and BSS have been illustrated in Tables 4–6, respectively. The BSS is assumed to be fully charged and produces its rated output power.

Step 2: APL, RPL, and bus voltages are obtained by integrating WTG, WTG+SPV, and WTG+SPV+BSS (referred as Case 1, Case 2, and Case 3, respectively) in the EDS to analyze the results obtained in Step 1.

Step 3: Reliability indices are estimated for EDS considering two different WTG and SPV reliability data, including λ_p and RT (for Scenario 1 to Scenario 6).

Step 4: Furthermore, the reliability improvement is analyzed by adding BSS (considering 100% reliable) to the EDS in the presence of WTG and SPV. All related reliability data used are mentioned in Table A2 of the Appendix A.

5.1. DG Location and DG Rating

The bus number is obtained to allocate WTG, SPV, and BSS. The two indexes are implemented to obtain the locations as described in Section 2.1. It is observed from the analysis that Index-1 provides the location suitable for a single DG. This index examines the effective apparent power injection to the buses. The Load Factor value of lth line depends on whether the lth is in the path of the ith bus to the source node or not. The multiplication of Load Factor and injected apparent power provides the

APL and RPL of *lth* line due to the *ith* bus AP_a injection. Thus, the maximum value of Index-1 at *ith* bus indicates the candidate bus to place one DG. The first six values of this index for respective buses are provided in Table 3.

The optimal siting for several DGs are found by implementing the Index-2. This index provides the hierarchy of weak buses in the EDS. It shows the sensitivity of the bus towards voltage collapse. The value of Index-2 must be greater than or equals to zero for ensuring the stable operation of the distribution system. The minimum value of this index depicts more sensitivity to the voltage collapse and thus, referred to as the weak bus. The optimal location(s) and size(s) of WTG, SPV, and BSS for three cases are obtained and reproduced in Table 7. The DGs are accommodated according to the locations obtained from the indexes, as mentioned in Section 2.1. The DG size and minimum APL are then evaluated, implementing CF-PSO as described in Section 2.

Table 7. DG location and DG size obtained.

	Case 1	Case 2	Case 3
Location (bus no.)	6	30, 13	30, 13, 24
size@upf (MW)	2.564	1.148, 0.843	1.048, 0.801, 1.105

5.2. APL, RPL, and Bus Voltages

The accommodation of DG at an optimal location with optimal size reflects in VP improvement and minimization of APL and RPL. Most of the research has concentrated on APL minimization because of the dominance of I^2R losses in the EDS. In contrast, the RPL minimization for overall voltage improvement of all the 32 buses of the EDS has also been observed. The estimation of APL, RPL, and VP is considered before analyzing the system's reliability. This is done to analyze the system's reliability with the optimal DG size, DG location, minimum power loss, and better VP.

Several kinds of research are performed to obtain APL minimization, which is cited in Table 2 for single and multiple DGs, respectively. It is observed from Table 2 that the authors of the mentioned literature have not dealt with the cases considering various DGs combination. Therefore, it is vital to observe that the results obtained considering several types of DGs are compared with the conventional DGs [77]. The output results obtained for the APL are tabulated in Table 8. This table shows that the APL value is better for Case 1, and comparable for Case 2 and Case 3. The slight variations in APL for Case 2 and Case 3 are observed because the authors have considered the pf of WTG only. Furthermore, the bus voltage profile with one DG and multiple DGs is drawn in Figure 7. The voltages at all the buses vary according to the real power loss and reactive power loss in the distribution system. Therefore, the system requires real power support for active power loss minimization, which improves the bus voltages by compensating the I^2R losses. Furthermore, it is concluded from Table 8 that the active power loss minimization is not reduced significantly for Case 3 as compared to Case 2. Thus, there is a marginal improvement in bus voltage profile for Case 3 as compared to Case 2, which is depicted in Figure 7. Also, the improvement in bus voltages is observed when multiple DGs are placed. This VP is further improved at 0.85 and 0.82 pfs. This is because of increment in reactive power support at system buses. It is the point of interest to know about the two voltage peaks when the system is operated with single DG. The two voltage peaks appear at bus number 7 and 26 because these buses are directly connected to bus 6, at which single DG is placed optimally. Also, the size of the single DG is greater than the sum of the size of two DG and slightly lesser than the sum of the size of three DG, as obtained in Table 8. A comparison between present work and the best available method is made for ELM. Simultaneously, from Table 9 it can be inferred that the minimum bus voltage is improved and RPL is minimized with the implementation of multiple DGs at different pfs as illustrated. The graphical representation of APL and RPL for without DG, one DG, and multiple DGs are represented in Figure 8.

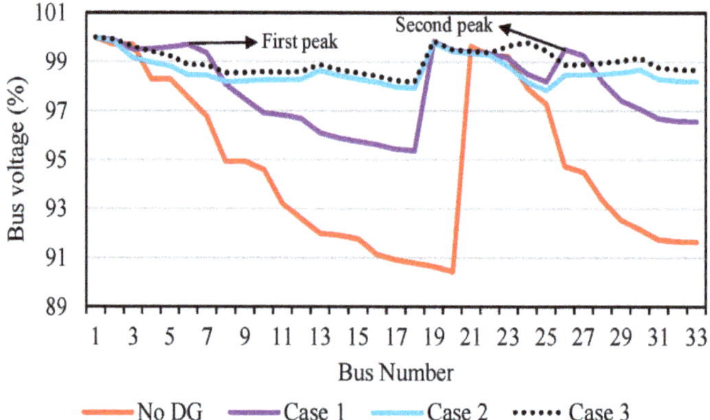

Figure 7. Voltage profile for 33 bus system (WTG at 0.9 pf).

Table 8. APL (MW) obtained considering WTG power factor.

	pf	No DG	Case 1	Case 2	Case 3
	Unity		0.11104	0.0727	0.05148
Present work	0.85	0.21101	0.06831	0.04539	0.02795
	0.82		0.06831	0.0444	0.02702
APL obtained considering power factor of all Conventional DGs					
	pf	No DG	Single DG	Two DG	Three DG
	Unity		0.11107	0.087172	0.072787
EA [77]	0.85	0.211	0.068170	0.03119	0.01552
	0.82		0.067870	0.03041	0.01514

Table 9. Minimum voltage, DG location, and RPL (MVAr) obtained.

	Minimum Voltage (%)				RPL			
pf	No DG	Case 1	Case 2	Case 3	No DG	Case 1	Case 2	Case 3
Unity		94.26	96.88	96.86		0.08168	0.05121	0.03848
0.85	90.44	95.74	98.12	98.15	0.14306	0.05504	0.03257	0.02185
0.82		96.0	98.20	98.22		0.05504	0.03195	0.02119

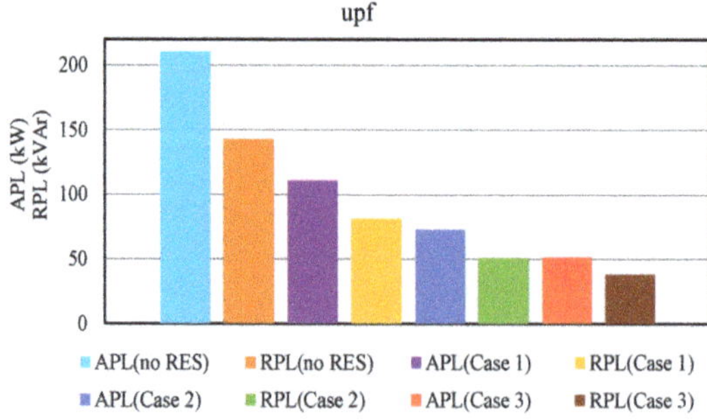

Figure 8. Active and Reactive power losses (WTG at UPF).

5.3. Reliability Assessment

The indices obtained show the improvement in the system's reliability. The indices are calculated for two different reliability data of DGs. It is observed that the indices are dependent on two reliability data, namely

λ_p and RT of the system's elements. The present work has considered different reliability data for DG only. The best reliability improvement is observed for 0.2 of λ_p and 12 h of RT. A detailed description of the DG reliability data effect on indices is given as per the following Scenarios.

- Scenario 1: 0.2 f/yr and 12 h (as provided in Table A2 of Appendix A.2)
- Scenario 2: 0.4 f/yr and 12 h
- Scenario 3: 0.6 f/yr and 12 h
- Scenario 4: 0.2 f/yr and 24 h
- Scenario 5: 0.2 f/yr and 48 h
- Scenario 6: No failure

Scenarios 1, 2, and 3 are considered by fixing the RT and varying λ_p. The appropriate case from the first three cases is then considered for variable RT to extract the best case from the top Five Scenarios. The values of these reliability data are fed manually in the optimization technique to get the values of indices for the system's reliability improvement. Furthermore, the following key assumptions are considered to assess the reliability of the EDS.

- Circuit breakers, distribution lines, and potential transformers are available throughout with 100% reliability.
- The λ_p and RT of DG, Buses, feeders, and substations are given in Table A2.
- RT for each distribution branch = 10 h.

5.3.1. Effect on Load-Oriented Indices

The EENS and AENS are obtained and tabulated in Tables 10 and 11 for all cases as illustrated in Figure 9a–d, respectively. It is important to note that the EENS and AENS decrease with the number of DGs, and these are also decreased with decreasing values of λ_p and RT. As the increasing number of DGs are integrated into an EDS, the supplied energy is improved in the EDS, and thus, the indices related to the energy not supplied are reduced. This reduction is more while integrating the DGs with lesser λ_p and RT values. The reducing EENS and AENS are desirable, and thus, the EDS reliability enhances with the integration of DGs with appropriate reliability data values.

Table 10. EENS (MWh per year) evaluated for different Scenarios.

	Scenario 1	Scenario 2	Scenario 3	Scenario 4	Scenario 5	Scenario 6
No DG	82.763	82.763	82.763	82.763	82.763	82.763
Case 1	65.533	68.465	73.397	68.465	78.329	58.601
Case 2	31.817	29.249	31.817	29.249	34.385	24.113
Case 3	30.135	27.567	30.135	27.567	32.703	22.431

Table 11. AENS (MWh per customer per year) evaluated for different Scenarios.

	Scenario 1	Scenario 2	Scenario 3	Scenario 4	Scenario 5	Scenario 6
No DG	0.0255	0.0255	0.0255	0.0255	0.0255	0.0255
Case 1	0.0196	0.0203	0.0226	0.0211	0.0241	0.0181
Case 2	0.0082	0.009	0.0098	0.009	0.0106	0.0074
Case 3	0.0077	0.0085	0.0093	0.0085	0.0101	0.0069

5.3.2. Effect on System-Oriented Indices

The SAIDI, and SAIFI are obtained and tabulated in Tables 12 and 13 for all cases considering all scenarios as shown in Figure 10a–d, respectively. The important point to be noted here that the SAIDI, and SAIFI, decreases with the increasing number of DGs; SAIDI is also decreased with the decreasing values of λ_p and RT. It is worthy to note that the SAIFI is not affected by the RT of the DG. It is because this index is independent of RT as shown in Equation (43). As the increasing number of DGs are incorporated into the EDS, the duration of the interruption and the number of interruptions occurred are reduced in the system. Thus, the SAIDI, and SAIFI are reduced. Customer Average Interruption Duration Index can be determined using the ratio of SAIDI, and SAIFI as shown in Equation (45). The reduction in the value of indices is more while integrating the DGs with lesser λ_p and RT values. Moreover, reducing SAIDI and SAIFI are desirable for EDS reliability enhancement.

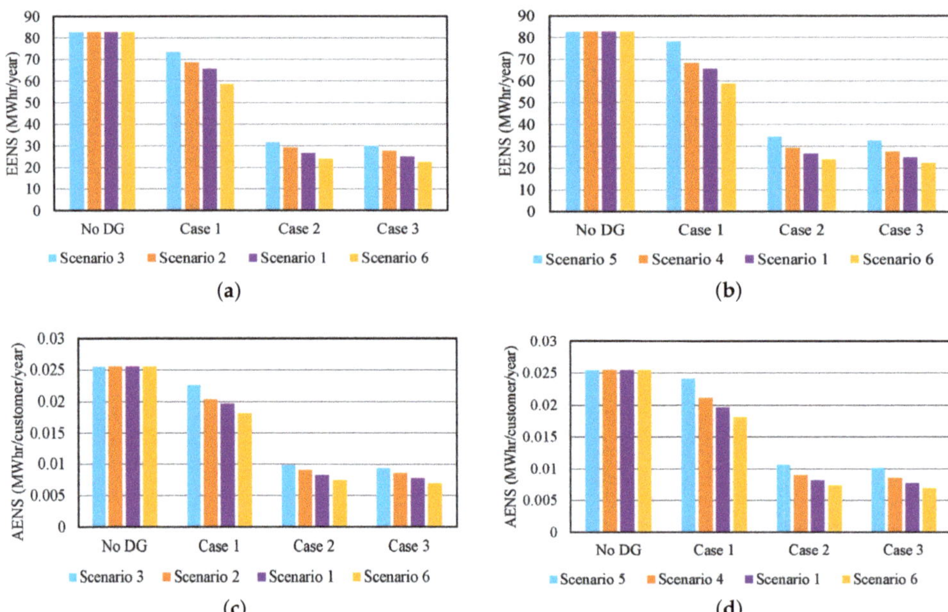

Figure 9. Load-oriented indices for different DG reliability data (**a**) EENS at different λ_p, (**b**) EENS at different RT, (**c**) AENS at different λ_p, and (**d**) AENS at different RT.

Table 12. SAIDI (hour per customer per year) evaluated for different Scenarios.

	Scenario 1	Scenario 2	Scenario 3	Scenario 4	Scenario 5	Scenario 6
No DG	24.012	24.012	24.012	24.012	24.012	24.012
Case 1	18.764	20.085	21.406	20.085	22.728	17.442
Case 2	7.388	8.053	8.719	8.053	9.385	6.722
Case 3	7.201	7.866	8.532	7.866	9.198	6.535

Table 13. SAIFI (failure per customer per year) evaluated for different Scenarios.

	Scenario 1	Scenario 2	Scenario 3	Scenario 4	Scenario 5	Scenario 6
No DG	3.179	3.179	3.179	3.179	3.179	3.179
Case 1	2.109	2.219	2.329	2.109	2.109	1.999
Case 2	0.915	0.970	1.026	0.915	0.915	0.859
Case 3	0.842	0.897	0.953	0.842	0.842	0.786

The ASAI is determined and tabulated in Table 14 for all cases considering six scenarios as illustrated in Figure 11a,b. The electrical power service availability for all loads increases with the integration of multiple DGs. This index is further increased when DGs have a lower λ_p and RT values. The increment in ASAI increases leads to the decrement in ASUI as given in Equations (46a) and (46c) which is desirable for the EDS reliability improvement.

Table 14. ASAI (pu) evaluated for different Scenarios.

	Scenario 1	Scenario 2	Scenario 3	Scenario 4	Scenario 5	Scenario 6
No DG	0.99726	0.99726	0.99726	0.99726	0.99726	0.99726
Case 1	0.99786	0.99771	0.99756	0.99771	0.99741	0.99801
Case 2	0.99916	0.99908	0.99900	0.99908	0.99893	0.99923
Case 3	0.99918	0.99910	0.99903	0.99910	0.99895	0.99925

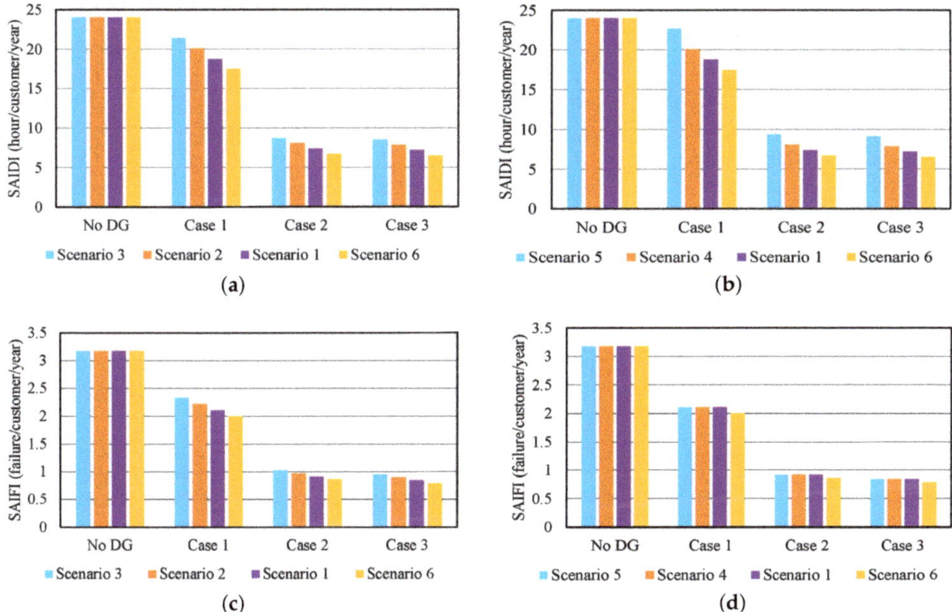

Figure 10. System-oriented indices for different DG reliability data (**a**) SAIDI at different λ_p, (**b**) SAIDI at different RT, (**c**) SAIFI at different λ_p, and (**d**) SAIFI at different RT.

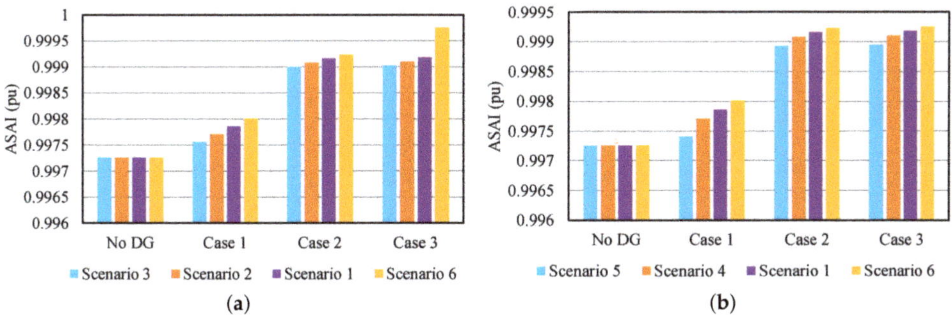

Figure 11. System-oriented indices for different DG reliability data (**a**) ASAI at different λ_p, (**b**) ASAI at different RT.

6. Conclusions and Scope for Future Work

The optimal location(s) and size(s) of distributed generations derived for three cases ensure an enhancement in the electrical loss minimization and also, improves the bus voltage profile when compared to a system without distributed generations. While analyzing all the cases, it is observed that Case 3 has provided the best result. It was observed that the APL value is reduced at UPF by 0.00003 MW, 0.014472 MW, and 0.021307 MW for Case 1, Case 2, and Case 3, respectively when compared to [77]. The minimum value of bus voltage is improved by 6.42% for Case 3 at UPF when compared to No DG Case. The bus voltage profile was further improved at 0.85 and 0.82 power factors by 7.71% and 7.78%, respectively. After achieving the acceptable results, the reliability analysis is performed for the distributed generation integrated distribution system. It is concluded that the combination of Case 3 with Scenario 6 provides the best system's reliability results. However, Scenario 6 being an ideal one, Case 3 with Scenario 1 is considered to yield better results for the improvement in the system's reliability. This research analysis can be extended in the future for a greater number of standard distribution systems considering the following aspects.

- Reliability assessment of larger EDS.
- Inclusion of reliability data of subsystems.
- System reconfiguration.
- Considering CO_2 emission.
- Economical aspects related to the system's reliability, including net present value, Levelized cost of energy, and many other aspects.
- Security.

The future aspects can be dealt with large radial distribution systems, including IEEE 69 and 118 bus systems. Inclusion of reliability data in terms of electrical, mechanical, and structural subsystems which would be favorable in obtaining accurate reliability of the system. Distribution system reliability improvement can also be achieved by adjusting the number of branches which is termed as a system reconfiguration technique. Furthermore, the dependency of reliability analysis on CO_2 emission, economics, system protection, and system security can also be considered for further research works.

Author Contributions: Conceptualization, Methodology, and Draft writing: S.K.; Software Implementation and Review: K.S.; Mathematical formulations: A.S.S.V.; Investigation, Results validation, and Data curation: R.M.E.; Supervision, Results validation, and Review: R.K.S.; Review and Editing: N.D. All authors have read and agreed to the published version of the manuscript.

Funding: This research received no external funding.

Acknowledgments: The authors wholeheartedly thank the Department of Electrical Engineering, Indian Institute of Technology (Banaras Hindu University), Varanasi for providing the laboratory facilities and technical support to accomplish the research work promptly. The first author expresses his gratitude towards Govind Ballabh Pant Institute of Engineering and Technology, Pauri Garhwal, Uttarakhand, India for allowing him to pursue PhD from IIT(BHU) Varanasi, Uttar Pradesh, India.

Conflicts of Interest: The authors declare no conflict of interest.

Abbreviations

This work has used the following abbreviations:

AENS	Average Energy Not Supplied (MWh per customer per year)
APL	Active Power Loss (MW)
ASAI	Average Service Availability Index (pu)
BSS	Battery Storage System
CF-PSO	Constriction Factor-based Particle Swarm Optimization
DG	Distributed Generation
DGen	Diesel Generator
EDS	Electrical Distribution System
EENS	Expected Energy Not Supplied (MWh per year)
EIR	Energy index of reliability (pu)
ELM	Electrical Loss Minimization
ENS	Energy Not Supplied (MWh)
GA	Genetic Algorithm
GE	General Electric
IEEE	Institution of Electrical and Electronics Engineers
LOLE	Loss of Load Expectation (hour)
LOLP	Loss of Load Probability (pu)
MOGA	Multi-objective Genetic Algorithm
MW	Megawatt
OPF	Optimal Power Flow
pf	Power Factor (pu)
RA	Reliability Assessment
RDS	Radial Distribution System

RES	Renewable Energy Source
RPL	Reactive Power Loss (MW)
SAIDI	System Average Interruption Duration Interruption (hour per customer per year)
SAIFI	System Average Interruption Frequency Interruption (failure per customer per year)
SPR	Surface Plasmon Resonance
SPV	Solar photovoltaic
WTG	Wind Turbine Generator
VP	Voltage Profile

Appendix A

Appendix A.1

The IEEE 33 bus EDS as shown in Figure A1 has been selected for analysis. As mentioned in Section 5, it is inferred that the 33-bus distribution system is analyzed for obtaining the locations and sizes. The branch and load data have been taken from [76].

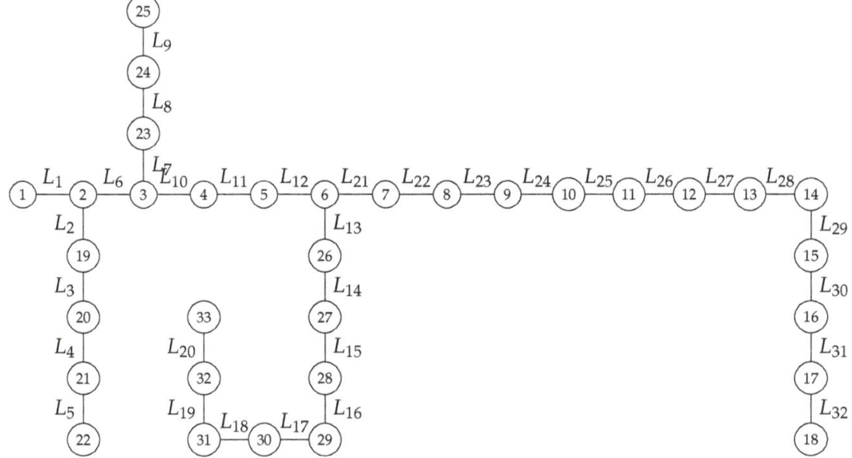

Figure A1. The IEEE 33 bus EDS.

Appendix A.2

Table A1. Load distribution for 33 bus [52].

Bus Number (or Load Point)	Number of Loads	Type of Load Mixed	Same Type of Loads		
			C	I	R
2–5	148	Industrial (I)			
6–9	10	Commercial (C)	"	"	"
11, 12	132	"	"	"	"
13–15	110	Residential (R)	"	"	"
16	2	"	"	"	"
17–20	118	"	"	"	"
21–26	126	"	"	"	"
27–31	108	"	"	"	"
32, 33	58	"	"	"	"

Table A2. Reliability data for 33 bus [52].

Bus, Feeder, etc.	Reliability Data for All Loads, Feeders, etc.	
	λ_p **(f/yr)**	**RT (h)**
Load@4	0.321	11.04
Load@(5, 7–12, 29, 30, 14, 16, 18–22, 25–28)	0.301	11.44
Load@(13, 15)	0.314	11.17
Load@(17, 23, 24)	0.208	1.75
Load@(31–33)	0.327	10.96
substation	0.1	5
feeder (2, 3, 6)	0.2	3
DG	0.2	12

From Table A2, average failure rate, and unavailability can be determined as given in Equations (A1) and (A2). The system reliability indices are thus evaluated for all six scenarios using the values of λ_p and U_p.

$$\text{average failure rate, } \lambda_p = (\text{no. of loads} \times \text{failure rates}) + (\text{no. of substations} \times \text{failure rate}) \\ + (\text{no. of feeders} \times \text{failure rate}) + (\text{no. of DG} \times \text{failure rate}) \tag{A1}$$

$$\text{outage duration or unavailability, } U_p = (\text{no. of loads} \times \text{failure rates} \times RT) + (\text{no. of substations} \times \text{failure rate} \times RT) \\ + (\text{no. of feeders} \times \text{failure rate} \times RT) + (\text{no. of DG} \times \text{failure rate} \times RT) \tag{A2}$$

$$\text{EENS} = \sum[(\text{Demand or load at } P^{th} \text{ load point}) \times (\text{Annual outage duration at } P^{th} \text{ load point})] \tag{A3}$$

$$\text{AENS} = \frac{\sum(\text{EENS at } P^{th} \text{ load point})}{\text{Total number of customers at all load points}} \tag{A4}$$

$$\text{SAIDI} = \frac{\sum[(\text{annual outage duration at } P^{th} \text{ load point}) \times (\text{Number of customers at } P^{th} \text{ load point})]}{\text{Total number of customers at all load points}} \tag{A5}$$

$$\text{SAIFI} = \frac{\sum[(\text{average failure rate at } P^{th} \text{ load point}) \times (\text{Number of customers at } P^{th} \text{ load point})]}{\text{Total number of customers at all load points}} \tag{A6}$$

$$\text{ASAI} = 1 - \frac{SAIDI}{8760} \tag{A7}$$

Table A3. Cost per kilowatt for Reliability worth estimation [78].

Type of Load	Interruption Duration (minutes)	Cost ($/kW)
Commercial	1	0.38
	20	2.97
	60	8.55
	240	31.32
	480	83.01
Industrial	1	1.63
	20	3.87
	60	9.09
	240	25.16
	480	55.81
Residential	1	0
	20	0.09
	60	0.48
	240	4.91
	480	15.69

References

1. Raju, K.; Madurai Elavarasan, R.; Mihet-Popa, L. An Assessment of Onshore and Offshore Wind Energy Potential in India Using Moth Flame Optimization. *Energies* **2020**, *13*, 3063.
2. Elavarasan, R.M. The motivation for renewable energy and its comparison with other energy sources: A review. *Eur. J. Sustain. Dev. Res.* **2019**, *3*, em0076. [CrossRef]
3. Kumar, N.M.; Chopra, S.S.; Chand, A.A.; Elavarasan, R.M.; Shafiullah, G. Hybrid renewable energy microgrid for a residential community: A techno-economic and environmental perspective in the context of the SDG7. *Sustainability* **2020**, *12*, 3944. [CrossRef]
4. Elavarasan, R.M.; Shafiullah, G.; Padmanaban, S.; Kumar, N.M.; Annam, A.; Vetrichelvan, A.M.; Mihet-Popa, L.; Holm-Nielsen, J.B. A comprehensive review on renewable energy development, Challenges, and policies of leading Indian states with an international perspective. *IEEE Access* **2020**, *8*, 74432–74457. [CrossRef]
5. Elavarasan, R.M.; Shafiullah, G.; Manoj Kumar, N.; Padmanaban, S. A State-of-the-Art review on the drive of renewables in Gujarat, State of India: Present situation, barriers and future initiatives. *Energies* **2020**, *13*, 40. [CrossRef]
6. Madurai Elavarasan, R.; Selvamanohar, L.; Raju, K.; Vijayaraghavan, R.R.; Subburaj, R.; Nurunnabi, M.; Khan, I.A.; Afridhis, S.; Hariharan, A.; Pugazhendhi, R. A Holistic Review of the Present and Future Drivers of the Renewable Energy Mix in Maharashtra, State of India. *Sustainability* **2020**, *12*, 6596. [CrossRef]
7. Tan, X.; Li, Q.; Wang, H. Advances and trends of energy storage technology in microgrid. *Int. J. Electr. Power Energy Syst.* **2013**, *44*, 179–191. [CrossRef]
8. Elavarasan, R.M.; Afridhis, S.; Vijayaraghavan, R.R.; Subramaniam, U.; Nurunnabi, M. SWOT analysis: A framework for comprehensive evaluation of drivers and barriers for renewable energy development in significant countries. *Energy Rep.* **2020**, *6*, 1838–1864.
9. Elavarasan, R.M. Comprehensive review on India's growth in renewable energy technologies in comparison with other prominent renewable energy based countries. *J. Sol. Energy Eng.* **2020**, *142*, 030801. [CrossRef]
10. Sharma, S.; Bhattacharjee, S.; Bhattacharya, A. Grey wolf optimisation for optimal sizing of battery energy storage device to minimise operation cost of microgrid. *IET Gener. Transm. Distrib.* **2016**, *10*, 625–637. [CrossRef]
11. Energy Storage System: Roadmap for India 2019–32. 2019. Available online: http://www.indiasmartgrid.org/reports/ISGF_Report_Energy_Storage_System_RoadmapforIndia_2019\to2032_11July2019_Draft.pdf (accessed on 29 September 2020).
12. Hassan, T.; Abbassi, R.; Jerbi, H.; Mehmood, K.; Tahir, M.F.; Cheema, K.M.; Elavarasan, R.M.; Ali, F.; Khan, I.A. A Novel Algorithm for MPPT of an Isolated PV System Using Push Pull Converter with Fuzzy Logic Controller. *Energies* **2020**, *13*, 4007. [CrossRef]
13. Ebrahimi, H.; Marjani, S.R.; Talavat, V. Optimal planning in active distribution networks considering nonlinear loads using the MOPSO algorithm in the TOPSIS framework. *Int. Trans. Electr. Energy Syst.* **2020**, *30*, e12244. [CrossRef]
14. Liu, L.; Zhang, Y.; Da, C.; Huang, Z.; Wang, M. Optimal allocation of distributed generation and electric vehicle charging stations based on intelligent algorithm and bi-level programming. *Int. Trans. Electr. Energy Syst.* **2020**, *30*, e12366. [CrossRef]
15. da Silva Seta, F.; de Oliveira, L.W.; de Oliveira, E.J. Comprehensive approach for distribution system planning with uncertainties. *IET Gener. Transm. Distrib.* **2019**, *13*, 5467–5477. [CrossRef]
16. Ahmadi, M.; Lotfy, M.E.; Howlader, A.M.; Yona, A.; Senjyu, T. Centralised multi-objective integration of wind farm and battery energy storage system in real-distribution network considering environmental, technical and economic perspective. *IET Gener. Transm. Distrib.* **2019**, *13*, 5207–5217. [CrossRef]
17. Deb, G.; Chakraborty, K.; Deb, S. Spider monkey optimization technique–based allocation of distributed generation for demand side management. *Int. Trans. Electr. Energy Syst.* **2019**, *29*, e12009. [CrossRef]
18. Dehnavi, E.; Aminifar, F.; Afsharnia, S. Congestion management through distributed generations and energy storage systems. *Int. Trans. Electr. Energy Syst.* **2019**, *29*, e12018. [CrossRef]
19. Chedid, R.; Sawwas, A. Optimal placement and sizing of photovoltaics and battery storage in distribution networks. *Energy Storage* **2019**, *1*, e46. [CrossRef]

20. Babu, K.B.; Maheswarapu, S. A solution to multi-objective optimal accommodation of distributed generation problem of power distribution networks: An analytical approach. *Int. Trans. Electr. Energy Syst.* **2019**, *29*. [CrossRef]

21. Hesaroor, K.; Das, D. Annual energy loss reduction of distribution network through reconfiguration and renewable energy sources. *Int. Trans. Electr. Energy Syst.* **2019**, *29*, e12099. [CrossRef]

22. BiazarGhadikolaei, M.; Shahabi, M.; Barforoushi, T. Expansion planning of energy storages in microgrid under uncertainties and demand response. *Int. Trans. Electr. Energy Syst.* **2019**, *29*, e12110. [CrossRef]

23. Amir, V.; Azimian, M.; Razavizadeh, A.S. Reliability-constrained optimal design of multicarrier microgrid. *Int. Trans. Electr. Energy Syst.* **2019**, *29*, e12131. [CrossRef]

24. RaguRaman, L.; Ravindran, M. MFLRS-RDF technique for optimal sizing and performance analysis of HRES. *Int. J. Numer. Model. Electr. Netw. Devices Fields* **2020**, *33*, e2675. [CrossRef]

25. Gholami, M.; Zakariazadeh, A. Olympic ranking–based allocation of distributed generation units in distribution networks. *Int. Trans. Electr. Energy Syst.* **2020**, *30*, e12220. [CrossRef]

26. Muhammad, M.A.; Mokhlis, H.; Amin, A.; Naidu, K.; Franco, J.F.; Wang, L.; Othman, M. Enhancement of simultaneous network reconfiguration and DG sizing via Hamming dataset approach and firefly algorithm. *IET Gener. Transm. Distrib.* **2019**, *13*, 5071–5082. [CrossRef]

27. Manna, D.; Goswami, S.K. Optimum placement of distributed generation considering economics as well as operational issues. *Int. Trans. Electr. Energy Syst.* **2020**, *30*, e12246. [CrossRef]

28. Pandey, A.K.; Kirmani, S. Optimal location and sizing of hybrid system by analytical crow search optimization algorithm. *Int. Trans. Electr. Energy Syst.* **2020**, *30*, e12327. [CrossRef]

29. Kiani, A.T.; Nadeem, M.F.; Ahmed, A.; Khan, I.; Elavarasan, R.M.; Das, N. Optimal PV Parameter Estimation via Double Exponential Function-Based Dynamic Inertia Weight Particle Swarm Optimization. *Energies* **2020**, *13*, 4037. [CrossRef]

30. Kumar, S.; Saket, R.; Dheer, D.K.; Holm-Nielsen, J.; Sanjeevikumar, P. Reliability enhancement of electrical power system including impacts of renewable energy sources: A comprehensive review. *IET Gener. Transm. Distrib.* **2020**, *14*, 1799–1815. [CrossRef]

31. Samrout, M.; Yalaoui, F.; Châtelet, E.; Chebbo, N. New methods to minimize the preventive maintenance cost of series–parallel systems using ant colony optimization. *Reliab. Eng. Syst. Saf.* **2005**, *89*, 346–354. [CrossRef]

32. Shahzad, M.; Ahmad, I.; Gawlik, W.; Palensky, P. Load concentration factor based analytical method for optimal placement of multiple distribution generators for loss minimization and voltage profile improvement. *Energies* **2016**, *9*, 287. [CrossRef]

33. Parihar, S.S.; Malik, N. Optimal allocation of renewable DGs in a radial distribution system based on new voltage stability index. *Int. Trans. Electr. Energy Syst.* **2020**, *30*, e12295. [CrossRef]

34. Hassan, A.A.; Fahmy, F.H.; Nafeh, A.E.S.A.; Abu-elmagd, M.A. Hybrid genetic multi objective/fuzzy algorithm for optimal sizing and allocation of renewable DG systems. *Int. Trans. Electr. Energy Syst.* **2016**, *26*, 2588–2617. [CrossRef]

35. Atteya, I.I.; Ashour, H.; Fahmi, N.; Strickland, D. Radial distribution network reconfiguration for power losses reduction using a modified particle swarm optimisation. *CIRED Open Access Proc. J.* **2017**, *2017*, 2505–2508. [CrossRef]

36. Wazir, A.; Arbab, N. Analysis and optimization of IEEE 33 bus radial distributed system using optimization algorithm. *JETAE J. Emerg. Trends Appl. Eng.* **2016**, *1*, 2518–4059.

37. Paliwal, N.K.; Singh, A.K.; Singh, N.K.; Kumar, P. Optimal sizing and operation of battery storage for economic operation of hybrid power system using artificial bee colony algorithm. *Int. Trans. Electr. Energy Syst.* **2019**, *29*, e2685. [CrossRef]

38. Saric, M.; Hivziefendic, J.; Konjic, T.; Ktena, A. Distributed generation allocation considering uncertainties. *Int. Trans. Electr. Energy Syst.* **2018**, *28*, e2585. [CrossRef]

39. Baran, M.; Wu, F. Network reconfiguration in distribution systems for loss reduction and load balancing. *IEEE Trans. Power Deliv.* **1989**, *4*, 1401–1407. [CrossRef]

40. Chong, B.; Zhang, X.; Godfrey, K.; Yao, L.; Bazargan, M. Optimal location of unified power flow controller for congestion management. *Eur. Trans. Electr. Power* **2010**, *20*, 600–610. [CrossRef]

41. Aghajani, A.; Kazemzadeh, R.; Ebrahimi, A. Optimal energy storage sizing and offering strategy for the presence of wind power plant with energy storage in the electricity market. *Int. Trans. Electr. Energy Syst.* **2018**, *28*, e2621. [CrossRef]

42. Charfi, S.; Atieh, A.; Chaabene, M. Optimal sizing of a hybrid solar energy system using particle swarm optimization algorithm based on cost and pollution criteria. *Environ. Prog. Sustain. Energy* **2019**, *38*, e13055. [CrossRef]

43. Wu, C.; Lou, Y.; Lou, P.; Xiao, H. DG location and capacity optimization considering several objectives with cloud theory adapted GA. *Int. Trans. Electr. Energy Syst.* **2014**, *24*, 1076–1088. [CrossRef]

44. Duong, M.Q.; Pham, T.D.; Nguyen, T.T.; Doan, A.T.; Tran, H.V. Determination of optimal location and sizing of solar photovoltaic distribution generation units in radial distribution systems. *Energies* **2019**, *12*, 174. [CrossRef]

45. Madurai Elavarasan, R.; Ghosh, A.; K Mallick, T.; Krishnamurthy, A.; Saravanan, M. Investigations on performance enhancement measures of the bidirectional converter in PV–wind interconnected microgrid system. *Energies* **2019**, *12*, 2672. [CrossRef]

46. Natarajan, M.; Ramadoss, B.; Lakshmanarao, L. Optimal location and sizing of MW and MVAR based DG units to improve voltage stability margin in distribution system using a chaotic artificial bee colony algorithm. *Int. Trans. Electr. Energy Syst.* **2017**, *27*, e2287. [CrossRef]

47. Nawaz, S.; Bansal, A.K.; Sharma, M.P. Allocation of DG and capacitor units for power loss reduction in radial distribution system. In Proceedings of the 2016 International Conference on Recent Advances and Innovations in Engineering (ICRAIE), Jaipur, India, 23–25 December 2016; pp. 1–6.

48. Vita, V. Development of a decision-making algorithm for the optimum size and placement of distributed generation units in distribution networks. *Energies* **2017**, *10*, 1433. [CrossRef]

49. Basso, T. *IEEE 1547 and 2030 Standards for Distributed Energy Resources Interconnection and Interoperability with the Electricity Grid*; Technical report; National Renewable Energy Lab. (NREL): Golden, CO, USA, 2014.

50. Huang, W.; Zhang, N.; Yang, J.; Wang, Y.; Kang, C. Optimal configuration planning of multi-energy systems considering distributed renewable energy. *IEEE Trans. Smart Grid* **2017**, *10*, 1452–1464. [CrossRef]

51. Bahramirad, S.; Daneshi, H. Optimal Sizing of Smart Grid Storage Management System in a Microgrid. In Proceedings of the 2012 IEEE PES Innovative Smart Grid Technologies, ISGT 2012, Washington DC, USA, 16-20 January 2012; pp. 1–6.

52. Kumar, D.; Samantaray, S.; Kamwa, I.; Sahoo, N. Reliability-constrained based optimal placement and sizing of multiple distributed generators in power distribution network using cat swarm optimization. *Electr. Power Compon. Syst.* **2014**, *42*, 149–164. [CrossRef]

53. Soliman, S.A.H.; Mantawy, A.A.H. *Modern Optimization Techniques with Applications in Electric Power Systems*; Springer Science & Business Media: New York, NY, USA, 2011.

54. Bhumkittipich, K.; Phuangpornpitak, W. Optimal placement and sizing of distributed generation for power loss reduction using particle swarm optimization. *Energy Procedia* **2013**, *34*, 307–317. [CrossRef]

55. Aman, M.; Jasmon, G.; Mokhlis, H.; Bakar, A. Optimal placement and sizing of a DG based on a new power stability index and line losses. *Int. Trans. Electr. Energy Syst.* **2012**, *43*, 1296–1304. [CrossRef]

56. Shukla, T.; Singh, S.; Srinivasarao, V.; Naik, K. Optimal sizing of distributed generation placed on radial distribution systems. *Electr. Power Compon. Syst.* **2010**, *38*, 260–274. [CrossRef]

57. Abdel-Aal, H.; Bassyouni, M.; Abdelkreem, M.; Abdel-Hamid, S.; Zohdy, K. Feasibility Study for a Solar-Energy Stand-Alone System:(SESAS). *Smart Grid Renew. Energy* **2012**, *3*, 204. [CrossRef]

58. Hamouda, Y.A. Wind energy in Egypt: Economic feasibility for Cairo. *Renew. Sustain. Energy Rev.* **2012**, *16*, 3312–3319. [CrossRef]

59. Zhao, B.; Guo, C.; Cao, Y. A multiagent-based particle swarm optimization approach for optimal reactive power dispatch. *IEEE Trans. Power Syst.* **2005**, *20*, 1070–1078. [CrossRef]

60. Kusiak, A.; Zhang, Z.; Xu, G. Minimization of wind farm operational cost based on data-driven models. *IEEE Trans. Sustain. Energy* **2013**, *4*, 756–764. [CrossRef]

61. Wang, L.; Singh, C. Multicriteria design of hybrid power generation systems based on a modified particle swarm optimization algorithm. *IEEE Trans. Energy Convers.* **2009**, *24*, 163–172. [CrossRef]

62. Ramezani, M.; Haghifam, M.R.; Singh, C.; Seifi, H.; Moghaddam, M.P. Determination of capacity benefit margin in multiarea power systems using particle swarm optimization. *IEEE Trans. Power Syst.* **2008**, *24*, 631–641. [CrossRef]

63. Yang, H.; Xie, K.; Tai, H.M.; Chai, Y. Wind farm layout optimization and its application to power system reliability analysis. *IEEE Trans. Power Syst.* **2015**, *31*, 2135–2143. [CrossRef]

64. Sunil Joseph, P.; DineshBalaji, C. Transmission loss minimization using optimization technique Based on PSO. *2011 IEEE Symp. Ind. Electron. Appl.* **2013**, *1*, 1–5. [CrossRef]

65. Clerc, M.; Kennedy, J. The particle swarm-explosion, stability, and convergence in a multidimensional complex space. *IEEE Trans. Evol. Comput.* **2002**, *6*, 58–73. [CrossRef]

66. El-Zonkoly, A. Optimal placement of multi-distributed generation units including different load models using particle swarm optimisation. *IET Gener. Transm. Distrib.* **2011**, *5*, 760–771. [CrossRef]

67. Kansal, S.; Kumar, V.; Tyagi, B. Optimal placement of different type of DG sources in distribution networks. *Int. J. Electr. Power Energy Syst.* **2013**, *53*, 752–760. [CrossRef]

68. Kansal, S.; Sai, B.; Tyagi, B.; Kumar, V. Optimal placement of distributed generation in distribution networks. *Int. J. Eng. Sci. Technol.* **2011**, *3*, 47–55. [CrossRef]

69. Subcommittee, D. IEEE Guide for Electric Power Distribution Reliability Indices. *Distribution* **2012**, *1997*, 1–43.

70. Heier, S. *Grid Integration of Wind Energy: Onshore and Offshore Conversion Systems*; John Wiley & Sons: Sussex, UK, 2014.

71. Koutroulis, E.; Kolokotsa, D.; Potirakis, A.; Kalaitzakis, K. Methodology for optimal sizing of stand-alone photovoltaic/wind-generator systems using genetic algorithms. *Sol. Energy* **2006**, *80*, 1072–1088. [CrossRef]

72. van Sark, W.G. Teaching the relation between solar cell efficiency and annual energy yield. *Eur. J. Phys.* **2007**, *28*, 415. [CrossRef]

73. Singh, R.; Sharma, M.; Rawat, R.; Banerjee, C. An assessment of series resistance estimation techniques for different silicon based SPV modules. *Renew. Sustain. Energy Rev.* **2018**, *98*, 199–216. [CrossRef]

74. Yang, J.; Sun, Y.; Xu, Y. Modeling impact of environmental factors on photovoltaic array performance. *Int. J. Energy Environ.* **2013**, *4*, 955–968.

75. Gunawardana, A. Proper sizing of energy storage for grid connected photovoltaic system. Master's Thesis, Department of Engineering Science Faculty of Engineering and Science, University of Agder, Kristiansand, Norway, 2014.

76. Sultana, S.; Roy, P.K. Multi-objective quasi-oppositional teaching learning based optimization for optimal location of distributed generator in radial distribution systems. *Int. J. Electr. Power Energy Syst.* **2014**, *63*, 534–545. [CrossRef]

77. Mahmoud, K.; Yorino, N.; Ahmed, A. Optimal distributed generation allocation in distribution systems for loss minimization. *IEEE Trans. Power Syst.* **2015**, *31*, 960–969. [CrossRef]

78. Billington, R.; Allan, R.N. *Reliability Evaluation of Power Systems*; Springer: New York, NY, USA, 1996.

Publisher's Note: MDPI stays neutral with regard to jurisdictional claims in published maps and institutional affiliations.

Article

High-Performance 3-Phase 5-Level E-Type Multilevel–Multicell Converters for Microgrids

Marco di Benedetto [1,*], Alessandro Lidozzi [1], Luca Solero [1], Fabio Crescimbini [1] and Petar J. Grbović [2]

[1] C-PED, Center for Power Electronics and Drives, Roma Tre University, 00146 Roma, Italy; alessandro.lidozzi@uniroma3.it (A.L.); luca.solero@uniroma3.it (L.S.); fabio.crescimbini@uniroma3.it (F.C.)

[2] Innsbruck Power Electronics Lab Innsbruck, University of Innsbruck, 6020 Innsbruck, Austria; petar.grbovic@uibk.ac.at

* Correspondence: marco.dibenedetto@uniroma3.it

Abstract: This paper focuses on the analysis and design of two multilevel–multicell converters (MMCs), named 3-phase 5-Level E-Type Multilevel–Multicell Rectifier (3Φ5L E-Type MMR) and 3-phase 5-Level E-Type Multilevel–Multicell Inverter (3Φ5L E-Type MMI) to be used in microgrid applications. The proposed 3-phase E-Type multilevel rectifier and inverter have each phase being accomplished by the combination of two I-Type topologies connected to the T-Type topology. The two cells of each phase of the rectifier and inverter are connected in interleaving using an intercell transformer (ICT) in order to reduce the volume of the output filter. Such an E-Type topology arrangement is expected to allow both the high efficiency and power density required for microgrid applications, as well as being capable of providing good performance in terms of quality of the voltage and current waveforms. The proposed hardware design and control interface are supported by the simulation results performed in Matlab/Simulink. The analysis has been then validated in terms of an experimental campaign performed on the converter prototype, which presented a power density of 8.4 kW/dm^3 and a specific power of 3.24 kW/kg. The experimental results showed that the proposed converter can achieve a peak efficiency of 99% using only silicon power semiconductors.

Keywords: multilevel–multicell converter; wide bandgap devices; high performance; interleaved topology; power density; specific power; microgrid

Citation: di Benedetto, M.; Lidozzi, A.; Solero, L.; Crescimbini, F.; Grbović, P.J. High-Performance 3-Phase 5-Level E-Type Multilevel–Multicell Converters for Microgrids. *Energies* **2021**, *14*, 843. https://doi.org/10.3390/en14040843

Academic Editor: Massimiliano Luna
Received: 12 January 2021
Accepted: 2 February 2021
Published: 5 February 2021

Publisher's Note: MDPI stays neutral with regard to jurisdictional claims in published maps and institutional affiliations.

1. Introduction

The electrical power demand has increased in all applications, such as transportation, industry, household, and the commercial sectors [1]. We are in fact becoming more and more hungry for electric power: on one hand, this is provoked by the constant increase in the world population, and consequently the urban centers are increasing in size with more people living in cities; on the other hand, the hunger for energy, especially in developing countries always in need of new infrastructure such as hospitals, schools, and transport, comes from the middle class. Thus, the increased demand of energy should be addressed by both significantly increasing the use of renewable energy sources and improving the efficiency of energy systems [2]. Currently, the most abundant renewable energy sources are wind and solar [3,4]; for both of them, adequate conversion equipment is needed. For example, wind energy requires turbines and generators, which leads to variable frequency and variable voltage electricity. As a result, between the generator and the grid, a power conversion system is needed in order to meet the grid requirement of a fixed frequency of 50 or 60 Hz at certain standard voltage levels. Concerning solar energy, a photovoltaic (PV) cell provides a DC source at unregulated low voltage, and thus power electronics systems must be able to adjust this DC voltage level to one suitable for supplying the load [5]. In the 21st century, therefore, modern solutions based on different energy sources coming from wind, solar, energy storage and micro-turbines have been used to feed the load or different loads [6,7]. Particularly, in a stand-alone microgrid, more energy sources are connected to a single DC-bus to supply the load, as shown in Figure 1. The photovoltaic system and the battery system are connected to the DC-bus thanks to a DC/DC power converter [8,9], while the wind system and the micro-turbine are connected to the DC-bus through an AC/DC converter. The power flows from the DC-bus to the load thanks to the DC/AC converter connected to a power filter.

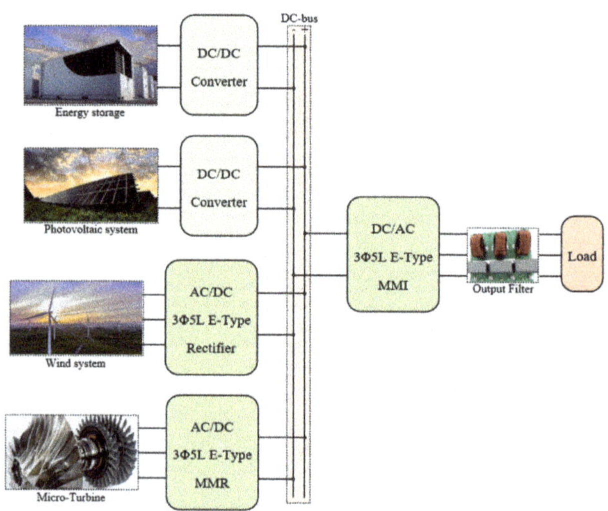

Figure 1. Structure of stand-alone microgrid with the integrated renewable energy source.

It is evident that the modern renewable energy system could not operate, and the microgrid could not be realized, without power electronics. In fact, power electronics converters and controllers are the most important elements in this application, and the efforts of power conversion designers consistently focuses on improving the efficiency and reducing the volumes of the conversion system [10,11]. These goals can be reached thanks to the continuous improvement of power electronics technologies and power converter topologies. Indeed, on one hand the industrial manufacturing of power components keeps introducing new high-performance modules and power semiconductors on the market [12,13]; on the other hand, newer and emerging power converter topologies are constantly proposed in literature. For example, in high power and/or high voltage applications, the limit of the power conversion system is attributed to the power semiconductors capable of withstanding limited voltage and current stresses. To overcome this problem, academia and industry are working on new solutions of power converters [14–16]. Thus, the use of multilevel configurations allows the arrangement of power conversion systems with power semiconductors that are required to withstand only a fraction of both the overall DC-bus voltage and the converter output current. This usually allows higher switching frequency f_{sw} and, therefore, multilevel configurations lead to significant improvements in terms of quality of both voltage and current waveforms without giving up the benefit of high efficiency and high-power density. Furthermore, the weight and volume of the converter passive components are likely to be reduced [14–17], which is very appealing for several applications, where the overall size and weight of the electrical generating system needs to be minimized. Given the high number of benefits, more and more multilevel converter topologies are proposed in the literature [17–21]. It is understood that multilevel topology can act on the voltage stress of the power semiconductors, leaving their current ratings unchanged. To reduce the current stress flowing through the power semiconductors, more parallel cell converters can be used. Thus, the power semiconductors with low current rating can be chosen, leading to an improvement in the conduction losses. As another advantage, parallel converters enhance the current ripple, resulting in a further improvement of the waveform quality. Obviously, the multilevel converters present some disadvantages, such as the reliability and cost. In fact, the number of power semiconductors in multilevel converters increases as the number of the voltage levels increases, and, therefore, we could think that cost and reliability worsen. However, some multilevel topologies proposed in the literature have fault tolerance capability [22–24]. Additionally, even if the number of power devices increases, there is a reduction in passive devices, which leads to less use of copper and iron. These materials, actually, are more expensive than power semiconductors. Naturally, the real disadvantage of having many devices could lie in the driver circuits used for switching the devices [25]. Moreover, tuning the control algorithm could also be more complicated, given the difficulty in finding an analytical model of the multilevel converter [26,27]. Nevertheless, apart from the analytical effort required to find the control law, the continuous trend of increasing the controller performance and memory size along with the reduction in the cost has now reached a point where the increase in the number of power semiconductors is not such a disadvantage. Multilevel converters based on Neutral Point Clamped (NPC) and T-Type topologies have been proposed using Super-junction MOSFETs [17,18]. Here, the peak efficiency is estimated to be above 99%. A five-level T-Type converter able to reach an efficiency of 99.2% using only SiC technology is presented in [19]. In this paper, the confirmed power density and specific power are 1.4 kW/dm^3 and 2.5 kW/kg. In [20] a three-phase T-Type converter, called Swiss Rectifier, has been designed using SiC power devices. The declared peak efficiency and power density were 99.26% and 4 kW/dm^3, respectively. A hybrid Five-Level Active NPC Inverter that uses SiC and is able to achieve a peak efficiency above 98% has been proposed in [21].

One of the goals of the proposed paper is analyzing, designing, and testing the 3Φ5L E-Type MMR and the 3Φ5L E-Type Multilevel–Multicell Inverter (MMI) to obtain high, efficiency, power density and specific power by using only silicon (Si) power semiconductors. For this purpose, the operation modes of the proposed converters are clearly explained and a solution to balance the DC-bus voltages is discussed. Then, the investigation focused on the design of the proposed converters; an analytical approach to calculate the device stress is presented to select the suitable power semiconductors. Starting from this analysis, the power semiconductors have been selected and the efficiency as a function of the power has been analytically calculated for both converters. The theoretical investigation has been supported by the model of the converters created in the Matlab/Simulink and Plexim/Plecs environments. A prototype of the converters has been built and the proposed MMR and MMI are integrated on the same power board to reduce the power density. Furthermore, the control structures for both MMR and MMI to be used in microgrid applications have been implemented and verified through preliminary simulations. Finally, experimental tests have been performed in order to confirm the obtained theoretical analysis. This paper is organized as follows: the topology, the operation principle, and the voltage unbalancing issue of the 3Φ5L E-Type MMR and 3Φ5L E-Type MMI are presented in Section 2. The hardware aspect design of the proposed rectifier and inverter are illustrated in Section 3. Based on the proposed rectifier and inverter, the control strategies regarding stand-alone microgrid applications have been discussed in Section 4. Simulation results and experimental results from a laboratory prototype are shown in Sections 5 and 6, respectively. Conclusions are presented in Section 7.

2. Operation Structure of 3-Phase 5-Level E-Type MMC

2.1. 3Φ5L E-Type MMR and MMI

The circuit of the 3Φ5L E-Type MMR is illustrated in Figure 2. A single-phase of the rectifier has two cells: cell 1 and cell 2. This converter is based on both *I-Type* and *T-Type* topology [28–30]. In fact, each cell is composed of two I-Type legs connected to the T-Type leg. The power flows in one direction in the 3Φ5L E-Type MMR due to the presence of the diode into the T-Type cell. The two cells are connected in an interleaving manner, using an intercell transformer (ICT). The advantages of paralleling the cells using the ICT lies in the fact that the phase current is equally shared between the cells, the amplitude of the total current ripple is reduced, and the harmonic contents of the voltage at high frequency is shifted at twice the switching frequency. The 3Φ5L E-Type MMI is also composed of the I-Type and T-Type topologies like the rectifier, as illustrated in Figure 3. Each phase has two cells connected through the ICT.

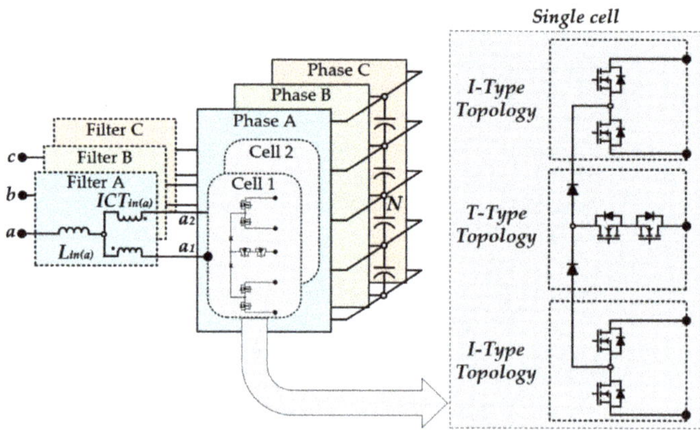

Figure 2. Circuit of 3Φ5L E-Type Multilevel–Multicell Rectifier (MMR).

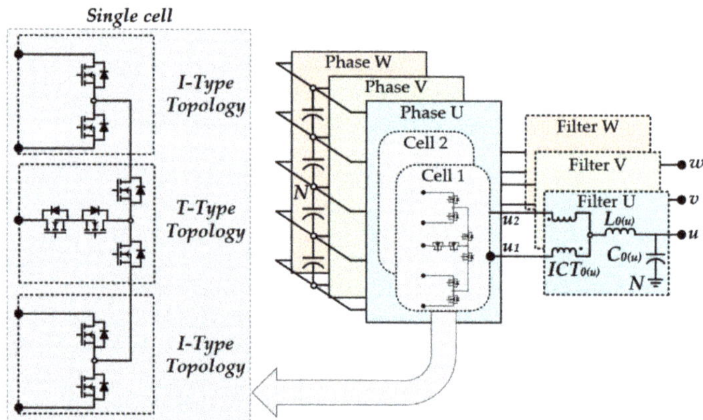

Figure 3. Circuit of 3Φ5L E-Type Multilevel–Multicell Inverter (MMI).

According to the modulation control scheme, both converters show five voltage levels on a single cell, while each phase has nine voltage levels, as shown in Figures 4 and 5. Thus, the line-to-line voltage shows seventeen voltage levels. A carrier-based pulse width modulation (PWM) method has been implemented, taking into account the multiple power semiconductors of the converters. The gate signals of the power devices are generated by the comparison of the modulating signals with the carriers, as shown in Figure 6. Considering the interleaving concept, a phase displacement is applied between the parallel cells in order to achieve highest quality of the output waveforms. Thus, four carrier signals, c_{t11}, c_{t12}, c_{t13} and c_{t14} (solid line) control the power devices in cell 1, and the other carrier signals c_{t21}, c_{t22}, c_{t23} and c_{t24} (dashed line) in the opposite phase control the power semiconductors in cell 2. Furthermore, as can be seen from Figure 6, two devices are controlled by a single carrier signal. The switching frequency of each power semiconductors is f_{sw} while the output waveform effective switching frequency is twice f_{sw}. This phenomenon, present in the interleaved converters, is usually called the multiplicative effect of the effective switching frequency, and leads to a drastic reduction in the output filter.

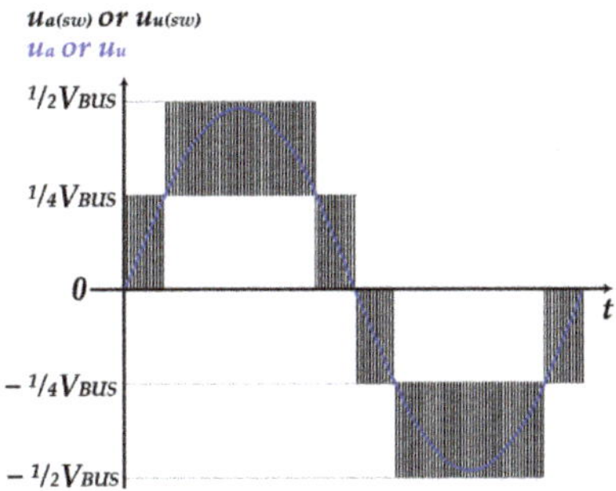

Figure 4. Phase-to-neutral switching voltage $u_{a1(sw)}$ (or $u_{a2(sw)}$).

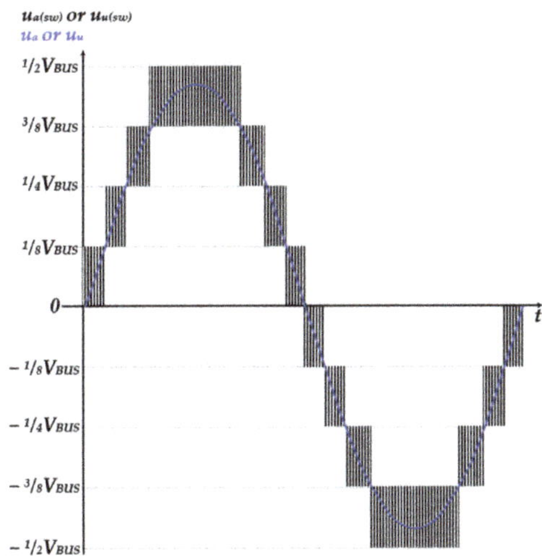

Figure 5. Line-to-line switching voltage $u_{a(sw)}$ (or $u_{u(sw)}$).

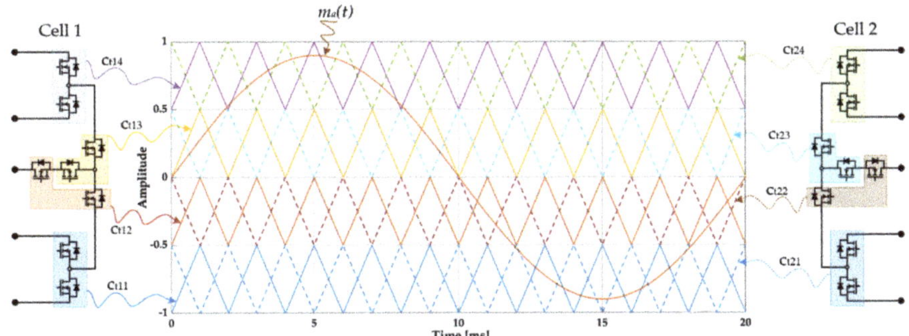

Figure 6. Modulation control scheme for power semiconductors located in a single-phase of the inverter.

2.2. Balancing Circuit

The main problem of the multilevel converter based on T-Type or E-Type topologies is the unequal voltage across the DC-bus capacitors [31–33]. This problem cannot be solved in a simple way with a control algorithm due to the uncontrollable current flow through the internal nodes [28] of the capacitors. The only chance to balance the DC-bus capacitors of the 3Φ5L E-Type converters is to use an external circuit. The focus of this paper was not to study the unbalanced voltage capacitor problem. Here, two series resonant balancing circuits (SRBCs) have been used to solve the voltage unbalancing problem. Figure 7 illustrates the circuit and the implemented prototype of the SRBCs.

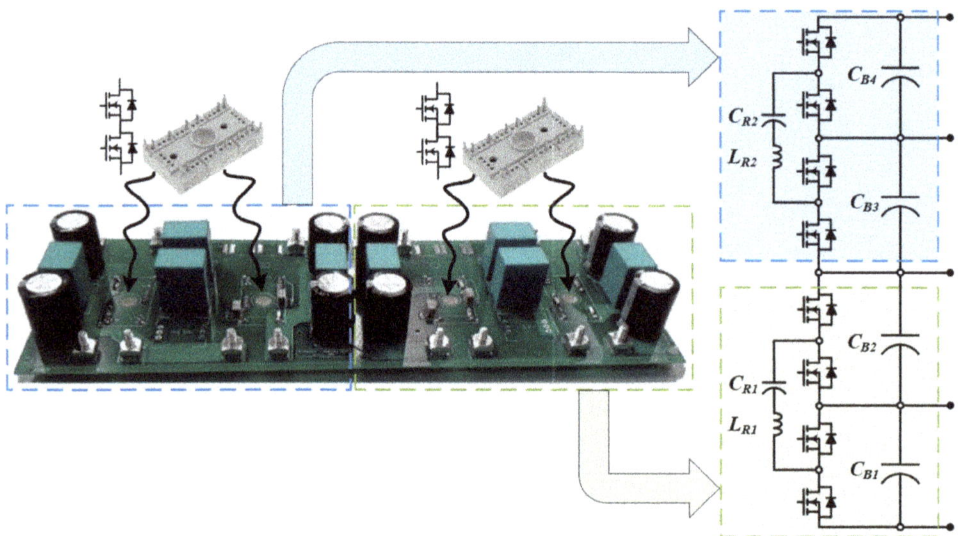

Figure 7. Series resonant balancing circuit prototype.

The SRBCs were built with four Semitop3 SK75GB066T modules (rated 60 A—600 V), 4 µH and 16 µF as resonant total inductance (L_{R1}/L_{R2}) and total capacitance (C_{R1}/C_{R2}), respectively. Furthermore, one DC-bus film capacitor and two DC-bus electrolytic capacitors were used as a DC link. The energy was transferred from the capacitor C_{B1} to C_{B2} and from the capacitor C_{B3} to C_{B4} through the capacitors C_{R1} and C_{R2}. The auxiliary inductors L_{R1} and L_{R2} were used to achieve a zero-current switching (ZCS) condition. The power semiconductors were driven with complementary control signals with a constant duty cycle at 50%; no control loops and sensors were required with the system being self-balanced. These two SRBCs were used to balance the voltage across the capacitors to ensure equal DC currents $i_{C1} = i_{C2} = i_{C3} = i_{C4}$.

3. Hardware Design and Prototype of E-Type Topology

The 3Φ5L E-Type MMR and MMI have been designed to maximize the efficiency, power density, and specific power, without sacrificing the quality of the voltage and current waveforms. To accomplish these tasks, different actions have been carefully carried out.

The capacitors of the DC-bus have been chosen considering the maximum peak-to-peak voltage ripple ΔV_{BUS} equal to 100 V and the estimated Root Mean Square (RMS) current flow through the capacitors in the case of an asymmetric load condition. The DC-bus current harmonics were compensated by the SRBCs except the 100 Hz component, which had to be compensated by the capacitors. For this reason, the DC-bus capacitors were selected according to (1), where N_S and N_P were the numbers of series and parallel capacitors, and I_{CBUS} was the RMS current at double fundamental frequency defined in Equation (2).

$$C_{BUS} = \frac{2\sqrt{2}N_S N_P}{(2\pi 100)\Delta V_{BUS}} I_{CBUS} \tag{1}$$

$$I_{CBUS}\big|_{100Hz} = \frac{U_0 I_0}{\sqrt{2}V_{BUS}} \tag{2}$$

In (2), U_0 is the RMS voltage, I_0 is the RMS load current, and V_{BUS} is the total voltage across DC-bus capacitors. According to (1) and (2), six parallel and four series electrolytic capacitors, each one equal to 220 µF, 220 V were chosen as DC-bus capacitor tanks.

The power semiconductors were carefully selected considering the maximum voltage and current stress that the power components are able to withstand. Figure 8 shows the maximum voltage stress across the power semiconductors of a single cell of the inverter and rectifier.

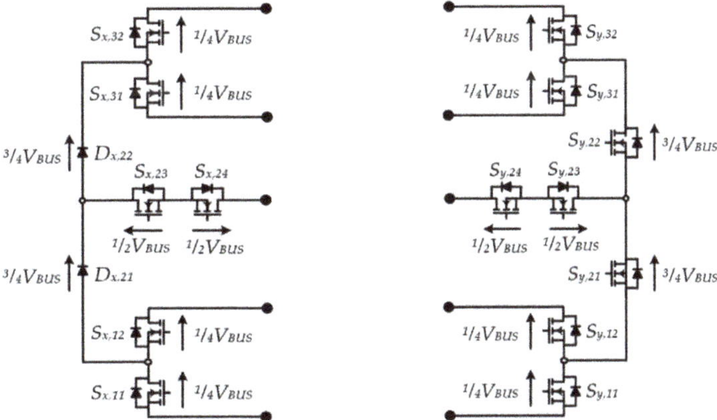

Figure 8. Maximum voltage stress that each cell of the E-Type Rectifier and Inverter can withstand at a steady state.

As can be seen, the switches $S_{x,21}$, $S_{x,22}$ and the diodes $D_{x,21}$, $D_{x,22}$, with $x \in \{1R, 2R\}$ and $x \in \{1I, 2I\}$, have the maximum voltage stress equal to $^3/_4 V_{BUS}$ compared to the other power semiconductors. Naturally, the overvoltage related to the commutated current must be added to this blocking voltage but, as explained in [34], the overvoltage commutation only appears when the blocking voltage at a steady state is equal to $\frac{1}{4}V_{BUS}$. Concerning the current stress, the use of parallel cells helps to reduce the current stress of the power semiconductors. To obtain the current stress of each power semiconductors, an analytical procedure has been performed. Particularly, the average (AVG) and the root mean square (RMS) currents flowing in the power semiconductors located in the rectifier and inverter are expressed in Equation (3), where M_{0R} is the modulation depth of the rectifier, I_{in} is the RMS phase current of the rectifier, M_{0I} is the modulation depth of the inverter, I_0 is the RMS phase current inverter, and $a_{RMS,i}$, $b_{RMS,i}$, $a_{AVG,i}$, $b_{AVG,i}$, $a_{RMS,j}$, $b_{RMS,j}$, $a_{AVG,j}$, $b_{AVG,j}$ are the coefficients related to the switches of the rectifier (i-index) and inverter (j-index). The derivation of Equation (3) requires very complex analysis, and it is beyond the scope of this discussion. A simplified discussion to obtaining the Equation (3), including the coefficients, is discussed in detail in Appendix A.

$$
\begin{aligned}
i_{RMS,R}(t) &= \sqrt{\frac{I_{in}{}^2 M_{0R}}{24\pi}\left(\frac{a_{RMS,i}}{M_{0R}} + b_{RMS,i}\right)} \\
|i_{AVG,R}(t)| &= \frac{\sqrt{2}I_{in}M_{0R}}{4\pi}\left(\frac{a_{AVG,i}}{M_{0R}} + b_{AVG,i}\right) \\
i_{RMS,I}(t) &= \sqrt{\frac{I_{out}{}^2 M_{0I}}{24\pi}\left(\frac{a_{RMS,j}}{M_{0I}} + b_{RMS,j}\right)} \\
|i_{RMS,I}(t)| &= \frac{\sqrt{2}I_{out}M_{0I}}{4\pi}\left(\frac{a_{AVG,j}}{M_{0I}} + b_{AVG,j}\right)
\end{aligned}
\tag{3}
$$

Starting from this analysis, the selected power semiconductors of the 3Φ5L E-Type Rectifier and Inverter are listed in Table 1. To improve the power density and the specific power, the 3Φ5L E-Type MMR and MMI were integrated on the same power board. The power switches of the rectifier and inverter were driven by three boards, each one controlling a single-phase of the rectifier and inverter.

Table 1. Power semiconductors used to build the 3Φ5L E-Type MMR and MMI with $x \in \{1R, 2R\}$ and $y \in \{1I, 2I\}$.

Device	Part Number	Voltage Rating	Current Rating	Technology	Manufacturer
		3Φ5L E-Type MMR			
$S_{x,11}, S_{x,12}$ $S_{x,31}, S_{x,32}$	IPT210N25NFD	250 V	69 A	OptiMOS^{TM} 3	Infineon
$D_{x,21}, D_{x,22}$	IDP30E120	1200 V	30 A	Si Diode	Infineon
$S_{x,23}, S_{x,24}$	IPL60R104C7	650 V	20 A	CoolMOS^{TM}C7	
		3Φ5L E-Type MMI			
$S_{y,11}, S_{y,12},$ $S_{y,31}, S_{y,32}$	IPT210N25NFD	250 V	69 A	OptiMOS^{TM} 3	Infineon
$S_{y,21}, S_{y,22}$	IKW40N120H3	1200 V	40 A	IGBT H3	Infineon
$S_{y,23}, S_{y,24}$	IKW20N60T	600 V	20 A	Trenchstop^{TM} IGBT	Infineon

The Infineon integrated circuit (IC) (part number 1EDI60I12AF) was employed as a gate driver chip. The printed circuit boards (PCBs) of the power board and the gate driver board have been optimized to reduce the current path during the commutations; in this way, the commutation inductance, i.e., the resulting inductance in the commutation circuit, has been reduced, and with it also the overvoltage commutation.

Additionally, because the high gate driver signals result from the high number of the power switches located in the 3Φ5L E-Type MMR and MMI, the interconnecting board which routes all the signals between the diver boards to the control board has been built.

Finally, the input and output filters have been designed to reduce the volume and to obtain high quality of the input currents in the rectifier side and high quality of the output voltages in the inverter side. Thus, the input and output ICTs and the input and output inductors have been built according to the analysis proposed in [30]. The complete prototype of the 3Φ5L E-Type MMR and MMI is illustrated in Figure 9, and features a power density of 8.4 kW/dm^3 and a specific power of 3.24 kW/kg.

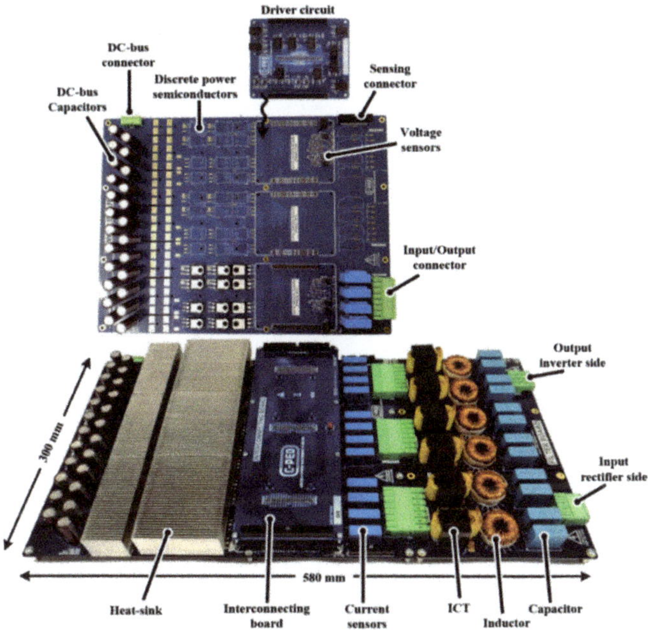

Figure 9. Prototype of the 3Φ5L E-Type Rectifier and Inverter including the input and output power filters, measuring 580 mm × 300 mm × 45 mm. The power density is 8.4 kW/dm^3 and the specific power is 3.24 kW/kg.

4. Control Interface

Figure 10 shows the block scheme of the 3Φ5L E-Type MMI. This converter must be provided with sinusoidal three-phase voltage waveforms with reduced total harmonic distortion (THD_v). To meet this target, a multi-resonant controller [35] has been carefully chosen. Figure 11 illustrates the control algorithm of the 3Φ5L E-Type MMR connected to the wind source or to the micro-turbine source. As can be seen, the external speed loop provides the current reference to the q-controller inner loop. The task of this controller is to regulate the phase current to reduce the THD. The ICT circulating currents are regulated by adding an offset into the modulating signals with an additional loop.

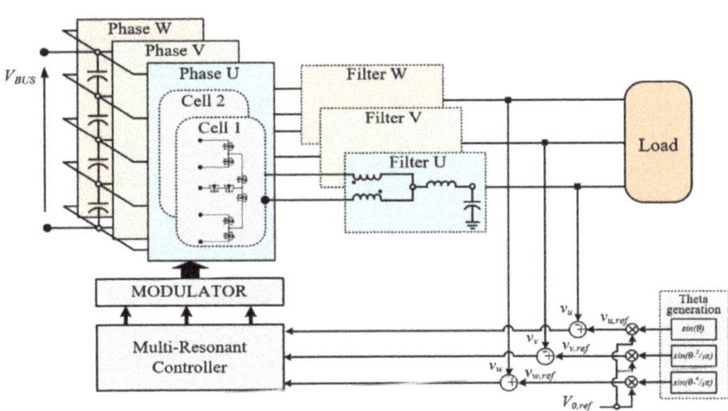

Figure 10. Block diagram of the grid-side 3Φ5L E-Type MMI control algorithm.

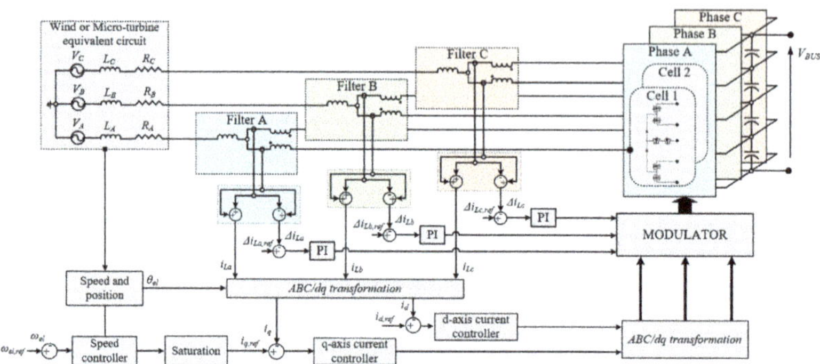

Figure 11. Block diagram of the 3Φ5L E-Type MMR control algorithm for wind or micro-turbine sources.

5. Simulation Results

The hardware design previously addressed has been verified using a simulation model realized in the Matlab/Simulink and Plexim environments. Particularly, the power converter models have been implemented in Plecs, while the control structures have been implemented in Simulink. The stress of the power semiconductors and the losses distribution have been evaluated in the Plecs environment, which has a specific domain for modeling power semiconductors. Moreover, based on the manufacturer of the power semiconductors, the 2D look-up tables and 3D look-up tables have been created in Plecs to evaluate the loss distribution. Figures 12 and 13 show the AVG and RMS current flowing into power semiconductors of the E-Type MMI and E-Type MMR for different values of the output power. Given the symmetry of the circuit, only the current flows through the power switches located in the bottom side of the E-Type MMI and E-Type MMR are illustrated. Both the analytical and experimental approaches provide the same results.

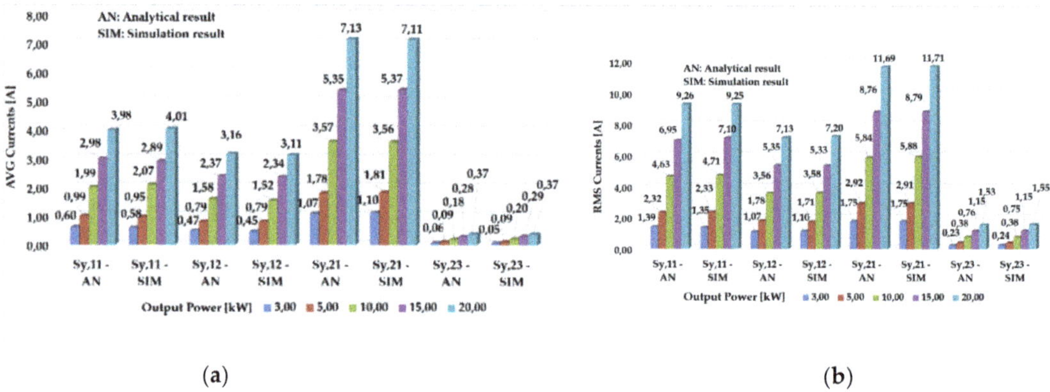

(**a**) (**b**)

Figure 12. Comparison between the analytical and simulation approaches of the currents flowing in the power semiconductors located in the 3Φ5L E-Type MMI: (**a**) average (AVG); (**b**) root mean square (RMS).

As can be seen, in the inverter side the most stressed switches are $S_{y,21}$, $S_{y,22}$, while the least stressed switches are located in the middle leg $S_{y,23}$, $S_{y,24}$. In the rectifier side, the most stressed and the least stressed power semiconductors are $D_{x,21}$, $D_{x,22}$ and $S_{x,23}$, $S_{x,24}$, respectively.

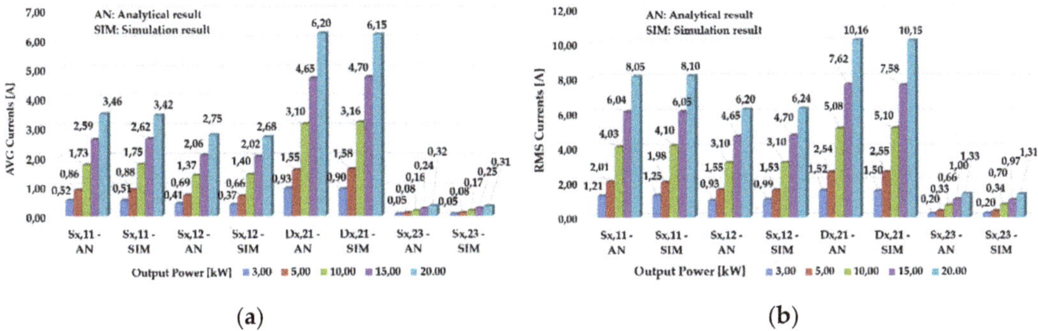

(a)

(b)

Figure 13. Comparison between the analytical and simulation approaches of the currents flowing in the power semiconductors located in the 3Φ5L E-Type MMR: (**a**) AVG; (**b**) RMS.

Based on the datasheet provided by the power semiconductor manufacturers, it has been possible to estimate the efficiency distribution of the proposed converters using an analytical approach and a simulation approach. Particularly, starting from the achieved AVG and RMS currents, analytical equations to estimate the losses of the converters have been obtained according to the method proposed in [36]. Numerical efficiency results from the obtained analytical equations have been compared with simulation results. Figure 14 shows the total efficiency of the 3Φ5L E-Type MMR and MMI, including the passive components, as a function of the power. These results have been obtained based on the selected power semiconductors and the operating parameters listed in Tables 1 and 2. The peak efficiency occurs when the power is close to 10 kW, while the efficiency at a nominal point is above 98%. As can be seen, the simulation results closely match the analytical results. Operation modes and characteristic waveforms of the 3Φ5L E-Type MMR and MMI have been evaluated according to the operating point listed in Table 2.

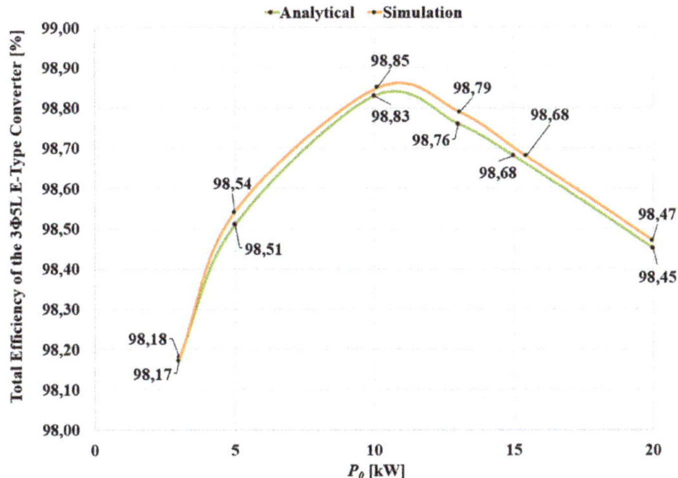

Figure 14. Total efficiency of the 3Φ5L E-Type MMC as a function of the power P_0: analytical result (green line), simulation result (orange line).

Table 2. Operating parameters of the 3Φ5L E-Type MMR and MMI.

	3Φ5L E-Type MMR	**3Φ5L E-Type MMI**
DC-Bus voltage	V_{BUS} = 600 V	V_{BUS} = 600 V
Switching frequency	f_{sw} = 20 kHz	f_{sw} = 20 kHz
Fundamental frequency	f_{in} = 100 Hz	f_0 = 50 Hz
Modulation depth	M_{OR} = 0.93	M_{OI} = 0.93

Figure 15 shows the output phase voltages u_u, u_v, u_w, the phase-to-neutral switching voltages $u_{a(sw)}$, $u_{b(sw)}$, $u_{c(sw)}$, and the inductor phase currents i_u, i_u, i_u of the 3Φ5L E-Type MMI under resistive three-phase loads. As can be seen, the voltage waveforms show a sinusoidal trend with very low total harmonic distortion.

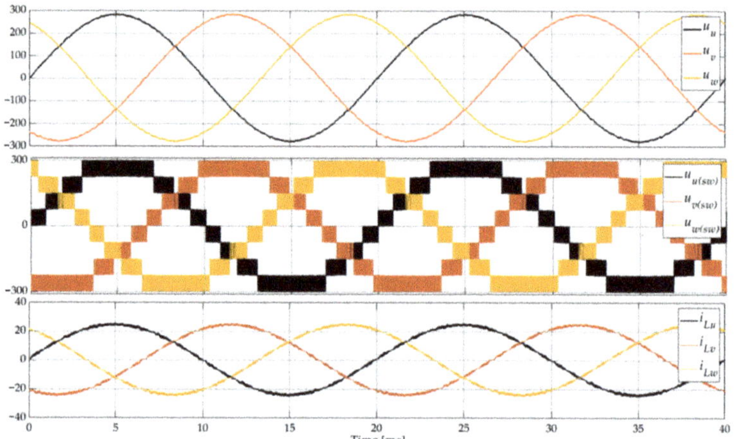

Figure 15. Waveforms of the 3Φ5L E-Type MMI, from top to bottom: output phase voltages u_u, u_v, u_w, phase-to-neutral switching voltages $u_{u(sw)}$, $u_{u(sw)}$, $u_{u(sw)}$, inductor phase currents i_{Lu}, i_{Lv}, i_{Lw}.

The waveforms of the 3Φ5L E-Type MMR are illustrated in Figure 16, where it is possible to notice, from the top to bottom, the phase back electromotive force (EMF) u_a, u_b, u_c, the phase-to-neutral switching voltage $u_{a(sw)}$, $u_{b(sw)}$, $u_{c(sw)}$, the cell-to-neutral switching voltage $u_{a1(sw)}$, $u_{b1(sw)}$, $u_{c1(sw)}$, and the electrical machine phase current i_a, i_b, i_c.

Here, the cell-to-neutral switching voltages also show five voltage levels, while the phase-to-neutral switching voltages exhibit nine voltage level. Thanks to the combination of the proposed topology and the control algorithm, the phase currents are regulated as three-phase sinusoidal waveforms with low total harmonic distortion.

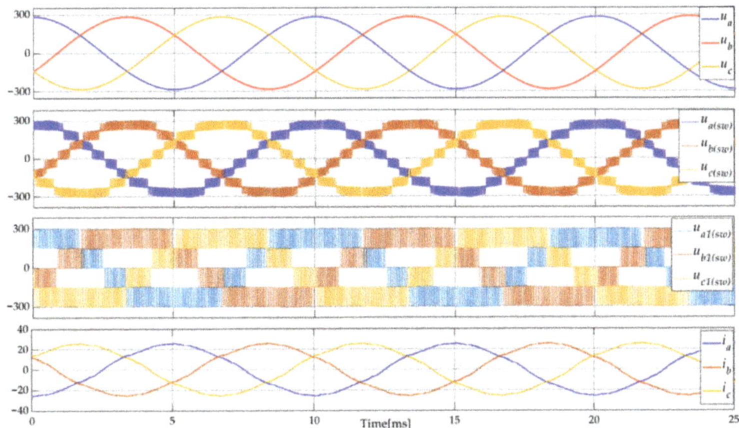

Figure 16. Waveforms of the 3Φ5L E-Type MMR, from top to bottom: phase back electromotive force (EMF) u_a, u_b, u_c, phase-to-neutral switching voltages $u_{a(sw)}$, $u_{b(sw)}$, $u_{c(sw)}$, cell-to-neutral switching voltages $u_{a1(sw)}$, $u_{b1(sw)}$, $u_{c1(sw)}$, electrical machine phase currents i_a, i_b, i_c.

6. Experimental Results

Experimental results have been carried out on 20 kVA 3Φ5L E-Type MMC prototypes previously described to support the proposed analysis. The DC-bus voltage was kept at 600 V by one port of the multi-port Dual Active Bridge (DAB) converter available in the laboratory [9]. The 3Φ5L E-Type MMR was connected to a permanent magnet synchronous motor (PMSM) to emulate the wind source and the 3Φ5L E-Type MMI was connected to the resistive load bench. Figure 17 shows the experimental setup of the multilevel converter including the SRBCs.

Figure 17. Experimental setup of the 3Φ5L E-Type MMR and MMI.

As can be seen from the Figure 17, the 3Φ5L E-Type MMR and MMI were controlled using two different control boards, which were based on the National Instruments sbRIO-9651 System on Module (SoM), as shown in Figure 18. The SoM is equipped with both a microprocessor (μP) and a field-programmable gate array (FPGA), the control loop of the voltages or, in case of the rectifier side, the control loop of the currents and speed, run on the FPGA using 32-bit floating point arithmetic, while system managements and communication infrastructure are managed by the μP.

Figure 18. Control architecture with the National Instruments sbRIO-9651 System on Module (SoM).

Figure 19 shows the phase-to-neutral switching voltage $u_{u(sw)}$, the cell-to-neutral switching voltage $u_{u1(sw)}$, the voltage waveform after the filter u_u, and the phase current i_u under resistive load, when the fundamental frequency f_0 was equal to 50 Hz, the switching frequency f_{sw} was equal to 20 kHz, and the modulation depth M_{0I} was equal to 0.93. Figure 19 shows the five voltage levels across the single cell converter $u_{u1(sw)}$ and nine voltage levels across the single phase $u_{u(sw)}$ for a fixed modulation index. Figure 20 shows the output voltage waveforms under resistive load. These results prove the good capability of the multi-resonant controller to perfectly track the voltage references and to compensate the harmonics introduced by dead component time.

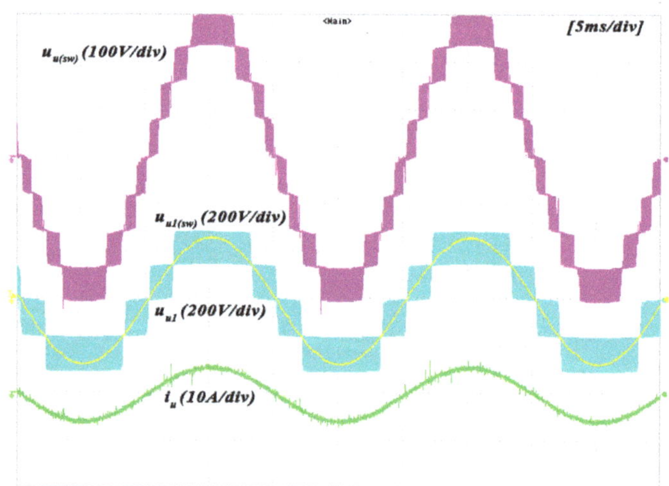

Figure 19. 3Φ5L E-Type MMI waveforms, from top to bottom: phase-to-neutral switching voltage $u_{u(sw)}$, cell-to-neutral switching voltage $u_{u1(sw)}$, output voltage u_u and phase current i_u.

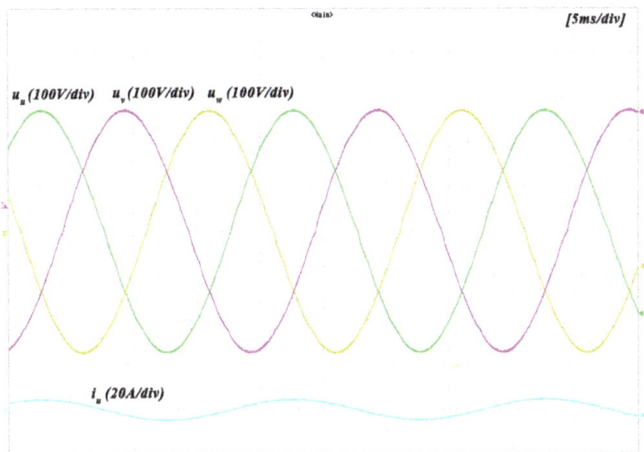

Figure 20. Phase-to-neutral voltages u_u, u_v, u_w of the 3Φ5L E-Type MMI side under resistive load.

Figure 21 illustrates the normalized harmonic spectrum of the phase-to-neutral voltage u_u. The amplitude was normalized with respect to the fundamental. The harmonics magnitude from the 15th to 50th order exhibited an amplitude less than 0.1%. The THD_v valuated up to the 50th order was close to 0.88%.

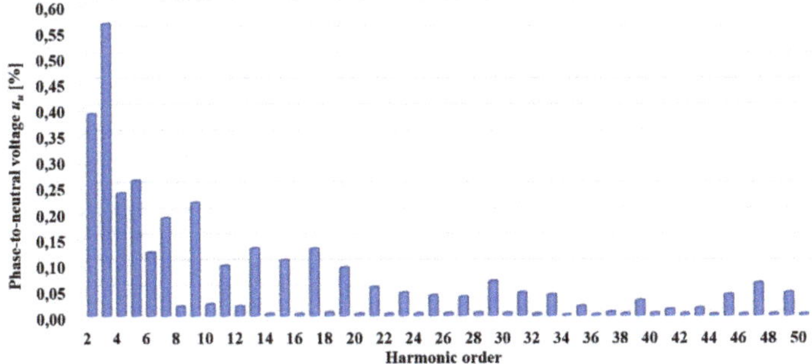

Figure 21. Harmonic spectrum of the phase-to-neutral voltage u_u.

Figure 22 shows the phase-to-neutral switching voltage $u_{a(sw)}$, the extracted fundamental component, the electrical machine phase current i_a, and angular position θ_{el}, when the fundamental frequency f_0 was equal to 100 Hz, the switching frequency f_{sw} was equal to 20 kHz, and modulation depth M_{0R} was equal to 0.93.

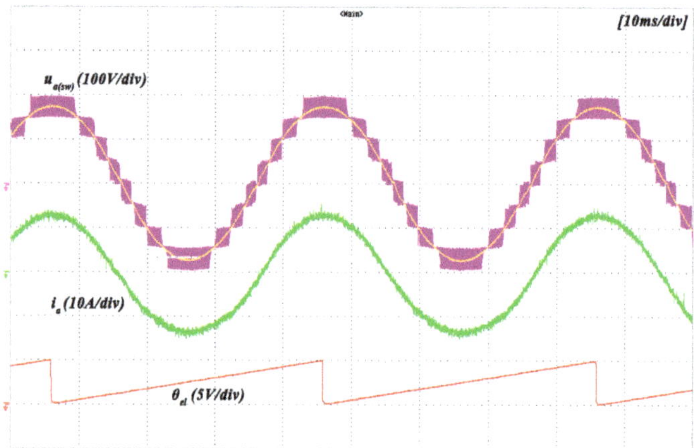

Figure 22. 3Φ5L E-Type MMR waveforms, from top to bottom: phase-to-neutral switching voltage $u_{a(sw)}$, fundamental component (extracted), electrical machine phase current i_a, and angular position θ_{el}.

Here, the nine voltage levels are also clearly visible, and the control algorithm provided good tracking capability, making the machine current almost a pure sinusoidal waveform. The normalized harmonic spectrum of the phase current normalized with respect to the fundamental is shown in Figure 23. The THD of the electrical machine phase current estimated up to the 50th order was equal to 1.95%. The efficiency of the 3Φ5L E-Type Rectifier and Inverter have been evaluated by using the PM3000A wattmeter, where one channel has been used to measure the input power at the DC-bus and two channels have been used to measure the output power through Aron's insertion. Figure 24 illustrates the experimental efficiency (blue line) of the 3Φ5L E-Type MMR plus 3Φ5L E-Type MMI including filters. As can be seen, the peak efficiency was equal to 98.81% by using only the Si power semiconductors, and at nominal power the efficiency was above 98%. Furthermore, the experimental results showed a good matching compared to the theoretical analysis. Consequently, the achieved experimental point validated the theoretical performance analysis of the 3Φ5L E-Type MMR and the 3Φ5L E-Type MMI.

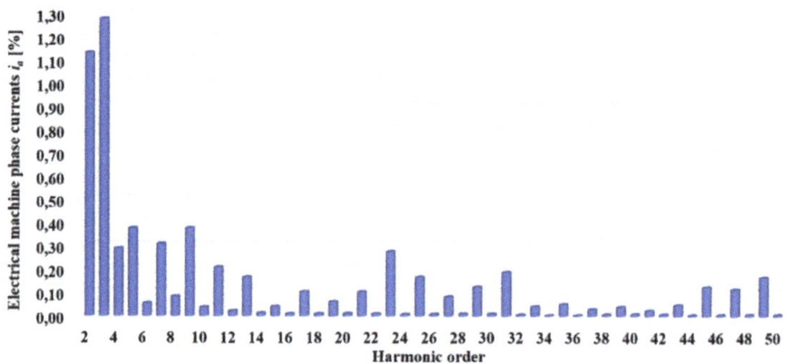

Figure 23. Harmonic spectrum of the electrical machine phase currents i_a.

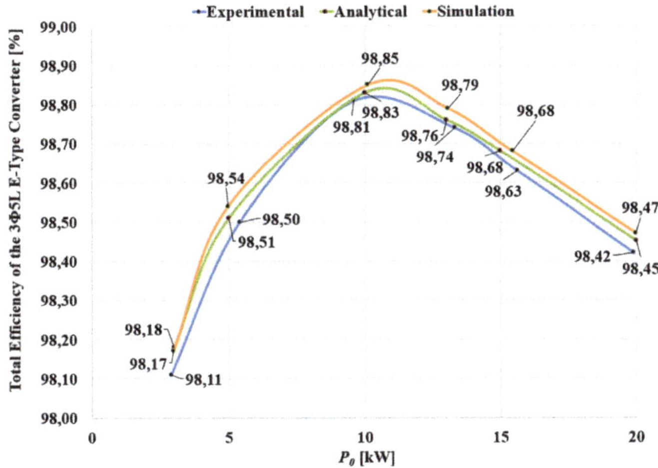

Figure 24. Total efficiency of the 3Φ5L E-Type MMC as a function of the power P_0: analytical result (green line), simulation result (orange line) and experimental result (blue line).

7. Conclusions

The multilevel–multicell 3Φ5L E-Type MMI and 3Φ5L E-Type MMR for stand-alone microgrid applications have been presented and discussed in this paper. The E-Type topology has been carefully studied with reference to the multicell interleaving configuration. The advantages and disadvantages of the proposed multilevel–multicell converters have been clearly explained. To build the prototype of the MMR and MMI, the hardware design process has been discussed. The prototype of the proposed multilevel-multicell has been built, aiming for improvements in the power density and specific power, as well as the power quality of the voltage and current waveforms. In fact, the complete prototype of the 3Φ5L E-Type MMR plus 3Φ5L E-Type MMI presented a power density of 8.4 kW/dm^3 and a specific power of 3.24 kW/kg. To evaluate the performance of the multilevel–multicell converters, the control strategies have been introduced with particular regard to stand-alone microgrid applications. Experimental results confirmed the effectiveness of the proposed multilevel–multicell converters, achieving a peak efficiency of 98.81% using Si power semiconductors, as well as a THD_v of 0.88% and a THD_i of 1.95%.

Author Contributions: Conceptualization, P.J.G., and L.S.; methodology, F.C. and L.S.; software, M.d.B. and A.L.; validation, M.d.B. and A.L.; formal analysis, M.d.B. and A.L.; investigation, M.d.B.; resources, L.S., F.C. and A.L.; data curation, L.S. and A.L.; writing—original draft preparation, M.d.B.; writing— review and editing A.L., L.S., F.C. and P.J.G.; visualization, M.d.B.; supervision, L.S., A.L., F.C. and P.J.G.; project administration, L.S. and P.J.G.; funding acquisition, L.S., A.L., F.C. All authors have read and agreed to the published version of the manuscript.

Funding: This research received no external funding.

Institutional Review Board Statement: Not applicable.

Informed Consent Statement: Not applicable.

Data Availability Statement: Not applicable.

Acknowledgments: The authors are thankful to "Sky Research doo" for building the prototype of the multilevel converters, including the power filter used in the experimental campaign.

Conflicts of Interest: The authors declare no conflict of interest.

Appendix A

The analytical approach is presented in this section to calculate the AVG and the RMS current flowing through the power semiconductors in the 3Φ5L E-Type MMR and MMI. In general, the AVG and RMS current over one fundamental period can be found by Equations (A1) and (A2), where $\theta = \omega t$, ω is the fundamental frequency, i is the sinusoidal phase current, and d_d is the duty cycle of the devices.

$$I_{RMS} = \sqrt{\frac{1}{2\pi} \int_0^\pi [i^2(\theta) \cdot d_d(\theta)] \, d(\theta)} \tag{A1}$$

$$I_{AVG} = \frac{1}{2\pi} \int_0^\pi [i(\theta) \cdot d_d(\theta)] \, d(\theta) \tag{A2}$$

To find the RMS and AVG currents, the duty cycles of the power semiconductors in both rectifier and inverter must be obtained. According to the modulation strategy illustrated in Figure 6, the duty cycle of the devices can be derived from Equation (A3), where $\theta_{in} = \omega_{in} t$, $\theta_0 = \omega_0 t$, ω_{in} and ω_0 the fundamental frequency of the rectifier and inverter, respectively, $A_{n,car}$ is the amplitude of the carriers and $m_{n,car}$ is the offset of the carriers, with $n = 1, 2, 3, 4$, and $m_x(\theta_{in})$, $m_y(\theta_0)$ are the modulation index of the rectifier and inverter, respectively, defined in (A4), with $z \in \{A, B, C\}$ and $w \in \{U, V, W\}$ and $k = 0, 1, 2$.

$$\begin{cases} d_{rect,devices}(\theta_{in}) = \frac{1}{A_{n,car}} \left[\left(\frac{A_{n,car}}{2} - m_{n,car} \right) + m_z(\theta_{in}) \right] \\ d_{inv,devices}(\theta_0) = \frac{1}{A_{n,car}} \left[\left(\frac{A_{n,car}}{2} - m_{n,car} \right) + m_w(\theta_0) \right] \end{cases} \tag{A3}$$

$$\begin{cases} m_z(\theta_{in}) = M_{0R} \sin\left(\theta_{in} - k\frac{2\pi}{3}\right) \\ m_w(\theta_0) = M_{0I} \sin\left(\theta_0 - k\frac{2\pi}{3}\right) \end{cases} \tag{A4}$$

Substituting (A4) into (A3), the duty cycles for each power semiconductors in the single cell rectifier and inverter can be expressed as (A5) and (A6), where $\alpha_{1R} = \arcsin(0.5/M_{0R})$ and $\alpha_{1I} = \arcsin(0.5/M_{0I})$ are the angles between the carrier signals and the modulating signals of the rectifier and inverter, respectively. Replacing (A5) and (A6) into (A1) and (A2), and performing some algebraic manipulations, the RMS and AVG currents in each power semiconductors can be written as in Equation (3), where the coefficients $a_{RMS,i}$, $b_{RMS,i}$, $a_{AVG,i}$, $b_{AVG,i}$, $a_{RMS,j}$, $b_{RMS,j}$, $a_{AVG,j}$, $b_{AVG,j}$ are listed in Tables A1 and A2.

$$\begin{aligned}
d_{S_{x,11}}(\theta_{in}) &= \begin{cases} 0 & \theta_{in} \in [0, \pi], \theta_{in} \in [\pi, \pi + \alpha_1], \theta_{in} \in [2\pi - \alpha_{1R}, 2\pi] \\ -1 - 2M_{0R}\sin(\theta_{in}) & \theta_{in} \in [\pi, 2\pi - \alpha_{1R}] \end{cases} \\
d_{S_{x,12}}(\theta_{in}) &= \begin{cases} 1 & \theta_{in} \in [0, \pi], \theta_{in} \in [\pi, \pi + \alpha_{1R}], \theta_{in} \in [2\pi - \alpha_{1R}, 2\pi] \\ 2[1 + M_{0R}\sin(\theta_{in})] & \theta_{in} \in [\pi + \alpha_{1R}, 2\pi - \alpha_{1R}] \end{cases} \\
d_{D_{x,21}}(\theta_{in}) &= \begin{cases} 0 & \theta_{in} \in [0, \pi] \\ 1 & \theta_{in} \in [\pi, 2\pi] \end{cases} \\
d_{D_{x,22}}(\theta_{in}) &= \begin{cases} 1 & \theta_{in} \in [0, \pi] \\ 0 & \theta_{in} \in [\pi, 2\pi] \end{cases} \\
d_{S_{x,23}}(\theta_{in}) &= \begin{cases} 1 - 2M_{0R}\sin(\theta_{in}) & \theta_{in} \in [0, \alpha_{1R}], \theta_{in} \in [\pi - \alpha_{1R}, \pi] \\ 0 & \theta_{in} \in [\alpha_{1R}, \pi - \alpha_{1R}] \\ 1 & \theta_{in} \in [\pi, 2\pi] \end{cases} \\
d_{S_{x,24}}(\theta_{in}) &= \begin{cases} 1 & \theta_{in} \in [0, \pi] \\ 1 + 2M_{0R}\sin(\theta_{in}) & \theta_{in} \in [\pi, \pi + \alpha_{1R}], \theta_R \in [2\pi - \alpha_{1R}, 2\pi] \\ 0 & \theta_{in} \in [\pi + \alpha_{1R}, 2\pi - \alpha_{1R}] \end{cases} \\
d_{S_{x,31}}(\theta_{in}) &= \begin{cases} 1 & \theta_{in} \in [0, \alpha_{1R}], \theta_R \in [\pi - \alpha_{1R}, \pi], \theta_{in} \in [\pi, 2\pi] \\ 2[1 - M_{0R}\sin(\theta_{in})] & \theta_{in} \in [\alpha_{1R}, \pi - \alpha_{1R}] \end{cases} \\
d_{S_{x,32}}(\theta_{in}) &= \begin{cases} -1 + 2M_{0R}\sin(\theta_{in}) & \theta_{in} \in [\alpha_{1R}, \pi - \alpha_{1R}] \\ 0 & \theta_{in} \in [0, \alpha_{1R}], \theta_{in} \in [\pi - \alpha_{1R}, \pi], \theta_{in} \in [\pi, 2\pi] \end{cases}
\end{aligned} \tag{A5}$$

$$d_{S_{y,11}}(\theta_0) = \begin{cases} 0 & \theta_0 \in [0, \pi], \theta_0 \in [\pi, \alpha_{1I} + \pi], \theta_0 \in [2\pi - \alpha_{1I}, 2\pi] \\ -1 - 2M_{0I}\sin(\theta_0) & \theta_0 \in [\alpha_{1I} + \pi, 2\pi - \alpha_{1I}] \end{cases}$$

$$d_{S_{y,11}}(\theta_0) = \begin{cases} 1 & \theta_0 \in [0, \pi], \theta_0 \in [\pi, \alpha_{1I} + \pi], \theta_0 \in [2\pi - \alpha_{1I}, 2\pi] \\ 2[1 + M_{0I}\sin(\theta_I)] & \theta_0 \in [\alpha_{1I} + \pi, 2\pi - \alpha_{1I}] \end{cases}$$

$$d_{S_{y,21}}(\theta_0) = \begin{cases} 0 & \theta_0 \in [0, \pi] \\ -2M_{0I}\sin(\theta_I) & \theta_0 \in [\pi, \alpha_{1I} + \pi], \theta_0 \in [2\pi - \alpha_{1I}, 2\pi] \\ 1 & \theta_0 \in [\alpha_{1I} + \pi, 2\pi - \alpha_{1I}] \end{cases}$$

$$d_{S_{y,22}}(\theta_0) = \begin{cases} 2M_{0I}\sin(\theta_I) & \theta_0 \in [0, \alpha_{1I}], \theta_0 \in [\pi - \alpha_{1I}, \pi] \\ 1 & \theta_0 \in [\alpha_{1I}, \pi - \alpha_{1I}] \\ 0 & \theta_0 \in [\pi, 2\pi] \end{cases}$$

$$d_{S_{y,23}}(\theta_0) = \begin{cases} 1 - 2M_{0I}\sin(\theta_I) & \theta_0 \in [0, \alpha_{1I}], \theta_0 \in [\pi - \alpha_{1I}, \pi] \\ 0 & \theta_0 \in [\alpha_{1I}, \pi - \alpha_{1I}] \\ 1 & \theta_0 \in [\pi, 2\pi] \end{cases}$$

$$d_{S_{y,24}}(\theta_0) = \begin{cases} 1 & \theta_0 \in [0, \pi] \\ 1 + 2M_{0I}\sin(\theta_I) & \theta_0 \in [\pi, \alpha_{1I} + \pi], \theta_0 \in [2\pi - \alpha_{1I}, 2\pi] \\ 0 & \theta_0 \in [\alpha_{1I} + \pi, 2\pi - \alpha_{1I}] \end{cases}$$

$$d_{S_{y,31}}(\theta_0) = \begin{cases} 1 & \theta_0 \in [0, \alpha_{1I}], \theta_0 \in [\pi - \alpha_{1I}, \pi], \theta_0 \in [\pi, 2\pi] \\ 2[1 - M_{0I}\sin(\theta_I)] & \theta_0 \in [\alpha_{1I}, \pi - \alpha_{1I}] \end{cases}$$

$$d_{S_{y,32}}(\theta_0) = \begin{cases} 0 & \theta_0 \in [0, \alpha_{1I}], \theta_0 \in [\pi - \alpha_{1I}, \pi], \theta_0 \in [\pi, 2\pi] \\ -1 + 2M_0\sin(\theta_I) & \theta_0 \in [\alpha_{1I}, \pi - \alpha_{1I}] \end{cases} \tag{A6}$$

Table A1. 3Φ5L E-Type MMR power semiconductor coefficients of the RMS and AVG currents.

i	Power Semiconductor	Coefficients
1	$D_{x,21}, D_{x,22}$	$a_{RMS,1} = 3\pi - 6\alpha_{1R} - 3\sin(2\alpha_{1R})$ $b_{RMS,1} = 8(\cos(\alpha_{1R}) + 2)(\cos(\alpha_{1R}) - 1)^2$ $a_{AVG,1} = 2\cos(\alpha_{1R})$ $b_{AVG,1} = 2\alpha_{1R} - \sin(2\alpha_{1R})$
2	$S_{x,11}, S_{x,32}$	$a_{RMS,2} = 3\pi - 6\alpha_{1R} + 3\sin(2\alpha_{1R})$ $b_{RMS,2} = -4\cos(\alpha_{1R})(\cos^2(\alpha_{1R}) - 3)$ $a_{AVG,2} = -2\cos(\alpha_{1R})$ $b_{AVG,2} = \sin(2\alpha_{1R}) - 2\alpha_{1R} + \pi$
3	$S_{x,12}, S_{x,31}$	$a_{RMS,3} = \frac{3}{2}(\pi - 2\alpha_{1R} + \sin(2\alpha_{1R}))$ $b_{RMS,3} = 2\left[\cos(\alpha_{1R})(\cos^2(\alpha_{1R}) - 3) + (\cos(\alpha_{1R}) + 2)(\cos(\alpha_{1R}) - 1)^2\right]$ $a_{AVG,3} = 4\cos(\alpha_{1R})$ $b_{AVG,3} = 4\alpha_{1R} - 2\sin(2\alpha_{1R}) - \pi$
4	$S_{x,23}, S_{x,24}$	$a_{RMS,4} = \frac{1}{2}[2\alpha_{1R} - \sin(2\alpha_{1R})]$ $b_{RMS,4} = 2(\cos(\alpha_{1R}) - 1)^2(-\cos(\alpha_{1R}) - 2)$ $a_{AVG,4} = 2(1 - \cos(\alpha_{1R}))$ $b_{AVG,4} = \sin(2\alpha_{1R}) - 2\alpha_{1R}$

Table A2. 3Φ5L E-Type MMI power semiconductor coefficients of the RMS and AVG currents.

i	Power Semiconductor	Coefficients
1	$S_{y,21}, S_{y,22}$	$a_{RMS,1} = 3\pi - 6\alpha_{1I} - 3\sin(2\alpha_{1I}) + 6\sin(2\alpha_{1I})\cos^2(\varphi_0)$ $b_{RMS,1} = -4\sin(\varphi_0)^2 + 6\sin\left(\frac{\alpha_{1I}}{2} + \varphi_0\right)^2 - 2\sin\left(\frac{3\alpha_{1I}}{2} + \varphi_0\right)^2 + \sin(3\alpha_{1I} - \varphi_0)$ $\quad - 6\cos(\alpha_{1I}) + 8\cos(\varphi_0) - 3\cos(\alpha_{1I} - 2\varphi_0)$ $a_{AVG,1} = 4\cos(\alpha_{1I})\cos(\varphi_0)$ $b_{AVG,1} = 2\sin(\alpha_{1I}) + \cos(\varphi_0)[4\alpha_{1I} - 2\sin(2\alpha_{1I}) - 2\varphi_0]$
2	$S_{y,11}, S_{y,32}$	$a_{RMS,2} = 6\alpha_{1I} - 3\pi + 3\sin(2\alpha_{1I}) - 3(1 + \cos(2\varphi_0))\sin(2\alpha_{1I})$ $b_{RMS,2} = 8\cos^3(\alpha_{1I}) + 24\cos^2(\varphi_0)\cos(\alpha_{1I}) - 12\cos^3(\alpha_{1I})\cos(\varphi_0)$ $a_{AVG,2} = -2\cos(\varphi_0)\cos(\alpha_{1I})$ $b_{AVG,2} = \cos(\varphi_0)[\sin(2\alpha_{1I}) - 2\alpha_{1I} + \pi]$
3	$S_{y,12}, S_{y,31}$	$a_{RMS,3} = \frac{3}{2}[\pi - 2\alpha_{1I} + \sin(2\alpha_{1I})\cos(2\varphi_0)]$ $b_{RMS,3} = \frac{1}{2}[3 - 12\cos(\alpha_{1I}) + 4\cos(\varphi_0) + \cos(2\varphi_0) - 6\cos(2\varphi_0)\cos(\alpha_{1I}) +$ $\quad + 2\cos(3\alpha_{1I})\cos(2\varphi_0)]$ $a_{AVG,3} = 8\cos(\alpha_{1I})\cos(\varphi_0)$ $b_{AVG,3} = 2\sin(\varphi_0) + 2\alpha_{1I}\cos(\varphi_0) - 2\varphi_0\cos(\varphi_0) - 3\sin(2\alpha_{1I})\cos(\varphi_0) - \pi\cos(\varphi_0)$
4	$S_{y,23}, S_{y24}$	$a_{RMS,4} = 3\alpha_{1I} - 3\cos(2\varphi_0)cos(\alpha_1)\sin(\alpha_{1I})$ $b_{RMS,4} = -6 + 6cos(\alpha_{1I}) - 2\cos(2\varphi_0) + 6\cos(2\varphi_0)\cos(\alpha_{1I}) - 4\cos(2\varphi_0)\cos^3(\alpha_{1I})$ $a_{AVG,4} = 2 - 2\cos(\varphi_0)\cos(\alpha_{1I})$ $b_{AVG,4} = -2\cos(\varphi_0) + \sin(2\alpha_{1I})\cos(\varphi_0) - 2\alpha_{1I}\cos(\varphi_0) + 2\varphi_0\cos(\varphi_0)$

References

1. Elavarasan, R.M.; Shafiullah, G.M.; Padmanaban, S.; Kumar, N.M.; Annam, A.; Vetrichelvan, A.M.; Mihet-Popa, L.; Holm-Nielsen, J.B. A Comprehensive Review on Renewable Energy Development, Challenges, and Policies of Leading Indian States with an International Perspective. *IEEE Access* **2020**, *8*, 74432–74457. [CrossRef]
2. Ganesan, S.; Subramaniam, U.; Ghodke, A.A.; Elavarasan, R.M.; Raju, K.; Bhaskar, M.S. Investigation on Sizing of Voltage Source for a Battery Energy Storage System in Microgrid with Renewable Energy Sources. *IEEE Access* **2020**, *8*, 188861–188874. [CrossRef]
3. Kesavan, T.; Sheebarani, S.; Gomathy, V.; Kavin, R.; Sivaranjani, S. Renewable Energy Based on Energy Conservation and Crossover System. In Proceedings of the 2020 6th International Conference on Advanced Computing and Communication Systems (ICACCS), Coimbatore, India, 6–7 March 2020; pp. 155–157.
4. Guo, Z.; Wei, W.; Chen, L.; Dong, Z.; Mei, S. Impact of Energy Storage on Renewable Energy Utilization: A Geometric Description. *IEEE Trans. Sustain. Energy* **2020**, 1. [CrossRef]
5. Panigrahi, R.; Mishra, S.K.; Srivastava, S.C.; Srivastava, A.K.; Schulz, N.N. Grid Integration of Small-Scale Photo-voltaic Systems in Secondary Distribution Network—A Review. *IEEE Trans. Ind. Appl.* **2020**, *56*, 3178–3195. [CrossRef]
6. Puchalapalli, S.; Tiwari, S.K.; Singh, B.; Goel, P.K. A Microgrid Based on Wind-Driven DFIG, DG, and Solar PV Ar-ray for Optimal Fuel Consumption. *IEEE Trans. Sustain. Energy* **2020**, *56*, 4689–4699.
7. Salman, U.T.; Al-Ismail, F.S.; Khalid, M. Optimal Sizing of Battery Energy Storage for Grid-Connected and Isolated Wind-Penetrated Microgrid. *IEEE Access* **2020**, *8*, 91129–91138. [CrossRef]
8. Mendola, M.L.; di Benedetto, M.; Lidozzi, A.; Solero, L.; Bifaretti, S. Four-Port Bidirectional Dual Active Bridge Converter for EVs Fast Charging. In Proceedings of the 2019 IEEE Energy Conversion Congress and Exposition (ECCE), Baltimore, MD, USA, 29 September–3 October 2019; pp. 1341–1347.
9. di Benedetto, M.; Lidozzi, A.; Solero, L.; Crescimbini, F.; Bifaretti, S. Hardware design of SiC-based Four-Port DAB Converter for Fast Charging Station. In Proceedings of the 2020 IEEE Energy Conversion Congress and Exposition (ECCE), Detroit, MI, USA, 11–15 October 2020; pp. 1231–1238.
10. Ebrahimzadeh, E.; Blaabjerg, F.; Wang, X.; Bak, C.L. Optimum Design of Power Converter Current Controllers in Large-Scale Power Electronics Based Power Systems. *IEEE Trans. Ind. Appl.* **2018**, *55*, 2792–2799. [CrossRef]
11. Xu, Q.; Jiang, W.; Blaabjerg, F.; Zhang, C.; Zhang, X.; Fernando, T. Backstepping Control for Large Signal Stability of High Boost Ratio Interleaved Converter Interfaced DC Microgrids with Constant Power Loads. *IEEE Trans. Power Electron.* **2019**, *35*, 5397–5407. [CrossRef]
12. Lazarević, V.Ž.; Zubitur, I.; Vasić, M.; Oliver, J.A.; Alou, P.; Patchin, G.; Eltze, J.; Cobos, J.A. High-Efficiency High-Bandwidth Four-Quadrant Fully Digitally Controlled GaN-Based Tracking Power Supply System for Linear Power Amplifiers. *IEEE J. Emerg. Sel. Top. Power Electron.* **2019**, *7*, 664–678. [CrossRef]
13. Yuan, X.; Laird, I.D.; Walder, S. Opportunities, Challenges, and Potential Solutions in the Application of Fast-Switching SiC Power Devices and Converters. *IEEE Trans. Power Electron.* **2021**, *36*, 3925–3945. [CrossRef]

14. Liang, G.; Tafti, H.D.; Farivar, G.G.; Pou, J.; Townsend, C.D.; Konstantinou, G.; Ceballos, S. Analytical Derivation of Intersubmodule Active Power Disparity Limits in Modular Multilevel Converter-Based Battery Energy Storage Systems. *IEEE Trans. Power Electron.* **2021**, *36*, 2864–2874. [CrossRef]

15. Ebrahimi, J.; Karshenas, H. N-Tuple Flying Capacitor Multicell Converter—A Generalized Modular Hybrid Topology. *IEEE Trans. Ind. Electron.* **2019**, *66*, 5004–5014. [CrossRef]

16. Meraj, M.; Bhaskar, M.S.; Iqbal, A.; Al-Emadi, N.; Rahman, S. Interleaved Multilevel Boost Converter with Minimal Voltage Multiplier Components for High-Voltage Step-up Applications. *IEEE Trans. Power Electron.* **2020**, *35*, 12816–12833. [CrossRef]

17. McNeill, N.; Yuan, X.; Anthony, S.P. High-Efficiency NPC Multilevel Converter Using Super-Junction MOSFETs. *IEEE Trans. Ind. Electron.* **2015**, *63*, 25–37. [CrossRef]

18. McNeill, N.; Yuan, X.; Jin, B. A Super-Junction MOSFET-Based 99%+ Efficiency T-Type Multilevel Converter. In Proceedings of the 2018 IEEE Energy Conversion Congress and Exposition (ECCE), Portland, OR, USA, 23–27 September 2018; pp. 5643–5650.

19. Shi, Y.; Shi, Y.; Wang, L.; Xie, R.; Li, H. A 50 kW high power density paralleled-five-level PV converter based on SiC T-type MOSFET modules. In Proceedings of the 2016 IEEE Energy Conversion Congress and Exposition (ECCE), Milwaukee, WI, USA, 18–22 September 2016; pp. 1–8.

20. Schrittwieser, L.; Leibl, M.; Haider, M.; Thöny, F.; Kolar, J.W.; Soeiro, T.B. 99.3% Efficient Three-Phase Buck-Type All-SiC SWISS Rectifier for DC Distribution Systems. *IEEE Trans. Power Electron.* **2019**, *34*, 126–140. [CrossRef]

21. Zhang, L.; Zheng, Z.; Li, C.; Ju, P.; Wu, F.; Gu, Y.; Chen, G. A Si/SiC Hybrid Five-Level Active NPC Inverter with Improved Modulation Scheme. *IEEE Trans. Power Electron.* **2019**, *35*, 4835–4846. [CrossRef]

22. Di Benedetto, M.; Lidozzi, A.; Solero, L.; Crescimbini, F.; Grbovic, P.J. Reliability and Real-Time Failure Protection of the Three-Phase Five-Level E-Type Converter. *IEEE Trans. Ind. Appl.* **2020**, *56*, 6630–6641. [CrossRef]

23. Deng, F.; Lu, Y.; Liu, C.; Heng, Q.; Yu, Q.; Zhao, J. Overview on submodule topologies, modeling, modulation, control schemes, fault diagnosis, and tolerant control strategies of modular multilevel converters. *Chin. J. Electr. Eng.* **2020**, *6*, 1–21. [CrossRef]

24. Pires, V.F.; Cordeiro, A.; Foito, D.; Pires, A.; Martins, J.; Chen, H. A Multilevel Fault-Tolerant Power Converter for a Switched Reluctance Machine Drive. *IEEE Access* **2020**, *8*, 21917–21931. [CrossRef]

25. Omer, P.; Kumar, J.; Surjan, B.S. A Review on Reduced Switch Count Multilevel Inverter Topologies. *IEEE Access* **2020**, *8*, 22281–22302. [CrossRef]

26. Benedetto, M.D.; Lidozzi, A.; Solero, L.; Crescimbini, F.; Grbovic, P.J. Low-Frequency State-Space Model for the Five-Level Unidirectional T-Rectifier. *IEEE Trans. Ind. Appl.* **2017**, *53*, 1127–1137. [CrossRef]

27. Benedetto, M.D.; Lidozzi, A.; Solero, L.; Crescimbini, F.; Grbovic, P.J. Small-Signal Model of the Five-Level Unidirec-tional T-Rectifier. *IEEE Trans. Power Electron.* **2017**, *32*, 5741–5751. [CrossRef]

28. Di Benedetto, M.; Solero, L.; Crescimbini, F.; Lidozzi, A.; Grbović, P.J. 5-Level E-type back to back power converters—A new solution for extreme efficiency and power density. In Proceedings of the 2017 13th Conference on Ph.D. Research in Microelectronics and Electronics (PRIME), Giardini Naxos, Italy, 12–15 June 2017; pp. 341–344.

29. Di Benedetto, M.; Lidozzi, A.; Solero, L.; Crescimbini, F.; Grbovic, P.J. Five-level back to back E-Type converter for high speed gen-set applications. In Proceedings of the IECON 2016—42nd Annual Conference of the IEEE Industrial Electronics Society, Florence, Italy, 23–26 October 2016; pp. 3409–3414.

30. Di Benedetto, M.; Lidozzi, A.; Solero, L.; Crescimbini, F.; Grbovic, P.J. Low Volume and Low Weight 3-Phase 5-Level Back to Back E-Type Converter. *IEEE Trans. Ind. Appl.* **2019**, *55*, 7377–7388. [CrossRef]

31. Di Benedetto, M.; Lidozzi, A.; Solero, L.; Grbovic, P.J.; Bifaretti, S. ISOP DC-DC converters equipped 5-level unidirec-tional T-Rectifier for aerospace applications. In Proceedings of the 2015 IEEE Energy Conversion Congress and Exposition (ECCE), Montreal, QC, Canada, 20–24 September 2015; pp. 1694–1700.

32. Cui, D.; Ge, Q. A Novel Hybrid Voltage Balance Method for Five-Level Diode-Clamped Converters. *IEEE Trans. Ind. Electron.* **2018**, *65*, 6020–6031. [CrossRef]

33. Lin, H.; Shu, Z.; Yao, J.; Yan, H.; Zhu, L.; Luo, D.; He, X. A Simplified 3-D NLM-Based SVPWM Technique with Voltage-Balancing Capability for 3LNPC Cascaded Multilevel Converter. *IEEE Trans. Power Electron.* **2020**, *35*. [CrossRef]

34. Di Benedetto, M.; Lidozzi, A.; Solero, L.; Crescimbini, F.; Grbovic, P.J. Five-Level E-Type Inverter for Grid-Connected Applications. *IEEE Trans. Ind. Appl.* **2018**, *54*, 5536–5548. [CrossRef]

35. Lidozzi, A.; Di Benedetto, M.; Bifaretti, S.; Solero, L.; Crescimbini, F. Resonant Controllers with Three Degrees of Freedom for AC Power Electronic Converters. *IEEE Trans. Ind. Appl.* **2015**, *51*, 4595–4604. [CrossRef]

36. Di Benedetto, M.; Lidozzi, A.; Solero, L.; Crescimbini, F.; Grbovic, P.J. Performance assessment of the 5-level 3-phase back to back E-type converter. In Proceedings of the 2017 IEEE Energy Conversion Congress and Exposition (ECCE), Cincinnati, OH, USA, 1–5 October 2017; pp. 2106–2113.

Article

Efficiency Comparison of 2-Level SiC Inverter and Soft Switching-Snubber SiC Inverter for Electric Motor Drives

Marco di Benedetto *, Luca Bigarelli, Alessandro Lidozzi and Luca Solero

C-PED, Center for Power Electronics and Drives, Roma Tre University, 00146 Roma, Italy;
luca.bigarelli@uniroma3.it (L.B.); alessandro.lidozzi@uniroma3.it (A.L.); luca.solero@uniroma3.it (L.S.)
* Correspondence: marco.dibenedetto@uniroma3.it

Abstract: This paper focuses on the investigation and implementation of a high-performance power conversion system to reduce the overvoltage phenomenon in variable speed electric drive applications. Particularly, the pros and cons of using Silicon Carbide power MOSFETs in the power converter when a long power cable is employed in electric motor drive systems has been addressed. The three-phase two level inverter with the addition of snubber circuits that consist of capacitors and diodes has been investigated, designed and tested in order to mitigate the overvoltage problems without sacrificing the conversion efficiency. Given that the snubber circuit added to the switches can increase losses, an additional circuit is used to recover the energy from the snubber circuit. The proposed analysis has been then validated through an experimental campaign performed on the converter prototype. The experimental results show that the proposed converter can reduce the overvoltage at the electric motor terminals with excellent conversion efficiency compared to the classical solution like the three-phase two level inverter.

Keywords: three-phase inverter; snubber circuit; electric motor overvoltage; a high-performance; power losses

Citation: di Benedetto, M.; Bigarelli, L.; Lidozzi, A.; Solero, L. Efficiency Comparison of 2-Level SiC Inverter and Soft Switching-Snubber SiC Inverter for Electric Motor Drives. *Energies* **2021**, *14*, 1690. https://doi.org/10.3390/en14061690

Academic Editor: Massimiliano Luna

Received: 12 February 2021
Accepted: 14 March 2021
Published: 18 March 2021

1. Introduction

Nowadays, thanks to continuous improvement of power electronics technologies, Wide-Band-Gap (WBG) switching devices are now reaching a satisfactory level of maturity and their performance and capabilities are very useful and promising for very high efficiency applications [1,2]. Among WBG devices, the most technologically mature are Silicon Carbide (SiC) based switches [3]. Using the SiC power semiconductors allows the switching rate to increase so that small power losses can be achieved. Alternatively, keeping constant the power losses, it is possible to increase the switching frequency leading to a reduction of volume and weight of passive energy-storage components. For this reason, being able to integrate components like inductors and capacitors, it is now possible to maximize the power density of the power conversion system. Furthermore, SiC power semiconductors present higher voltage-blocking capability and lower on-state resistance compared to silicon based power semiconductors, as well as several benefits related to thermal management, so they are perfect for high temperature applications [4–6]. Particularly, given that the SiC power semiconductors can operate at extremely high junction temperatures, it is either possible to increase the power density of the whole converter system by decreasing the size of the heatsink, or to reach higher voltage applications without the need to enlarge the heatsink. However, in some cases, the high-power density and the high efficiency can be achieved using sophisticated power conversion topologies, avoiding SiC power semiconductors [7–10]. In these cases, the complexity of the conversion system control strategy increases, due to the large number of power devices [11,12]. The SiC power Metal-Oxide-Semiconductor Field-effect Transistors (MOSFETs) will outshine Insulated Gate Bipolar Transistors (IGBTs) in all drives applications where the use of long cables and short circuit capability is not needed [13]. For example, in the automotive applications SiC power MOSFETs are already a viable alternative to replace IGBT [14]. In this application, it is not requested to have short circuit capability for the semiconductors and, due the presence of really short length connection, the dV/dt should be limited only to prevent Electro Magnetic Interference (EMI) problems and Radio Frequency (RF) oscillations [15]. Unfortunately, as always happens, there are several drawbacks linked to the high switching rate of SiC devices. These drawbacks become more evident when SiC power semiconductors are used inside of the power conversion systems in which there is not a filter or, more generally, a component between the converter and the load that reduces the semiconductor switching dynamics, like in the case of inverter fed electric drives [16,17]. In variable frequency drive applications one of the biggest drawbacks in using the high dynamic devices like SiC power MOSFETs is linked to the overvoltage at the electric motor terminals. In fact, a higher switching rate leads to overvoltage at the motor terminals [17,18]. This additional value of the voltage applied at the motor terminals is repeated at every commutation, during both turn-on and turn-off. Thus, a high stress at motor insulation system can result in a sensible reduction of the motor life [17,19]. Overvoltage problems in Variable Frequency Drives (VFD) have been known since 1980s, when Bipolar Junction Transistors (BJTs) were replaced by the more performant IGBTs [20]. The overvoltage at the motor terminal may be described using the travelling wave and reflection wave theory [21,22], due to the impedance mismatch between the cable-motor and the cable-inverter. In addition to the impedance mismatch between the various element that compose the electrical drive, the magnitude of the overvoltage at the motor terminal depends on several factors, such as cable length, rise time of the semiconductor devices, modulation strategy of the converter and layout of the power converter [20,23]. In fact, longer length gives higher overvoltage, as well as higher dV/dt gives higher overvoltage. Furthermore, the control strategy of the converter combined with fast rise time and long cable lengths can lead to motor stress greater than two times the value of the DC-bus voltage [24]. Moreover, higher switching rate leads to excitation of capacitive parasitic components which become paths for common mode (CM) and differential mode (DM) noises (EMI), bringing additional motor insulation problems and high frequency current components flowing through the grounding path [25]. A further drawback is the excitation of shaft voltage due to the electromagnetic coupling between the stator and the rotor of the electric machine, which allows parasitic currents (bearing currents) to flow through bearings resulting in their premature damage [26].

There are several common techniques to limit and to reduce the switch dynamic in the motor drive applications [27–36]. The first technique is to use the passive filters at the output of the converter since they are easy to use and they can be found easily in the market [28,29]. The second one is to use the gate drivers to shape the dynamic of the semiconductor and the third one is to increase the gate resistance value [15,27]. These latter solutions, in some cases, can increase the switching losses of the power semiconductors, so that the performance of the converter gets worse. To limit and to reduce the overvoltage at the motor terminals some advanced techniques and/or topologies have been proposed in the literature [28–36]. In [18] two different solutions based on the Cascaded H-Bridge multilevel converter and two-level inverter equipped with active gate drivers are proposed, while in [28] an active terminal filter is presented, and it is compared with the common RLC and RC filters. A three-level neutral point clamped transformerless medium-voltage converter combined to the passive filter used to address the common-mode voltage issues has been presented in [29]. In [30,31] the soft-switching snubber inverter is used to reduce the overvoltage on the motor side when both the voltage source inverter equipped with high-speed semiconductor devices and the long power cables are employed. The solution proposed in [30] consists in a simple modification of a two-level inverter by adding a diode-capacitor snubber for each switch of the inverter (snubber two-level inverter). The diode-capacitor snubber is like a conventional resistor-capacitor-diode (RCD) snubber, but it is arranged in a different way. Furthermore, an energy recovery circuit is shared by all the snubber circuits in order to recover the energy dissipated by the passive devices.

In [32] a design procedure for a passive electromagnetic interference filter based on the parallel connection of the inductors and the resistors has been proposed for reducing the overvoltage at the motor terminals. In [33] the mitigation of the overvoltage for an open-end winding motor configuration has been presented using two parallel power converters without employing passive filters, whereas in [34] the overvoltage is reduced thanks to the active reflected wave canceller circuit. In [35] a modular multilevel converter (MMC) with a filter capacitor and the arm inductors has been proposed to reduce the overvoltage at motor terminal. A soft-switching circuit, entitled the auxiliary resonant soft-edge pole (ARSEP), has been proposed in [36] to mitigate the dV/dt.

The goals of this paper are to investigate, to analyze, and to implement a possible solution to reduce the overvoltage across the electrical motor when a long power cable and fast dynamic power semiconductors are used in variable speed electric drives, avoiding the presence of passive filters between the motor and the converter. To this purpose, a snubber two-level inverter (S2L Inverter) already present in the literature which makes use of SiC power MOSFETs has been designed and used to reduce the dV/dt of the SiC devices and to limit the overvoltage on the electric motor side. Furthermore, the performance of the S2L inverter is compared to the classical three-phase two level (2L) inverter. In the paper, following the introduction, the pros and cons of the SiC power MOSFETs in the motor drive application are presented in Section 2. Then the overvoltage phenomenon in motor drive application using SiC power MOSFETs is experimentally shown in Section 3. A case study of a power conversion system called S2L Inverter which is able to limit the overvoltage issue in motor drive application is discussed in Section 4. The experimental results from a S2L laboratory prototype are illustrated in Section 5. Conclusions are finally presented in Section 6.

2. SiC Power MOSFETs in Motor Drive Applications: Pros and Cons

In several industrial drive applications, the inverter and the electric motor are located in different places. Consequently, the use of long length feeding cable is necessary. Due to the presence of the long cable and motor, the inverter load is not anymore only resistive and inductive, but the high value of parasitic capacitance of the entire system changes the load in an inductive, resistive and capacitive load.

SiC MOSFETs can be excellent candidates in motor drive application. Compared to Si IGBTs they have better performance at low load conditions, lower switching losses, an intrinsic body diode with low recovery energy and the possibility to work in synchronous rectification helps a lot to reduce the total amount of power losses. These advantages of the SiC MOSFET over the IGBT have been supported by experimental tests. Figure 1 shows the experimental setup used to test the switching commutation of the power semiconductors when a long cable is connected to the electric motor. Looking at the figure, it is possible to recognize a 7.5 kW induction motor (part number: M2QA132S2B), the power cables of 2.5 m and 15 m with a section area equal to 4 mm^2 (Igus, part number: CF38.60.04) and the inverter. The inverter has been equipped with different switching components and the characteristics and operating parameters of each power semiconductor used as Device Under Test (DUT) are listed in Table 1.

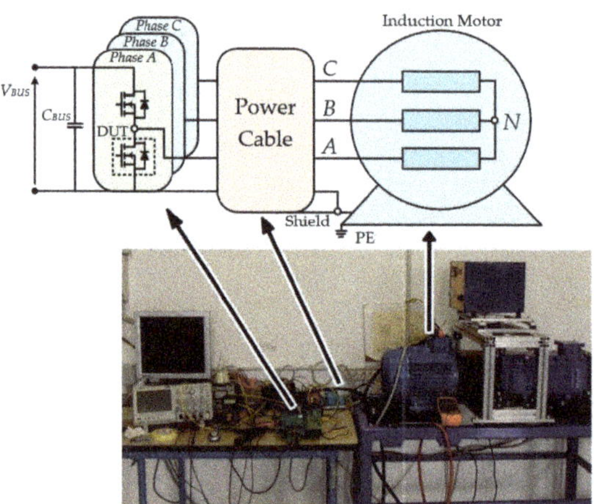

Figure 1. Experimental setup used for testing the effect of the long power cable connected to the motor on the power semiconductors.

Table 1. Operating parameters of each power semiconductors used as DUT.

Part Number	IKW40N120T2	IGBT A	IGBT B	IGBT C	IMW120R045T1
Package	TO 247-3	TO 247-3	TO 247-3	TO 247-3	TO 247-3
Technology	Si	Si	Si	Si	SiC
R_{Gon} [Ω]	20	47	20	10	33.2
R_{Goff} [Ω]	21	48	21	11	33.2
V_{GE} or V_{GS} on [V]	12	12	15	15	20
V_{GE} or V_{GS} off [V]	-15	-15	-5	-5	-5
V_{BUS} [V]	330	330	330	330	520
I_{Phase} RMS [A]	14.1	14.1	12.6	12.6	14.4
Modulation depth M_0	1	1	0.95	0.95	0.95
f_{sw} [kHz]	10	10	10	10	10
f_0 [Hz]	50	50	50	50	50
Dead-Time [μs]	1.8	1.8	1	1	1

To evaluate the thermal response of the different power switches, the case temperature of the device under test was checked every 10 s with a Negative Temperature Coefficient (NTC) sensor directly placed on the case. The power switches in all tests have been turned on and turned off every 60 s, as shown in Figure 2.

The thermal behavior demonstrates a huge improvement in terms of temperature performance of the SiC MOSFET (IMW120R045T1) compared to the IGBTs. This result shows that using the SiC MOSFET is possible to obtain one of these improvements: (1) higher efficiency of the power conversion system at the same switching frequency; (2) keeping constant the efficiency is possible to increase the switching frequency; (3) heatsink lower volume.

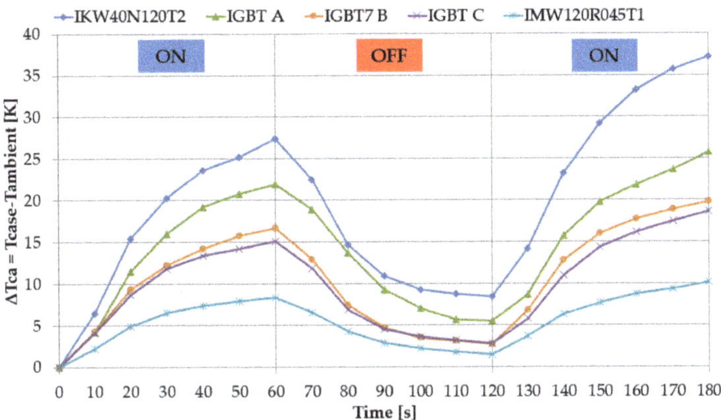

Figure 2. Thermal behavior of the power semiconductors in motor drive application.

The loss reduction of power semiconductors is the key point to increase the power density of the conversion system. Increasing the power density, it is possible either to shrink the system physical dimensions with the same power rating or to maintain constant the physical dimension constant and increase the power rating of the converter. The less effort in power dissipation is also an important sales point, given that less power losses of SiC MOSFETs can lead to a reduction of both the cooling effort per power unit and the cooling cost per power. Naturally, there are also drawbacks of using SiC MOSFETs in motor drives applications. The first one is related to the overload condition requested in this kind of application. Typical overload values are: (1) 150% to 180% of the rated current per 60 s for the standard drives, (2) 200% to 300% of the rated current per 3 s for the servo drives. At heavy loads, IGBTs have a better conduction performance compared to SiC MOSFETs, as well as easier short circuit detection. The second big drawback is the short circuit capability required in industrial drives; a short circuit time withstand of 8 µs to 10 µs is mandatory. Up to now only the commercial SiC MOSFETs from Infineon Technologies (IMW120RxxxM1) have a short circuit capability, however the time is 3 µs, which is far away from the requested ones. The last main drawback is high dV/dt when the SiC power MOSFETs are used in the VFD application.

3. Overvoltage Phenomenon in VFD Applications

In many industrial applications, the converter and the motor are at separate locations. Consequently, the use of long feeding power cables is needed. It has been demonstrated [20–22] that the voltage pulses travel with speed, which is approximately half the speed of light (150 m/µs). If the pulse takes longer than half the rise time to travel from the inverter to the motor, a full reflection occurs at the motor terminals, and the pulse amplitude is approximately doubled. So, if $t_r \leq 2t_d$, where t_r is the voltage rise time at the inverter output terminals and t_d is the cable propagation time, the line-to-line peak voltage V_m on the motor side is almost two times the DC-bus voltage V_{BUS}. Thus, the voltage reflection is a function of both the inverter pulse rise time and the length of the motor cables, as well as of the impedance mismatch between motor, cables and inverter. It is easy to understand that using new fast dynamic devices, line-to-line voltage V_m across the motor side is almost two times the DC bus voltage V_{BUS}. Furthermore, the fast switching produces complex transitions that highly stress the motor insulation; as a result, the machine winding can fail prematurely [37,38]. The influence of the length of power cables connected between the converter and the motor is shown in Figures 3 and 4.

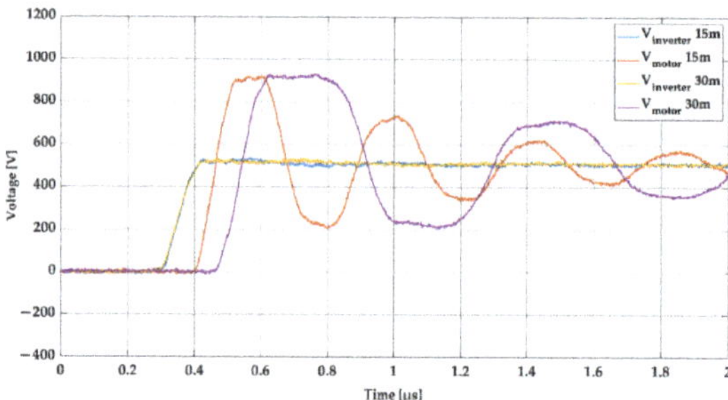

Figure 3. Voltage waveform across the inverter and the motor during the turn-off.

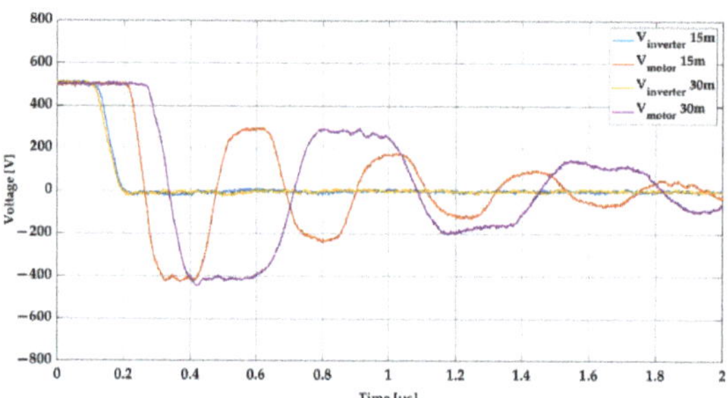

Figure 4. Voltage waveform across the inverter and the motor during the turn-on.

The converter is composed of the SiC power module (Semikron, part number: 26ACM12V17), while two Igus power cables (part number: CF270.UL.60.04.D) with different lengths, 15 m and 30 m, are used to connect the converter to the motor. The figures shown illustrate the turn-off and turn-on of the DUT. The blue waveform is the voltage between the output of the phase A (middle point) and minus DC-bus $V_{BOT,A}$ (voltage across the bottom device of the phase A) and the red waveform is the line-to-line voltage at the motor terminal V_m when the length of the power cable is equal to 15 m. The orange waveform is the inverter voltage $V_{BOT,A}$ and the violet waveform is the line-to-line voltage V_m across the motor when the length of the cable is equal to 30 m. As it can be seen, the voltage V_m is almost two times V_{BUS} in both the figures and the ringing expiration time (RET) increases when the length of the power cable increases too. If the RET is too long, the switching frequency f_{sw} is limited due to the overlap of the peaks, which can lead to an overvoltage higher than two times the DC-bus voltage V_{BUS}.

4. Case Study: Design of the Snubber 2-Level Inverter

A topology solution used to reduce the overvoltage issue at the motor side when the SiC MOSFETs power switches are used in the inverter is addressed in this section. Figure 5 shows the electrical circuit of the S2L inverter.

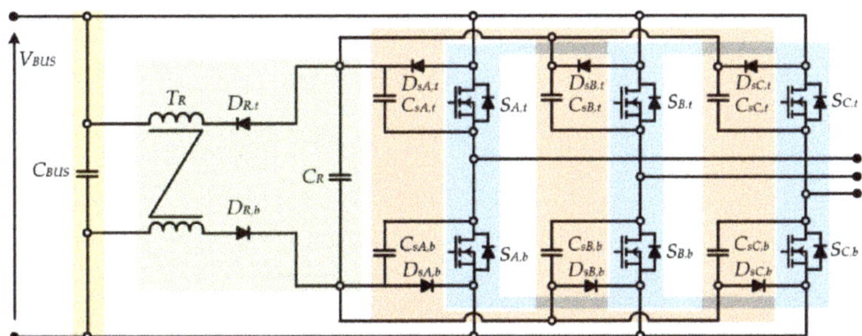

Figure 5. Electrical circuit of snubber two-level (S2L) inverter: (1) DC-bus capacitor in orange, (2) recovery circuit in green, (3) three-phase inverter circuit in blue, and (4) snubber circuit in pink.

The S2L Inverter is composed of four parts: (1) DC-bus capacitor, (2) recovery circuit, (3) three-phase inverter and (4) snubber circuit. Each switch has a snubber circuit that consists of one capacitor $C_{sx,t}$ (or $C_{sx,b}$) and one diode $D_{sx,t}$ (or $D_{sx,b}$), with $x \in \{A, B, C\}$. Since the snubber circuit added to the switches can increase the losses, an additional circuit is used to recovery energy from the snubber circuit. The energy recovery circuit, connected between the snubber circuits and the DC-bus capacitor, is composed by the capacitor C_R, two diodes $D_{R,t}$ and $D_{R,b}$, and the transformer T_R with unitary turns ratio, in order to assure the same current value in both the windings. The operation mode of the S2L Inverter is clearly explained in [31]; in the following the hardware aspect design is discussed.

4.1. Inverter Power Semiconductor Selection

The power semiconductors of the three-phase inverter have been selected according to the voltage and current stresses. The voltage rating of the switches is a function of the DC-bus voltage V_{BUS} and the overvoltage, as shown in (1).

$$V_{sw} = V_{BL} + \overbrace{L_\varsigma \frac{di_{sw}}{dt} + V_{FR}}^{\Delta V} \tag{1}$$

In (1) V_{BL} is the blocking voltage at steady state and ΔV is the switching overvoltage, function of different parameters such as the commutation inductance L_ς, the device current slope di_{sw}/dt and the forward recovery voltage V_{FR} of the complementary freewheeling diode. In the three-phase inverter the blocking voltage V_{BL} is equal to the DC-bus voltage V_{BUS}. Thus, considering 700 V as a nominal DC-bus voltage V_{BUS} the power semiconductors must withstand at least 700 V. However, the overvoltage ΔV must also be considered, especially in motor drive applications in which, as seen in the previous section, the overvoltage can be almost two times the DC-bus voltage V_{BUS} in case of long power cable. Thus, SiC power MOSFETs with at least 1200 V blocking voltage must be chosen. To select the current rating of the power semiconductors, the average (AVG) and Root Means Square (RMS) currents flowing into switches have been calculated according to Equation (2).

$$I_{AVG} = \frac{1}{T_0} \int_0^{T_0} [i_x(t)d_{sw}(t)]dt, \; I_{RMS} = \sqrt{\frac{1}{T_0} \int_0^{T_0} [i_x^2(t)d_{sw}(t)]dt} \tag{2}$$

$$d_{x,t} = \frac{1}{2}[1 + M_0 \sin(2\pi f_0 t)], \; d_{x,b} = \frac{1}{2}[1 - M_0 \sin(2\pi f_0 t)], \tag{3}$$

$$\begin{cases} I_{RMS,Sx} = \sqrt{2}I_{OUT}\sqrt{\frac{1}{24\pi}[8M_0\cos\varphi - 6\pi DTf_{sw} + 3\pi]} \\[2mm] I_{AVG,Sx} = \frac{\sqrt{2}I_{OUT}}{2}\left[\frac{1}{\pi} - \frac{2DTf_{sw}}{\pi} + \frac{M_0\cos\varphi}{4}\right] \\[2mm] I_{RMS,Dx} = \sqrt{2}I_{OUT}\sqrt{\frac{1}{24\pi}[-8M_0\cos\varphi - 6\pi DTf_{sw} + 3\pi]} \\[2mm] I_{AVG,Dx} = \frac{\sqrt{2}I_{OUT}}{2}\left[\frac{1}{\pi} - \frac{2DTf_{sw}}{\pi} - \frac{M_0\cos\varphi}{4}\right] \end{cases} \tag{4}$$

In (2), T_0 is the fundamental period, $i_x = I_0\sin(2\pi f_0 t - {}^2/_3 k\pi - \varphi)$ is the phase current, with φ being the phase displacement, $k = 0, 1, 2, x \in \{A, B, C\}$, and d_{sw} is the duty cycles of the switch. Using the Sinusoidal Pulse-Width-Modulation (S-PWM), the duty cycle of the top and bottom devices ($d_{x,t}$ and $d_{x,b}$) can be derived from the Equation (3), where M_0 is the modulation depth and f_0 is the fundamental frequency. Substituting the (3) into (2), the RMS and AVG currents of the switch and diode can be written as in (4), where DT is the dead-time. According to this analysis the power semiconductors have been selected.

4.2. Snubber Circuit Design

The dV/dt is a function of the snubber capacitor values $C_{sx,t}$ and $C_{sx,b}$, while the diodes $D_{sx,t}$ and $D_{sx,b}$ are used in order to block the current in one direction. During the turn-off of the switch $S_{x,t}$ and turn-on of the switch $S_{x,b}$, the capacitor $C_{sx,t}$ is charged and the capacitor $C_{sx,b}$ is discharged; on the other hand, during the turn-on of the switch $S_{x,t}$ and turn-off of the switch $S_{x,b}$ the capacitor $C_{sx,t}$ is discharged and the capacitor $C_{sx,b}$ is charged. The recap of the turn-off of the switch $S_{x,t}$ is illustrated in Figure 6, where the commutation inductance L_ζ is clearly highlighted.

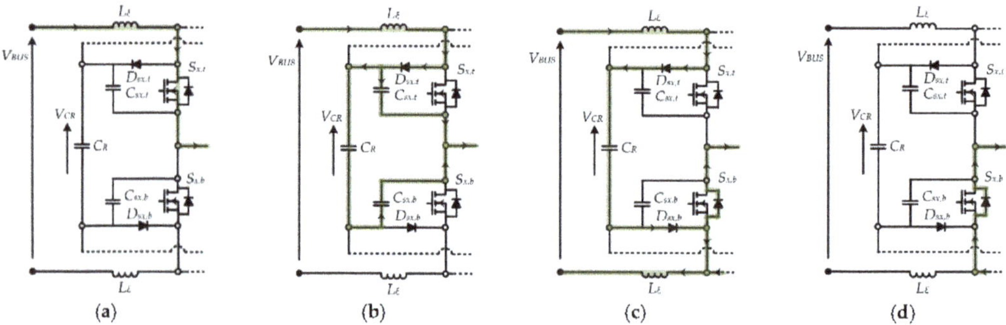

Figure 6. Current path highlighted in green during the turn-off of $S_{x,t}$: (a) step 1, (b) step 2, (c) step 3, (d) step 4.

The turn-off of $S_{x,t}$ occurs in four steps: after step 1, where the switch $S_{x,t}$ is "on" and the current flows through it, the swich $S_{x,t}$ is turned off, the snubber diode $D_{sx,t}$ is "on" and the snubber capacitors $C_{sx,t}$, $C_{sx,b}$ are respectively charged and discharged in step 2; in step 3, the snubber diode $D_{sx,b}$ and the antiparallel diode $D_{x,b}$ start to conduct the current; finally, in step 4, the current flows through the diode $D_{x,b}$ and the switch $S_{x,b}$ will be turned on under zero voltage. The turn-off of the bottom swich $S_{x,b}$ will occur in a similar way. In state two the commutation inductance L_ζ and snubber capacitor $C_{sx,t}$ form a resonant circuit, thus the second-order differential equation can be written in (5). The solution of the (5) is given in (6), where $\omega_r = 1/\sqrt{L_\zeta C_{sx,t}}$ is the resonant frequency, $Z_r = \sqrt{L_\zeta/C_{sx,t}}$ is the damping factor, $V_{Csx,t,0}$ is the initial voltage of the capacitor $C_{sx,t}$ and I_{L0} is the initial current of the commutation inductance L_ζ. From the Equation (6), the dV/dt across the power switch can be obtained as in (7).

$$\frac{dV^2_{Csx,t}}{dt} + \frac{V_{Csx,t}}{L_\zeta C_{sx,t}} = \frac{V_{BUS}}{L_\zeta C_{sx,t}} \tag{5}$$

$$V_{Csx,t} = V_{BUS} - (V_{BUS} - V_{Csx,t,0}) \cos(\omega_r t) + Z_r I_{L,0} \sin(\omega_r t) \tag{6}$$

$$\frac{dV_{Csx,t}}{dt} = \omega_r (V_{BUS} - V_{Csx,t,0}) \sin(\omega_r t) + \omega_r Z_r I_{L,0} \cos(\omega_r t) \tag{7}$$

It can be seen from (7) that the overvoltage is a function of the commutation inductance L_ξ, the snubber capacitance $C_{sx,t}$ and the initial current value I_{L0}. On one hand, high value of the inductance L_ξ provides low value of the resonant frequency ω_r and high value of the damping factor Z_r; consequently, the overvoltage will be high. The high initial current value also provides high overvoltage. On the other hand, the high value of the snubber capacitance $C_{sx,t}$ results in low resonant frequency ω_r and damping factor Z_r values; consequently the overvoltage during the commutation will be low. Finally, from (7) di/dt can be written as in (8). It can be noticed that di/dt is strictly dependent on the inductance L_ξ and the initial current value I_{L0}.

$$\frac{di}{dt} = \frac{1}{L_\xi}(V_{BUS} - V_{Csx,t,0}) \sin(\omega_r t) - \omega_r I_{L,0} \cos(\omega_r t) \tag{8}$$

The maximum value of dV/dt and di/dt occurs when the load current is equal to zero, that is when $\omega_r t = \pi/2$, $V_{Csx,t,0} = 0$, $I_{L,0} = 0$, and they are equal to (9).

$$\begin{cases} \dfrac{dV_{Csx,t}}{dt} = \omega_r V_{BUS} \\[2mm] \dfrac{di}{dt} = \dfrac{V_{BUS}}{L_\xi} \end{cases} \tag{9}$$

From the obtained equations it is possible to select the snubber capacitors, considering that the commutation inductance L_ξ depends on (1) the inductance L_σ introduced by the Printed Circuit Board (PCB) tracks that connect the switching devices to the DC-bus capacitor, (2) the inductance L_{ESL} related to the DC-bus capacitors and (3) the inductance L_{SW} associated with the die and wire bond of the power semiconductors.

4.3. Recovery Energy Circuit Design

As it can be seen from Figure 5, the energy recovery circuit is created using two SiC diodes, the capacitor C_R and the coupled inductor properly designed. The SiC diodes are used to have the unidirectional circulating current. The voltage stress of the diodes $D_{sx,t}$, $D_{sx,b}$, is equal to the DC-bus voltage V_{BUS} plus the overvoltage, whereas the current stress is very low, since they are in on-state for a short time. The transferred average power as energy recovery is given in (10), where $\Delta I_{L,0}$ is the current ripple related to the DC-bus during each switching event.

$$P_R = 3\left(\frac{1}{2}C_{sx,t} + \frac{1}{2}C_{sx,b}\right) V_{BUS}^2 f_{sw} + f_{sw} L_\xi \Delta I_{L,0}^2 \tag{10}$$

According to the analysis presented in [39], using the S-PWM with 3rd harmonic injection and assuming three balanced loads, the DC-bus current ripple over the interval $0\text{-}\pi/3$ is given in (11).

$$\Delta I_{L,0}^2 = \frac{3}{\pi} \int_{\frac{\pi}{3}}^{\frac{\pi}{2}} \left\{ 3\left[\sqrt{2}I_{OUT}\sin(\omega_0 t - \varphi)\right]^2 - \left[\sqrt{2}I_{OUT}\sin\left(\omega_0 t - \frac{2}{3}\pi\right)\right]^2 - \left[\sqrt{2}I_{OUT}\sin\left(\omega_0 t - \frac{4}{3}\pi\right)\right]^2 \right\} d\omega_0 t \tag{11}$$

Replacing (11) into (10) and performing some algebraic manipulations, the average power is given in (12).

$$P_R = 3\left(\frac{1}{2}C_{sx,t} + \frac{1}{2}C_{sx,b}\right) V_{BUS}^2 f_{sw} + f_{sw} L_\xi \frac{I_{OUT}^2}{2}\left[1 + \frac{6\sqrt{3}}{\pi}\right] \tag{12}$$

As it can be seen, the power managed by the energy recovery circuit is a function of switching frequency, load current, snubber capacitors and commutation inductance.

4.4. DC-bus Capacitor Design

The DC-bus capacitors are selected according to the DC-bus capacitor RMS current and DC-bus voltage ripple. It can be proven from [39] that the RMS current into DC-bus capacitor is given in (13).

$$I_{Crms} = I_{OUT}\sqrt{M_0\left[\frac{\sqrt{3}}{\pi} + \left(\frac{4\sqrt{3}}{\pi} - \frac{9}{4}M_0\right)\cos^2\varphi\right]}$$

(13)

This current is a function of the phase current I_{OUT}, modulation depth M_0, and phase displacement φ. Figure 7 shows the normalized RMS DC-bus current I_{Crms}/I_{OUT} as a function of the modulation depth M_0 when $\cos\varphi = 1$ and $\cos\varphi = 0$.

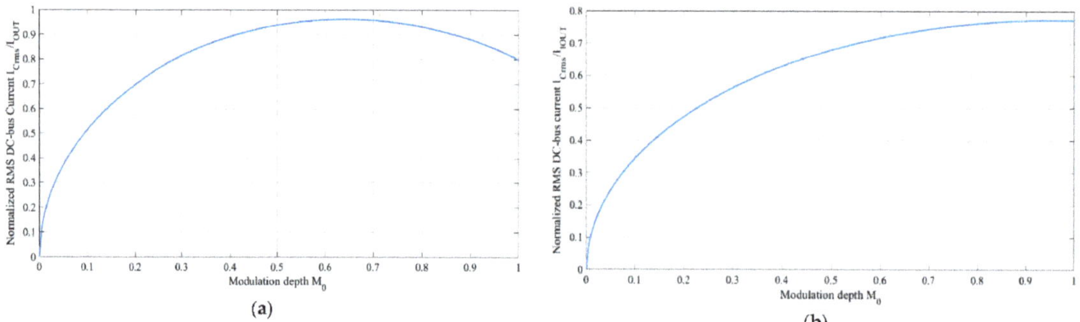

Figure 7. Normalized RMS DC-bus current as a function of the modulation depth: (**a**) $\cos\varphi = 1$, (**b**) $\cos\varphi = 0$.

As it can be seen, the maximum value of the RMS DC-bus current I_{Crms} occurs when the $\cos\varphi$ is equal to 1 and the modulation depth M_0 is between 0.55 and 0.6. The DC-bus capacitor must balance fluctuating instantaneous power on the DC-bus and its value can be selected according to the DC-bus voltage ripple. The required minimum DC-bus capacitor can be obtained from Equation (14), imposing the modulation depth M_0 equal to 0.5.

$$C_{BUS,min} = \frac{\sqrt{3}(1 - M_0)M_0 I_{OUT}}{\sqrt{2}f_{sw}\Delta V_{BUS}}$$

(14)

5. Experimental Results

According to the previous analysis, the three-phase S2L prototype converter has been realized. Particularly, given that the DC-bus is set at 700 V and the output power is set at 18 kW, 1700 V-45 mΩ SiC MOSFETs (part number C2M0045170D) and freewheeling parallel 1200 V-40 A SiC diodes (part number IDW40G120C5B) are selected. The snubber circuit is realized with one SiC diode and two Multilayer Ceramic Capacitors (MLCC). The chosen value of the snubber circuit capacitors $C_{Sx,t}$ and $C_{Sx,b}$ have been selected considering (7) and (8). The energy recovery circuit has been built using two SiC diodes (part number FFSD10120A) and a 40μF film capacitor (part number MKP1848). Based on (10), the coupled inductors have been designed in order to obtain 800 μH at the nominal load. The core selection has been based on the Area Product (AP) method. In particular, the number of turns is set at 41 and the material and the shape of the selected core are ferrite and EE 40/20, respectively. According to (13) and (14), four 40 μF film capacitors (part number MKP1848) are used as a DC-bus capacitor tank. All the components used to build up the S2L converter are listed in Table 2. To validate the proposed analysis and to obtain the overvoltage reduction at the motor side, using the SiC MOSFETs still keeping a high efficiency conversion, the experimental tests on the three-phase S2L inverter have been addressed. The experimental test setup is shown in Figure 8. It can be seen the three-phase S2L inverter, the synchronous machine, that is controlled by the S2L converter, mechanically coupled with an asynchronous machine can be controlled by an external test bench operating on 4-quadrants.

Table 2. Component description used to build the three-phase S2L prototype.

Reference	Part Number	Description
$S_{x,t}$, $S_{x,b}$	C2M0045170D	C2M0045170D: Cree SiC MOSFET 1700 V-45 mΩ-TO247-3
$D_{x,t}$, $D_{x,b}$	IDW40G120C5B	IDW40G120C5B: Infineon SiC Diode 1200 V-40 A-TO247-3
$D_{Sx,t}$, $D_{Sx,b}$	IDW40G120C5B	Infineon SiC Diode 1200 V-40 A-TO247-3
$C_{Sx,t}$, $C_{Sx,b}$	1812SC472KAT1A	AVX MLC Capacitor 4.7 nF-1500 V
$D_{R,t}$-$D_{R,b}$	FFSD10120A	On Semiconductor SiC Diode 1200 V-10 A-DPAK
C_R	MKP1848	MKP1848: Vishay Film Capacitor 40 µF 900V
T_R	N.A.	ICE Mutual Inductor 1:1-800 µH-4A-E42/20 Ferrite
C_{BUS}	5xMKP1848	Vishay Film Capacitor 40 µF 900 V
Current Sensor	LA100-TP	LEM Current Transducer I_{PN} = 100 A
Voltage Sensor	LV 20-P	LEM Voltage Transducer
Heatsink	SK 56/100 SA	Fischer Elektronik 300 mm × 100 mm × 40 mm

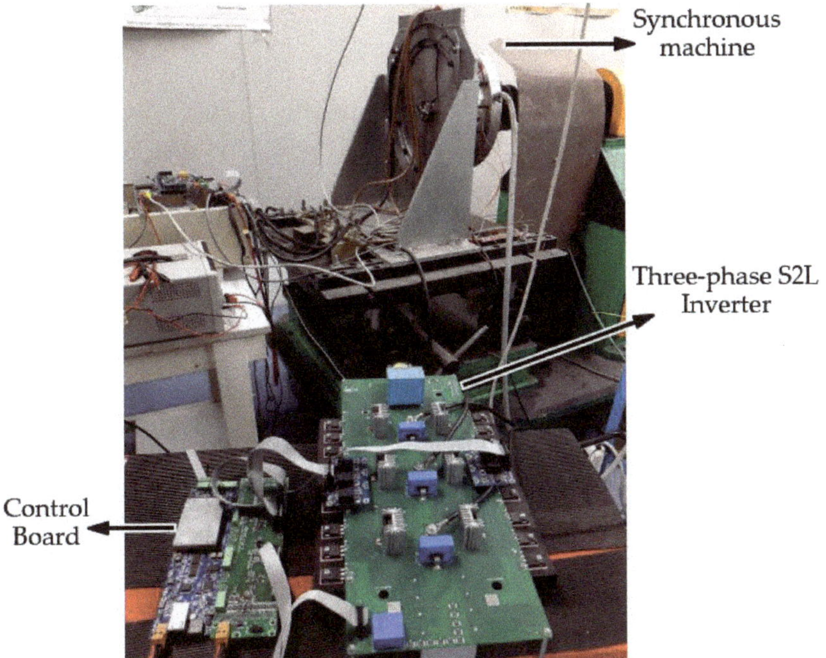

Figure 8. Experimental setup with S2L prototype including the motor drive.

The S2L inverter and the electric drive are connected to each other by 15 m power cables (CF38.60.04 from Igus). The S2L inverter is controlled by a dedicated board based on the National Instruments sbRIO-9651 System on Module (SoM).

The DC-bus voltage V_{BUS} is kept at 400 V using the DC Power Supply available in the laboratory. The achieved results from the three-phase S2L inverter have been compared to the three-phase 2L inverter with the same power devices. This latter inverter has been obtained by removing the snubber circuit and the energy recovery circuit form the three-phase S2L prototype converter. Thus, the power semiconductors and the parasitic inductor L_ζ are completely the same in both converters. Figure 9 shows the turn-off of $S_{A,t}$ located in the three-phase 2L inverter when the gate resistance $R_G = 4.7\,\Omega$, the dead-time $DT = 1\,\mu s$, the switching frequency $f_{sw} = 10$ kHz and the power cable length $l_C = 15$ m. When the phase current is equal to 4.75 A as in Figure 9a, the dV/dt in turn-on is really slow due to the total parasitic capacitances. The value of the voltage V_m at the motor side is only 1.285 times the V_{BUS}. When the phase current increases from 4.75 A to 31.88 A the turn-off behavior is drastically changed, as illustrated in Figure 9b. Here, the commutation inductance shows the negative influences when the 2-level inverter is considered, since the dV/dt is fast (9.95 V/ns), and the voltage overshoot at the drain-source voltage is equal to 131 V (yellow track). This overshoot leads to a voltage at the motor terminals two times the V_{BUS} (845 V). Figure 10 shows the turn-on and the turn-off of the $S_{A,t}$ located in the three-phase 2L inverter when the gate resistance $R_G = 22\,\Omega$, dead-time $DT = 1\,\mu s$, switching frequency $f_{sw} = 10$ kHz and the power cable length $l_C = 15$ m.

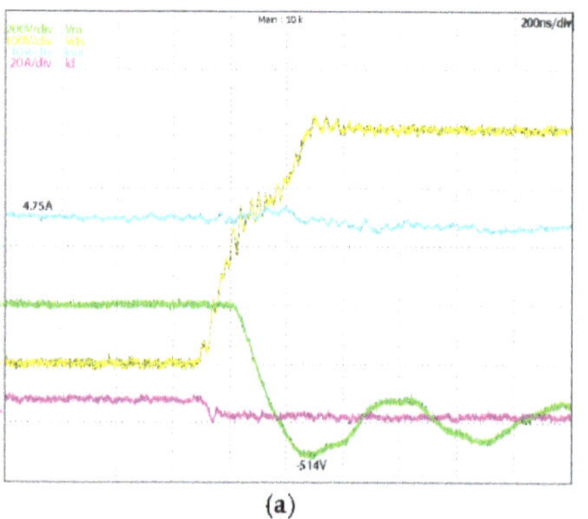

(a)

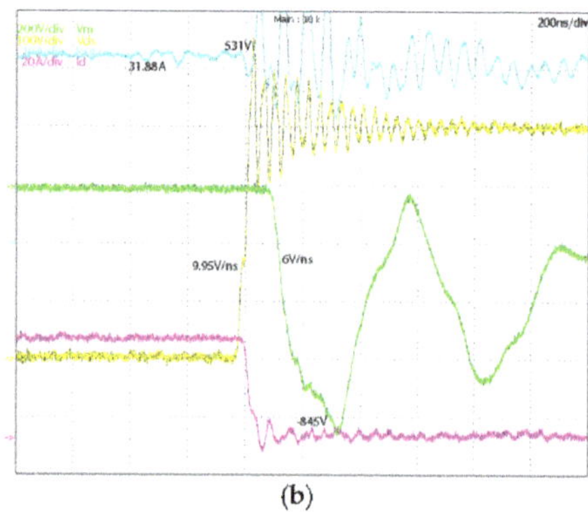

(b)

Figure 9. Turn-off of $S_{A,t}$ located in the three-phase 2L inverter with $R_G = 4.7\,\Omega$, dead-time $DT = 1\,\mu s$, switching frequency $f_{sw} = 10$ kHz and $l_C = 15$ m: (**a**) phase current of 4.75 A, (**b**) phase current of 31.88 A. Line-to-line voltage V_m at the motor side (green track), drain-source voltage V_{DS} of the $S_{A,t}$ (yellow track), drain current of $S_{A,t}$ (magenta track) and phase current I_A (ciano track).

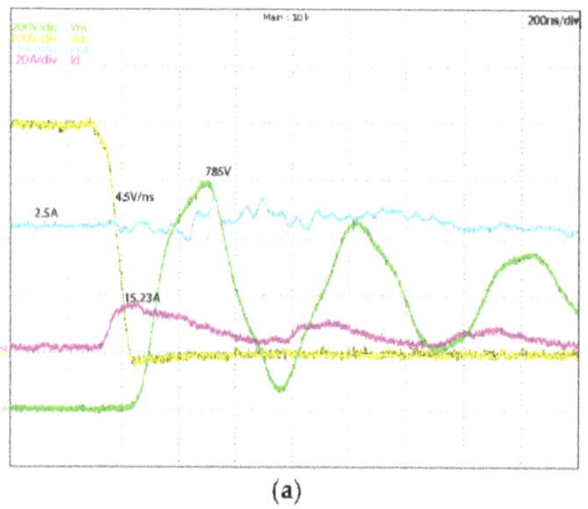

(a)

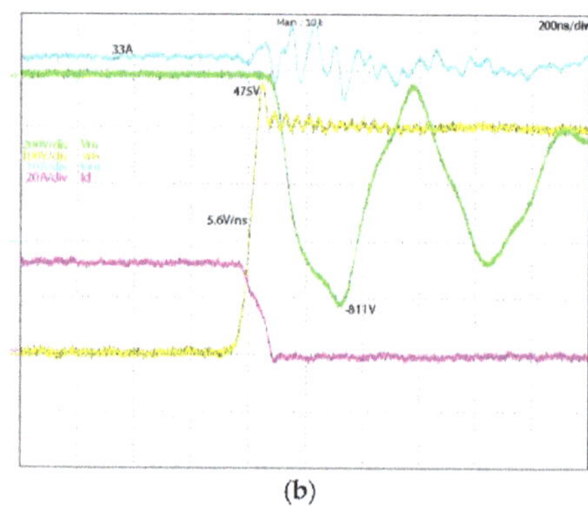

(b)

Figure 10. Commutation of $S_{A,t}$ located in the three-phase 2L inverter with $R_G = 22\,\Omega$, dead-time $DT = 1\,\mu s$, switching frequency $f_{sw} = 10$ kHz and $l_C = 15$ m: (**a**) turn-off, (**b**) turn-on. Line-to-line voltage V_m at the motor side (green track), drain-source voltage V_{DS} of the $S_{A,t}$ (yellow track), drain current of $S_{A,t}$ (magenta track) and phase current I_A (ciano track).

Increasing the gate resistance from 4.7 Ω to 22 Ω for both turn-on and turn-off, the overvoltage at the motor terminals is slightly reduced. As it can be seen from Figure 10a the phase current I_A is equal to 2.5 A, the dV/dt is equal to 4.5 V/ns and the voltage at the motor side V_m is equal to 785 V. Figure 10b shows the turn-off when the phase current I_A is equal to 33A. Thanks to the greater values of the R_G both the dV/dt and the voltage overshoot at the device side are reduced to 5.6 V/ns and 75 V, respectively. However, the overvoltage V_m is still more than double of the V_{BUS} (811 V is 2.02 times the V_{BUS}). Figure 11 shows the turn-on of the $S_{A,t}$ located in the three-phase S2L inverter when the gate resistance $R_G = 4.7$ Ω, snubber capacitors $C_{Sx,t} = 9.4$ μF, dead-time $DT = 1$ μs, switching frequency $f_{sw} = 10$ kHz and the power cable length $l_C = 15$ m. It is possible to notice that the overvoltage at the motor side is smaller than the previous case when the 2-level inverter is used. The values of the overvoltage at the motor terminal are 664 V and 660 V when the phase current I_A is 3.23 A and I_A is 25.6 A, respectively (almost 1.66 times the V_{BUS}). The dV/dt is equal to 2.24 V/ns and 2.06 V/ns when the phase current is 3.23 A and 26.6 A, respectively.

It is possible to notice that the important parameter in the overvoltage issue is not the dV/dt but the rise time (fall time in turn-off) of the drain source voltage. If the two times of the travelling time ($2t_d$) of the voltage wave is greater than the rise/fall time (t_f or t_r) of the drain source voltage, a double voltage appears at the motor terminals.

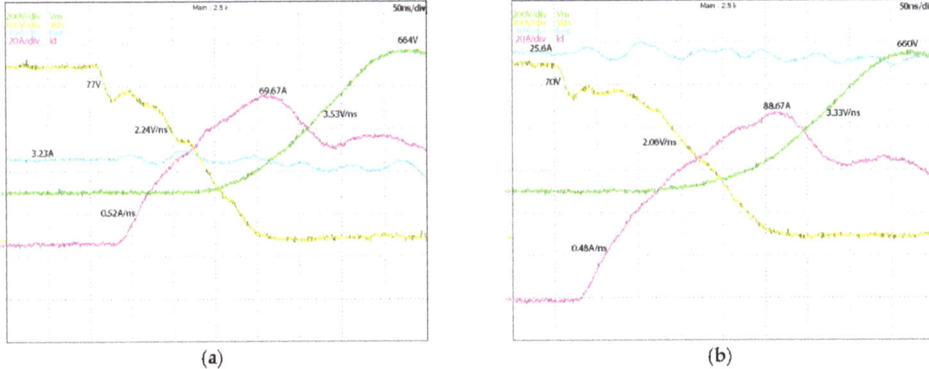

(a)   (b)

Figure 11. Turn-on of $S_{A,t}$ located in the three-phase S2L inverter with $R_G = 4.7\Omega$, $C_{SA,t} = 9.4$ μF dead-time $DT = 1$ μs, switching frequency $f_{sw} = 10$ kHz and $l_C = 15$ m: (**a**) phase current of 3.23 A, (**b**) phase current of 25.6 A. Line-to-line voltage V_m at the motor side (green track), drain-source voltage V_{DS} of the $S_{A,t}$ (yellow track), drain current of $S_{A,t}$ (magenta track) and phase current I_A (ciano track).

Figure 12 shows the turn-off of $S_{A,t}$ located in the three-phase S2L inverter when the phase current is equal to 5.7 A and 31.42 A. It is possible to recognize two different dV/dt (yellow line): the first one related to the snubber capacitor dynamic is equal to 0.3 V/ns and the second one is related to the turn-on of the bottom switch. The voltage at the motor terminals V_m is only 52 V higher than the V_{BUS}. When the phase current is 31.42 A (Figure 12b), the dV/dt is a little bit faster than the previous case (1.47 V/ns), but still no voltage overshoot occurs at the V_{DS} and the overvoltage value is equal to 562 V (1.4 times the V_{BUS}). The turn-off commutations of the three-phase S2L inverter occur at zero current; this means that all the losses are shifted to the turn-on commutations. The experimental efficiency of the three-phase 2L inverter and the three-phase S2L inverter is experimentally evaluated controlling the speed of the Permanent Magnet Synchronous Generators (PMSM) at constant torque. Figure 13 shows the efficiency of the three-phase 2L inverter versus modulations depth M_0 for different values of the switching frequency (10 kHz and 15 kHz) and gate resistance (4.7 Ω and 22 Ω) when the power cable length l_C is equal to 15 m, the DC-bus voltage V_{BUS} is equal to 400 V and the phase peak current I_x is equal to 30 A.

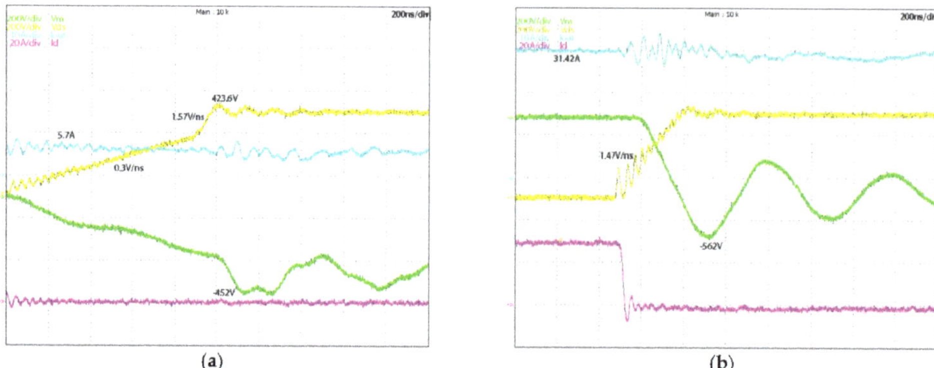

(a)   (b)

Figure 12. Turn-off of $S_{A,t}$ located in the three-phase S2L inverter with $R_G = 4.7\Omega$, $C_{SA,t} = 9.4$ μF dead-time $DT = 1$ μs, switching frequency $f_{sw} = 10$ kHz and $l_C = 15$ m: (**a**) phase current of 5.7 A, (**b**) phase current of 25.6 A. Line-to-line voltage V_m at the motor side (green track), drain-source voltage V_{DS} of the $S_{A,t}$ (yellow track), drain current of $S_{A,t}$ (magenta track) and phase current I_A (ciano track).

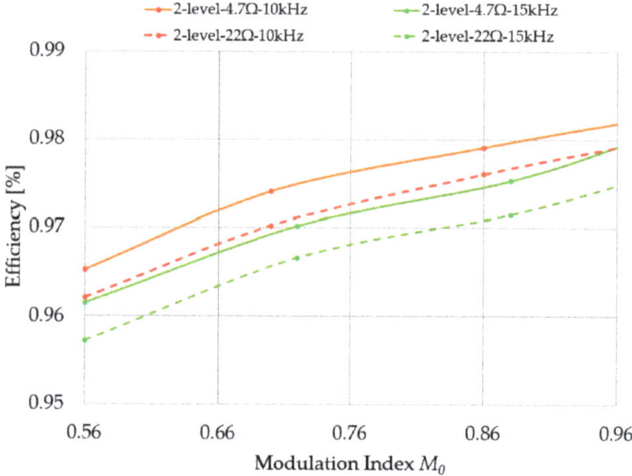

Figure 13. Efficiency of the three-phase 2L inverter as a function of modulation depth M_0 with $l_C = 15$ m $V_{BUS} = 400$ V, $I_x = 30$ A.

The efficiency shows the increasing trend when the modulation depth M_0 increases. However, when the gate drive resistance changes from 4.7 Ω to 22 Ω, the efficiency decreases. The efficiency trend at $f_{sw} = 10$ kHz, $R_G = 22$ Ω is close to the efficiency curve at $f_{sw} = 15$ kHz, $R_G = 4.7$ Ω. Figure 14 shows the efficiency of the three-phase S2L inverter versus modulations depth M_0 for different values of the switching frequency (10 kHz and 15 kHz), gate resistance (4.7 Ω and 22 Ω) and snubber capacitors (4.7 nF and 9.4 nF) when the power cable length l_C is equal to 15 m, the DC-bus voltage V_{BUS} is equal to 400 V and the phase peak current I_x is equal to 30 A. As it can be seen, the efficiency of the three-phase S2L inverter is lower than the efficiency of the three-phase 2L inverter. This is due to the snubber circuit losses. Furthermore, the efficiency of the S2L inverter highly depends on the snubber capacitor values and the switching frequency. When a capacitor value of 4.7 nF is used, the drop of efficiency from switching frequency of 10 kHz to switching frequency of 15 kHz is less than 0.5% (orange line). On the other hand, when a 9.4 nF snubber value is used, the efficiency drop increase to 0.7%. This happens due to the high value of the inrush current into snubber capacitors during the turn on, which is a function of the snubber capacitor values. Higher value of snubber capacitor led to a higher inrush current. From the overvoltage mitigation point of view, the best S2L inverter setup happens when R_G is equal to 4.7 Ω and the snubber capacitors $C_{Sx,t}$ and $C_{sx,b}$ are equal to 9.4 nF. In this condition, the performance of the S2L inverter is slightly lower than the 2L inverter.

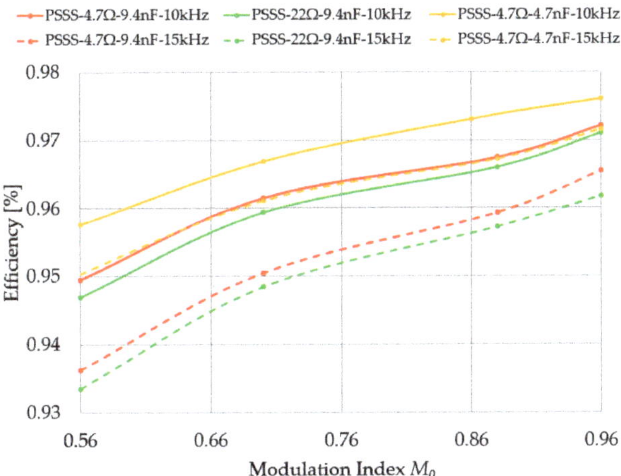

Figure 14. Efficiency of the three-phase S2L inverter as a function of modulation depth M_0 with $l_C = 15$ m $V_{BUS} = 400$ V, $I_x = 30$ A.

6. Conclusions

The benefits of using SiC MOSFETs in the motor drive applications have been discussed in the first part of this paper. The experimental tests performed on the power semiconductors in the first part of the paper show that SiC MOSFET (IMW120R045T1) allows a huge improvement in terms of temperature performance compared to the IGBTs. After that, the overvoltage issue when the electric drive is supplied by the converter with long power cables has been highlighted and supported by the experimental tests. The latter illustrate that the voltage at the motor terminal V_m is two times the DC bus voltage V_{BUS} when the length of the power cable is equal to 15 m or 30 m. A case study of a three-phase two level inverter with the addition of snubber circuits has been discussed to solve the overvoltage problem in the motor drive application. Particularly, the prototype of the S2L inverter has been built according to the proposed design guidelines. Experimental results confirm the effectiveness of the proposed converters in significantly reducing the overvoltage at motor terminals and mitigating the windings insulation stress. When the 2L inverter is used the peak voltage at the motor side is equal to 845 V and the dV/dt is equal to 9.95 V/ns, while in the S2L inverter the peak voltage is equal to 562 V and the dV/dt is equal to 1.47 V/ns. However, the conversion efficiency is slightly reduced compared to the three-phase 2L inverter. Considering the same operating conditions, the peak efficiency of the three-phase 2L inverter is equal to 0.982%, whereas the peak efficiency of the three-phase S2L inverter is equal to 0.976%. Consequently, the proposed analysis shows that by using the S2L inverter it is possible to find a good compromise between dV/dt, overvoltage and efficiency performance of the entire electric drive system.

Author Contributions: Conceptualization, L.S. and A.L.; methodology, L.S. and A.L.; software, M.d.B. and L.B.; validation, M.d.B. and L.B.; formal analysis, M.d.B. and L.B.; investigation, M.d.B.; resources, L.B. and M.d.B.; data curation, M.d.B. and L.B.; writing—original draft preparation, M.d.B. and L.B.; writing— review and editing L.S., and A.L.; visualization, M.d.B.; supervision, L.S. and A.L.; project administration, L.S. and A.L.; funding acquisition, L.S. All authors have read and agreed to the published version of the manuscript.

Funding: This research was funded by "ECPE—European Center for Power Electronics".

Institutional Review Board Statement: The study does not involve humans or animals.

Informed Consent Statement: The study does not involve humans or animals.

Data Availability Statement: The study does not report any data.

Conflicts of Interest: The authors declare no conflict of interest.

References

1. Jafari, A.; Nikoo, M.S.; Perera, N.; Yildirim, H.K.; Karakaya, F.; Soleimanzadeh, R.; Matioli, E. Comparison of Wide-Band-Gap Technologies for Soft-Switching Losses at High Frequencies. *IEEE Trans. Power Electron.* **2020**, *35*, 12595–12600. [CrossRef]
2. Shenai, K. High-Density Power Conversion and Wide-Bandgap Semiconductor Power Electronics Switching Devices. *Proc. IEEE* **2019**, *107*, 2308–2326. [CrossRef]
3. Tong, Z.; Gu, L.; Ye, Z.; Surakitbovorn, K.; Rivas-Davila, J. On the Techniques to Utilize SiC Power Devices in High- and Very High-Frequency Power Converters. *IEEE Trans. Power Electron.* **2019**, *34*, 12181–12192. [CrossRef]
4. La Mendola, M.; Di Benedetto, M.; Lidozzi, A.; Solero, L.; Bifaretti, S. Four-Port Bidirectional Dual Active Bridge Converter for EVs Fast Charging. In Proceedings of the 2019 IEEE Energy Conversion Congress and Exposition (ECCE), Baltimore, MD, USA, 29 September–3 Octomber 2019; pp. 1341–1347.
5. Hussain, M.W.; Elahipanah, H.; Zumbro, J.E.; Rodriguez, S.; Malm, B.G.; Mantooth, H.A.; Rusu, A. A SiC BJT-Based Negative Resistance Oscillator for High-Temperature Applications. *IEEE J. Electron Devices Soc.* **2019**, *7*, 191–195. [CrossRef]
6. Funaki, T.; Balda, J.C.; Junghans, J.; Kashyap, A.S.; Mantooth, H.A.; Barlow, F.; Hikihara, T. Power Conversion With SiC Devices at Extremely High Ambient Temperatures. *IEEE Trans. Power Electron.* **2007**, *22*, 1321–1329. [CrossRef]
7. Anderson, J.A.; Hanak, E.J.; Schrittwieser, L.; Guacci, M.; Kolar, J.W.; Deboy, G. All-Silicon 99.35% Efficient Three-Phase Seven-Level Hybrid Neutral Point Clamped/Flying Capacitor Inverter. *Cpss Trans. Power Electron. Appl.* **2019**, *4*, 50–61. [CrossRef]
8. Di Benedetto, M.; Solero, L.; Crescimbini, F.; Lidozzi, A.; Grbovic, P.J. 5-Level E-type back to back Power Converters—A New Solution for Extreme Efficiency and Power Density. In Proceedings of the 2017 13th Conference on Ph.D. Research in Microelectronics and Electronics (PRIME), Giardini Naxos, Italy, 12–15 June 2017; pp. 341–344.
9. Di Benedetto, M.; Lidozzi, A.; Solero, L.; Crescimbini, F.; Grbovic, P.J. Five-Level back to back E-Type Converter for High Speed Gen-Set Applications. In Proceedings of the IECON 2016—42nd Annual Conference of the IEEE Industrial Electronics Society, IEEE, Florence, Italy, 24–27 October 2016; pp. 3409–3414.
10. Di Benedetto, M.; Lidozzi, A.; Solero, L.; Grbovic, P.J.; Bifaretti, S. ISOP DC-DC Converters Equipped 5-Level Unidirection-al T-Rectifier for Aerospace Applications. In Proceedings of the 2015 IEEE Energy Conversion Congress and Exposition (ECCE), Montreal, QC, Canada, 20–24 September 2015; pp. 1694–1700.
11. Di Benedetto, M.; Lidozzi, A.; Solero, L.; Crescimbini, F.; Grbovic, P.J. Low-Frequency State-Space Model for the Five-Level Unidirectional T-Rectifier. *IEEE Trans. Ind. Appl.* **2017**, *53*, 1127–1137. [CrossRef]
12. Di Benedetto, M.; Lidozzi, A.; Solero, L.; Crescimbini, F.; Grbovic, P.J. Small-Signal Model of the Five-Level Unidirec-tional T-Rectifier. *IEEE Trans. Power Electron.* **2017**, *32*, 5741–5751. [CrossRef]
13. Klobucar, B.; Yuan, Z. 1200V Discrete CoolSiC(TM) MOSFETs in a Comparison with the Trenchstop HighSpeed IGBTs for High-Speed Spindles and Servo Drive System. In Proceedings of the PCIM Europe digital days 2020; International Exhibition and Conference for Power Electronics, Intelligent Motion, Renewable Energy and Energy Management, Nuremberg, Germany, 7–8 July 2020; pp. 1–6.
14. Han, D.; Sarlioglu, B. Comprehensive Study of the Performance of SiC MOSFET-Based Automotive DC–DC Converter under the Influence of Parasitic Inductance. *IEEE Trans. Ind. Appl.* **2016**, *52*, 5100–5111. [CrossRef]
15. Yang, Y.; Wen, Y.; Gao, Y. A Novel Active Gate Driver for Improving Switching Performance of High-Power SiC MOSFET Modules. *IEEE Trans. Power Electron.* **2019**, *34*, 7775–7787. [CrossRef]
16. Morya, A.K.; Gardner, M.C.; Anvari, B.; Liu, L.; Yepes, A.G.; Doval-Gandoy, J.; Toliyat, H.A. Wide Bandgap Devices in AC Electric Drives: Opportunities and Challenges. *IEEE Trans. Transp. Electrif.* **2019**, *5*, 3–20. [CrossRef]
17. Narayanasamy, B.; Sathyanarayanan, A.S.; Luo, F.; Chen, C. Reflected Wave Phenomenon in SiC Motor Drives: Conse-quences, Boundaries, and Mitigation. *IEEE Trans. Power Electron.* **2020**, *35*, 10629–10642. [CrossRef]
18. Leuzzi, R.; Monopoli, V.G.; Cupertino, F.; Zanchetta, P. Comparison of Two Possible Solution for Reducing Over-voltages at the Motor Terminals in High-Speed AC Drives. In Proceedings of the 2019 21st European Conference on Power Electronics and Applications (EPE '19 ECCE Europe), Genova, Italy, 3–5 Septemper 2019; pp. P.1–P.10.
19. Rodríguez, J.; Pontt, J.; Silva, C.; Musalem, R.; Newman, P.; Vargas, R.; Fuentes, S. Resonances and overvoltages in a medium-voltage fan motor drive with long cables in an under-ground mine. *IEEE Trans. Ind. Appl.* **2006**, *42*, 856–863. [CrossRef]
20. Skibinski, G.; Leggate, D.; Kerkman, R. Cable Characteristics and Their Influence on Motor over-Voltages. In Proceedings of the APEC 97—Applied Power Electronics Conference, Atlanta, GA, USA, 27 February 1997; Volume 1, pp. 114–121.
21. Persson, E. Transient effects in application of PWM inverters to induction motors. *IEEE Trans. Ind. Appl.* **2002**, *28*, 1095–1101. [CrossRef]
22. Pulsinelli, F.; Solero, L.; Chiola, D.; Sobe, K.; Brucchi, F. Overvoltages at Motor Terminals in SiC Electric Drives. In Proceedings of the 2018 International Symposium on Power Electronics, Electrical Drives, Automation and Motion (SPEEDAM), Amalfi, Italy, 20–22 June 2018; pp. 513–518.

23. Kerkman, R.J.; Leggate, D.; Schlegel, D.; Skibinski, G. PWM Inverters and Their Influence on Motor Overvoltage. In Proceedings of the APEC—Applied Power Electronics Conference, Atlanta, GA, USA, 27 February 1997; Volume 1, pp. 103–113.

24. Amarir, S.; Al-Haddad, K. Mathematical Analysis and Experimental Validation of Transient over-Voltage Higher than 2 per Unit along Industrial ASDM Long Cables. In Proceedings of the 2008 IEEE Power Electronics Specialists Conference, Rhodes, Greece, 15–19 June 2008; pp. 1846–1851.

25. Zhao, F.; Xiao, G.; Zhu, T.; Zhao, T.; Zheng, X.; Wu, Z. Harmonic Analysis and Suppression Method of Output Current Distortion for Medium-Voltage Motor Drives with Modular Multilevel Converter. *IEEE Trans. Power Electron.* **2019**, *35*, 744–759. [CrossRef]

26. Akagi, H.; Doumoto, T. An Approach to Eliminating High-Frequency Shaft Voltage and Ground Leakage Current from an Inverter-Driven Motor. *IEEE Trans. Ind. Appl.* **2004**, *40*, 1162–1169. [CrossRef]

27. Zhou, W.; Diab, M.S.; Yuan, X. Mitigation of Motor Overvoltage in SiC-Device-Based Drives using a Soft-Switching In-verter. In Proceedings of the 2020 IEEE Energy Conversion Congress and Exposition (ECCE), Detroit, MI, USA, 11–15 October 2020; pp. 662–669.

28. Yuen, K.K.-F.; Chung, H.S.-H.; Cheung, V.S.-P. An Active Low-Loss Motor Terminal Filter for Overvoltage Suppression and Common-Mode Current Reduction. *IEEE Trans. Power Electron.* **2011**, *27*, 3158–3172. [CrossRef]

29. Parreiras, T.M.; Prado, B.M.P.D.; Filho, B.D.C. Common-Mode Overvoltage Mitigation in a Medi-um-Voltage Pump Motor Transformerless Drive in a Mining Plant. *IEEE Trans. Ind. Appl.* **2018**, *54*, 848–857. [CrossRef]

30. Peng, F.; Su, G.-J.; Tolbert, L. A passive soft-switching snubber for PWM inverters. *IEEE Trans. Power Electron.* **2003**, *19*, 363–370. [CrossRef]

31. Pulsinelli, F.; Di Benedetto, M.; Lidozzi, A.; Solero, L.; Crescimbini, F. Power Losses Distribution in SiC Inverter Based Electric Motor Drives. *IEEE Trans. Ind. Appl.* **2019**, *55*, 7843–7853. [CrossRef]

32. Akagi, H.; Matsumura, I. Overvoltage Mitigation of Inverter-Driven Motors with Long Cables of Different Lengths. *IEEE Trans. Ind. Appl.* **2011**, *47*, 1741–1748. [CrossRef]

33. De Caro, S.; Foti, S.; Scimone, T.; Testa, A.; Scelba, G.; Pulvirenti, M.; Russo, S. Motor Overvoltage Mitigation on SiC MOSFET Drives Exploiting an Open-End Winding Configuration. *IEEE Trans. Power Electron.* **2019**, *34*, 11128–11138. [CrossRef]

34. Zhang, Y.; Li, H.; Peng, F.Z. A Low-Loss Compact Reflected Wave Canceller for SiC Motor Drives. *IEEE Trans. Power Electron.* **2021**, *36*, 2461–2465. [CrossRef]

35. Parida, N.; Das, A. A Modular Multilevel Converter with Filter Capacitor for Long-Cable-Fed Drive Application. *IEEE Trans. Ind. Appl.* **2019**, *55*, 7833–7842. [CrossRef]

36. Cai, M.; Wasynczuk, O.; Saeedifard, M. A Voltage-Edge-Rate-Limiting Soft-Switching Inverter Based on Auxiliary Reso-nant Pole. *IEEE J. Emerg. Sel. Top. Power Electron.* **2019**, *7*, 736–744. [CrossRef]

37. Bellomo, J.P.; Lebey, T.; Oraison, J.M.; Peltier, F. Influence of Rise Time on the Dielectric Behavior of Stator Insulation Mate-rials. In Proceedings of the Conference on Electrical Insulation and Dielectric Phenom—CEIDP '96, Millbrae, CA, USA, 23 October 1996; Volume 2, pp. 842–845.

38. Fabiani, D.; Montanari, G.C.; Cavallini, A.; Mazzanti, G. Relation between space charge accumulation and partial dis-charge activity in enameled wires under PWM-like voltage waveforms. *IEEE Trans. Dielectr. Electr. Insul.* **2004**, *11*, 393–405. [CrossRef]

39. Pei, X.; Zhou, W.; Kang, Y. Analysis and Calculation of DC-Link Current and Voltage Ripples for Three-Phase Inverter With Unbalanced Load. *IEEE Trans. Power Electron.* **2015**, *30*, 5401–5412. [CrossRef]

Article

Multi-Level Multi-Input Converter for Hybrid Renewable Energy Generators

Salvatore Foti [1], Antonio Testa [1,*], Salvatore De Caro [1], Luigi Danilo Tornello [2], Giacomo Scelba [2] and Mario Cacciato [2]

[1] Dipartimento di Ingegneria, University of Messina, 98166 Messina, Italy; sfoti@unime.it (S.F.); sdecaro@unime.it (S.D.C.)
[2] Dipartimento di Ingegneria Elettrica Elettronica e Informatica, University of Catania, 95125 Catania, Italy; luigi.tornello@unict.it (L.D.T.); giacomo.scelba@unict.it (G.S.); mario.cacciato@unict.it (M.C.)
* Correspondence: atesta@unime.it

Abstract: A three-phase multi-level multi-input power converter topology is presented for grid-connected applications. It encompasses a three-phase transformer that is operated on the primary side in an open-end winding configuration. Thus, the primary winding is supplied on one side by a three-phase N-level neutral point clamped inverter and, on the other side, by an auxiliary two-level inverter. A key feature of the proposed approach is that the N-level inverter is able to perform independent management of $N - 1$ input power sources, thus avoiding the need for additional dc/dc power converters in hybrid multi-source systems. Moreover, it can manage an energy storage system connected to the dc-bus of the two-level inverter. The N-level inverter operates at a low switching frequency and can be equipped with very low on-state voltage drop Insulated-Gate Bipolar Transistor (IGBT) devices, while the auxiliary inverter is instead operated at low voltage according to a conventional high-frequency two-level Pulse Width Modulation (PWM) technique and can be equipped with very low on-state resistance Metal-Oxide-Semiconductor Field-Effect Transistor (MOSFET) devices. Simulations and experimental results confirm the effectiveness of the proposed approach and its good performance in terms of grid current harmonic content and overall efficiency.

Keywords: multi-level inverter; multi-input converter; renewable energy sources; maximum power point tracking; solar power; wind power; open-end winding

Citation: Foti, S.; Testa, A.; De Caro, S.; Tornello, L.D.; Scelba, G.; Cacciato, M. Multi-Level Multi-Input Converter for Hybrid Renewable Energy Generators. *Energies* **2021**, *14*, 1764. https://doi.org/10.3390/en14061764

Academic Editors: Athanasios I. Papadopoulos and Massimiliano Luna

Received: 24 February 2021
Accepted: 19 March 2021
Published: 22 March 2021

Publisher's Note: MDPI stays neutral with regard to jurisdictional claims in published maps and institutional affiliations.

1. Introduction

The number of electricity generators powered by renewable energy sources (RESs) is continuously increasing because of concerns about environmental pollution and the limited reserves of fossil energy sources such as oil, coal, and gas [1]. Grid-connected photovoltaic (PV) and wind turbine (WT) generators are the most widely diffused types of RES power plants and their specific cost is continuously decreasing [2–4]. However, available solar and wind energy are affected by some factors, such as season cycle, daily cycle, temperature, and weather conditions, which make them intermittent and stochastic. Therefore, a power plant relying only on a single form of RES and without an energy storage capability can hardly cope with the requirements for a reliable electric power generation unit. Hybrid renewable energy systems (HRESs) combining more than one energy source are a viable solution to this problem [5] because they are effective not only in enhancing the reliability of power supply but also in reducing the size of energy storage systems [6,7]. However, in HRESs, a specific dc–dc power converter is normally used to manage each input power source, leading to a quite complex and expensive structure [8,9].

The multiple input power converter (MIPC) concept is a possible alternative to HRESs having to cope with sources with different power capacity and/or voltage levels, providing a well-regulated dc output voltage. Both isolated and non-isolated dc/dc multi-input converters find application in hybrid vehicles [10], the aerospace industry, and RES power

---

plants. A non-isolated double input dc/dc converter is proposed in [11] combining buck and buck-boost single-input topologies, while an *n*-input buck-boost topology is presented in [12], which however could not supply the load simultaneously from different sources. A bidirectional multi-input dc/dc converter was also developed in [13], which is burdened by high conduction losses. The efficiency of an MIPC can be increased by exploitation of zero voltage switching approaches, as in [14,15]. Some MIPC topologies have been purposely developed for application in HRESs [16]. Among them, three-port dc/dc converters are of major interest. They feature an input port, an output one, and a storage port, enabling a bidirectional power flow towards/from an energy storage system (ESS). Some non-isolated three-port converters are discussed in [17–21].

A different approach is proposed in this paper, where a particular kind of multiple-input multi-level converter (MMC) is exploited to connect photovoltaic and wind generators to an energy storage system and a three-phase ac grid. It is based on an open-end winding configuration, the asymmetrical hybrid multi-level inverter (AHMLI), whose applications on both motor drives and grid-connected generators are discussed in [22,23] and which has also been successfully exploited to reduce the overvoltage caused by long cables in PWM motor drives [24] as well as to realize a high-speed Gen-set [25,26]. In the present HRES application, the AHMLI topology encompasses an open-end winding three-phase transformer (OWT) whose primary winding operates in an open-end configuration. The primary winding is, in fact, supplied on one side by a three-phase neutral point clamped (NPC) multi-level inverter (MLI) and, on the other side, by a conventional two-level inverter (TLI). The MLI operates at a low switching frequency (<1 kHz), thus featuring very low switching power losses. It is tasked to control the active power supplied to the grid, while also managing $N - 1$ unidirectional input power flows, being N the number of the output voltage levels. Thus, $N - 1$ energy sources (ESs) such as photovoltaic (PV) strings or wind turbines (WTs) can be managed without the introduction of additional dc/dc power converters. Moreover, it can also accomplish a multi-channel maximum power point tracking (MPPT) function at the string level on PV arrays. Compared with the MLI, the TLI operates at a high switching frequency, but at a lower dc bus voltage. It is tasked to control the grid current and to compensate for low-order current harmonics and unbalanced components. Moreover, its dc-bus can act as the storage port of a three-port dc–dc converter, enabling the connection of an energy storage system (ESS) to the HRES. In other words, energy sources connected to the MLI are managed by the TLI, avoiding the introduction of dc/dc power converters.

The paper is organized as follows. Section 2 presents the proposed approach and its application to 5LI + TLI and 3LI + TLI topologies. In Section 3, the operations of these topologies are discussed. Simulation and experimental results are presented in Sections 4 and 5, respectively. Finally, Sections 7 and 8 concern the discussion and conclusion.

2. The Proposed MMC Topology

The proposed MMC for HRESs, tailored around the AHMLI topology, is shown in Figure 1. The primary winding of the three-phase transformer is connected to an N-level NPC inverter on one side and to an auxiliary TLI on the other side. The secondary winding is instead connected to a three-phase ac grid. The NPC inverter, which acts as an N-level multi-input converter, encompasses $N - 1$ DC-bus capacitors C_j, each one connected to a dc power source of voltage V_{ck} (k = 1, 2, 3, $N - 1$). According to the AHMLI topology and assuming that the two inverters are supplied by two independent power sources, $V_{DC}{}'$ and $V_{DC}{}''$, a phase voltage V_{pj} (j = 1,2,3,) of the primary winding of the OWT is given by:

$$V_{pj} = V_{NPCj} - V_{TLIj} - V_{O'O''} = \frac{2l'-2}{4} V'_{DC} - (2l'' - 1)V''_{DC}/2 - V_{O'O''}$$
$$l' = 0,1,2 \quad l'' = 0,1 \tag{1}$$

where V_{NPCj} is the NPC output j-phase voltage referred to the mid-point O' (Equation (2)), V_{TLIj} is the TLI output j-phase voltage referred to the mid-point O'' (Equation (3)), $V_{DC}{}''$ is the dc-bus voltage of the TLI, V_{DC}' is the total dc-bus voltage of the NPC, and $V_{O'O}''$ is the voltage between the mid points O' and O'' of the dc-buses of the two inverters (Equation (4)).

$$V_{NPCj} = \frac{2l' - 2}{4} V_{DC}' \quad l' = 0, 1, 2 \tag{2}$$

$$V_{TLIj} = \frac{2l'' - 1}{2} V_{DC}'' \quad l'' = 0, 1 \tag{3}$$

$$V_{O'O''} = \frac{1}{3} \sum_{j=1}^{3} (V_{MLIj} - V_{TLIj}) \tag{4}$$

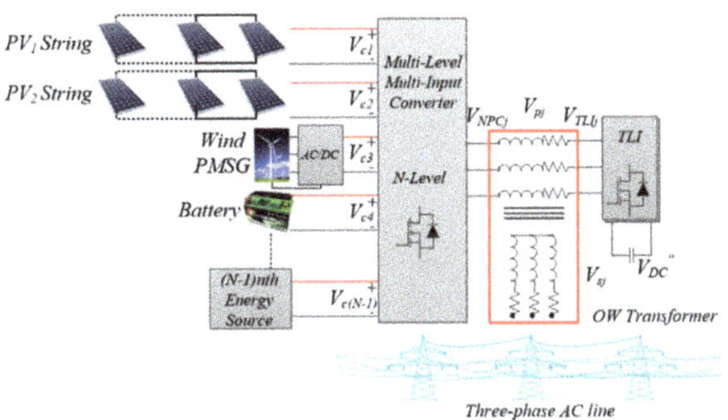

Figure 1. Proposed multi-level converter (MMC) configuration.

As V_{NPCj} may take N levels, while V_{TLIj} may take two, the transformer primary phase voltage V_{pj} may take $2N$ levels, whose amplitude is a function of V_{DC}' and V_{DC}''. Table 1 shows that in terms of phase voltage levels, the proposed configuration is equivalent to a conventional multi-level inverter with a larger number of power devices. From another point of view, a lower phase voltage THD (Total Harmonic Distortion) is obtained with the same number of switches.

Table 1. Basic multi-level inverter (MLI) topologies vs. the asymmetrical hybrid multi-level inverter (AHMLI).

MLI	NPC or Flying Capacitor		MMC (MLI+TLI) $V_{DC}'' = V_{DC}'/[(N-1)]$			
	Power Switches	Phase Voltage Levels	Power Switches			Phase Voltage Levels
			MLI	TLI	MLI + TLI	
3-L	12	9	12	6	18	17
5-L	24	17	24	6	30	25
7-L	36	25	36	6	42	33
9-L	48	33	48	6	54	41

As an example of an HRES application of the AHMLI topology, a six-level MMC is shown in Figure 2. PV strings, or groups of strings, are directly connected to the NPC's dc-bus capacitors while the permanent magnet synchronous generators of the wind turbines are connected through a three-phase controlled rectifier or a diode rectifier and an output dc-link capacitor. All dc sources must have about the same rated output voltage in order to prevent largely unbalanced dc-bus voltages. However, the TLI is able to compensate a NPC DC-bus capacitor voltage variation ΔV_c provided that it is lower than V_{DC}'', thus achieving a sinusoidal grid current. The dc-buses of the NPC and auxiliary inverters are isolated between them in order to prevent the circulation of zero-sequence currents [22]. Moreover, the TLI dc-bus is supplied through a floating capacitor; thus, an additional power source is not required. Another example is shown in Figure 3, where a five-level NPC inverter is used. In this case, an ESS is connected to the TLI dc-bus, and a bidirectional power flow can be managed towards/from the ESS. In both examples, the NPC provides the active power flow to the grid, while the TLI works as an active power filter, while also regulating the output current and the NPC dc-bus capacitor voltages V_{ck}.

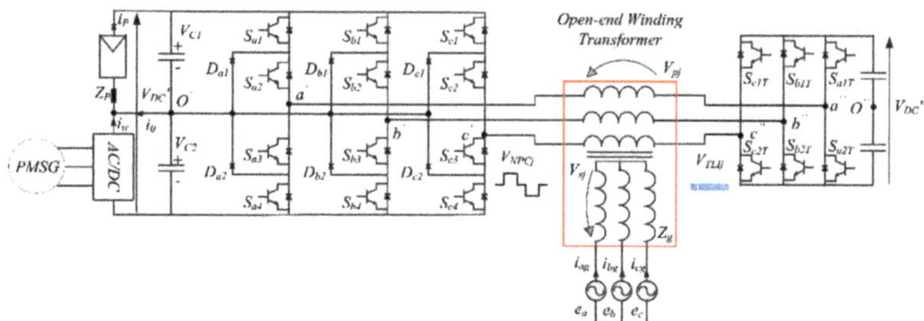

Figure 2. Six-level MMC (three-level inverter (3LI) + two-level inverter (TLI)) with two energy sources.

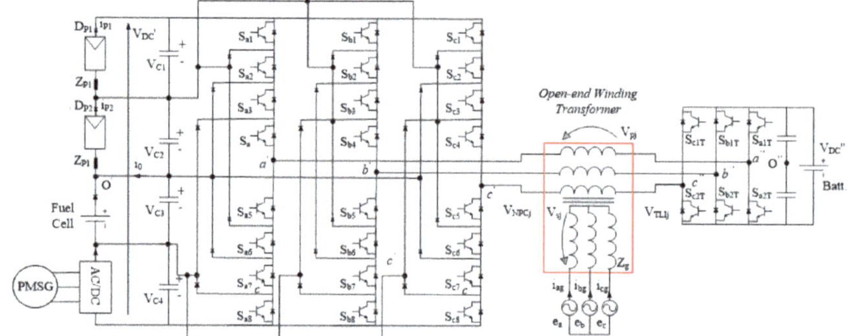

Figure 3. Ten-level MMC (5LI + TLI) with four energy sources and an energy storage device.

3. Proposed MMC Topology Operation

In order to manage multiple input sources while controlling the main power flow towards the grid, a suitable control system has been developed that is divided into two main parts: an MLI control subsystem and a TLI control subsystem.

3.1. MLI Control Subsystem

MLI switching power losses are kept low by taking advantage of low-frequency space vector modulation (SVM) or step modulation (ST). However, these modulation techniques must be suitably modified to allow for a direct periodical connection between the $N - 1$ energy sources connected to the MLI dc-bus and the TLI. This is necessary to enable independent voltage control on each of the $N - 1$ MLI dc-bus capacitors. According to (1), the space diagram of the transformer primary phase voltage V_{jp} is obtained by combining the voltage space vector diagrams of the two inverters. The simplest MMC configuration that can be obtained according to the proposed approach encompasses a three-level inverter (3LI + TLI), which provides six voltage levels if $V_{DC}'' = V_{DC}'/(N - 1)$. Such an MMC may take $3^3 = 27$ switching states; however, only 19 different voltage vectors can be generated, because some of the switching states are redundant. The MMC voltage space vector diagram can be obtained by adding the voltage space vector diagram of the TLI at the top of each voltage vector of the 3LI, as shown in Figure 4 [27]. Each input dc source can be independently managed by exploiting its periodical connection to the transformer's primary winding, which occurs when the 3LI generates one of the twelve possible low-voltage vectors (LVVs), namely PPO, OON, POO, ONN, ONO, POP, NNO, OOP, NOO, OPP, NON, and OPO, according to Figure 4 and Table 2. As an example,

Figure 5 shows that when the voltage vector PPO is generated, the capacitor C_1, which represents the output capacitor of the energy source ES_1, is directly connected to the TLI. Hence, when the voltage vector PPO is generated, V_{C1}, that is, the voltage across C_1, can be regulated by controlling the power stream between the 3LI and the TLI. This is accomplished through a closed loop voltage controller managing the power exchange between ES_1 and the TLI dc-bus capacitors through a specific set of components V_{kj} of the TLI reference voltage, as shown in Figure 5c. In practice, the voltage regulator, by processing the difference between the C_1 reference voltage V_{c1} * and the actual value V_{c1}, generates a coefficient I, which is multiplied by the actual values of the transformer's primary currents to obtain V_{kj}. Therefore, if I is positive, an additional power transfer is instated directed towards C_1, thus increasing V_{c1}. If I is negative, the additional power flow is directed from C_1 to the TLI dc-bus capacitors, thus discharging C_1 and reducing V_{c1}. Hence, it is possible to charge or discharge C_1 when the MLI generates the vector PPO. The same applies for the other dc bus capacitors when the MLI is generating the specific LVV.

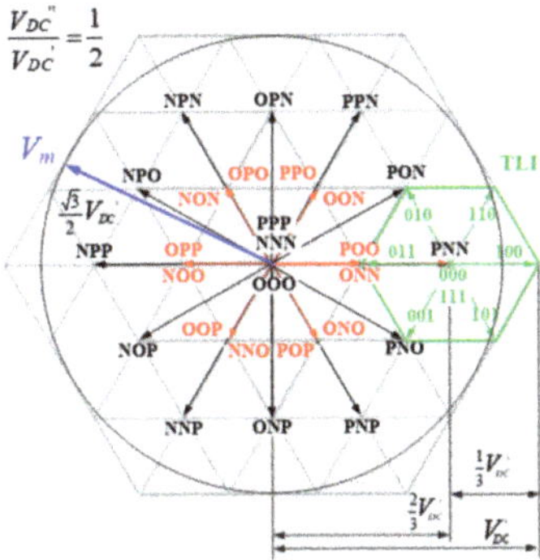

Figure 4. Three-level neutral point clamped (NPC) + TLI voltage space vector diagram ($V_{DC}''/V_{DC}' = 1/2$).

Table 2. Six-level MMC low voltage vectors.

Vector	S_{a1}	S_{a2}	S_{a3}	S_{a4}	S_{b1}	S_{b2}	S_{b3}	S_{b4}	S_{c1}	S_{c2}	S_{c3}	S_{c4}
PPO	1	1	0	0	1	1	0	0	0	1	1	0
OON	0	1	1	0	0	1	1	0	0	0	1	1
POO	1	1	0	0	0	1	1	0	0	1	1	0
ONN	0	1	1	0	0	0	1	1	0	0	1	1
ONO	0	1	1	0	0	0	1	1	0	1	1	0
POP	1	1	0	0	0	1	1	0	1	1	0	0
NNO	0	0	1	1	0	0	1	1	0	1	1	0
OOP	0	1	1	0	0	1	1	0	1	1	0	0
NOO	0	0	1	1	0	1	1	0	0	1	1	0
OPP	0	1	1	0	1	1	0	0	1	1	0	0
NON	0	0	1	1	0	1	1	0	0	0	1	1
OPO	0	1	1	0	1	1	0	0	0	1	1	0

In general, for a N-level NPC converter, $N - 1$ space vectors can be exploited for controlling $N - 1$ input dc sources. For instance, Figure 6 shows the space phasor diagram of a 10-level MMC composed of a five-level NPC and a TLI (5LI + TLI) as that shown in Figure 3, where V_{DC}''/V_{DC}' is set to 0.25. The energy sources that can be managed in this case are four, namely ES_1, ES_2, ES_3, and ES_4, each one connected to the MLI dc-bus through an output capacitor C_j ($j = 1 \ldots 4$). The voltage V_{Cj} across the output capacitor C_j can be regulated by acting on the six LVVs shown in Figure 6, being P_1, P_2, O, N_1, and N_2 the possible states of the j-leg, as shown in Table 3. When the vector R_2 ($P_2P_2P_1$) is generated, the capacitor C_1 is directly connected to the TLI through the primary windings of the transformer, as shown in Figure 7, making possible the regulation of the voltage V_{c1} by controlling the power stream between the two converters.

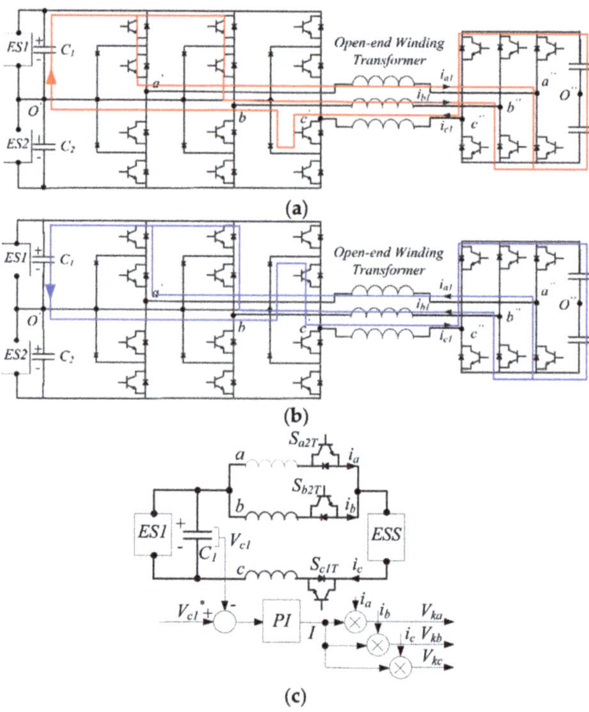

Figure 5. Three-level inverter (3LI) + TLI: C_1 voltage control when the PPO vector is generated: (**a**) discharging; (**b**) charging; (**c**) equivalent voltage control loop during PPO.

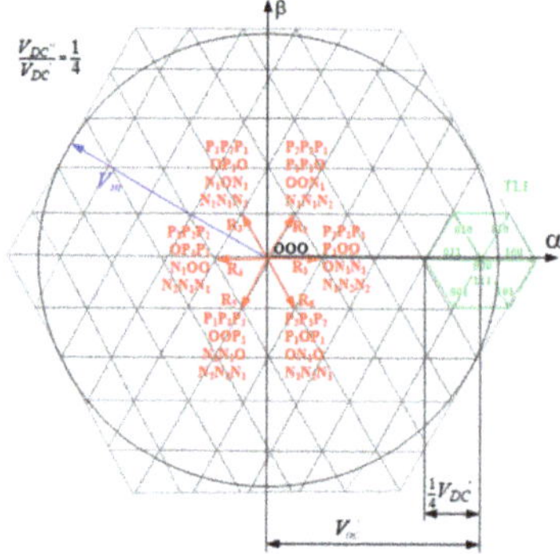

Figure 6. Five-level inverter (5LI) + TLI voltage space vector diagram ($V_{DC}'' / V_{DC}' = 1/4$).

Table 3. Five-level NPC j-leg states.

Vector	S_{j1}	S_{j2}	S_{j3}	S_{j4}	S_{j5}	S_{j6}	S_{j7}	S_{j8}
P_2	1	1	1	1	0	0	0	0
P_1	0	1	1	1	1	0	0	0
O	0	0	1	1	1	1	0	0
N_1	0	0	0	1	1	1	1	0
N_2	0	0	0	0	1	1	1	1

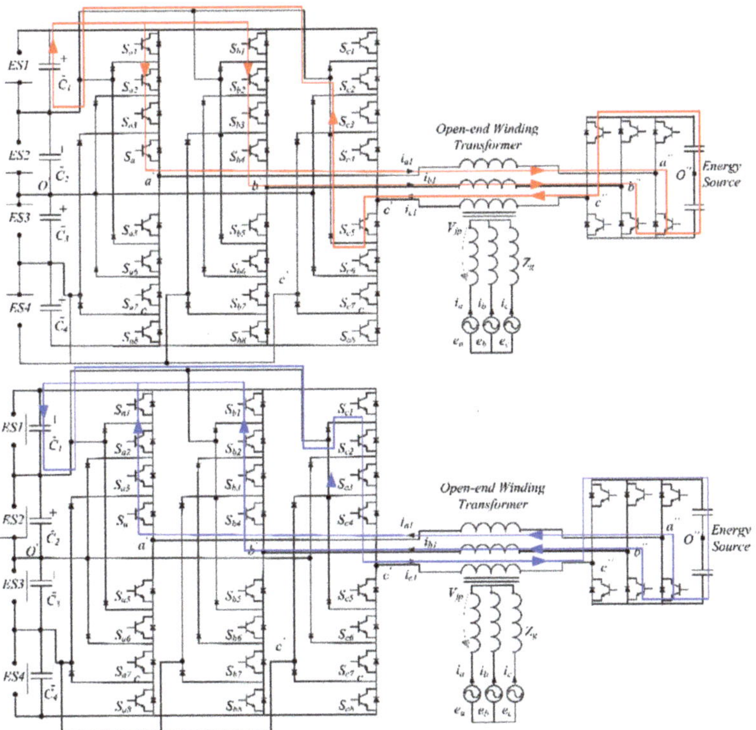

Figure 7. Five-level inverter (5LI) + TLI: C_1 voltage control when the $P_2P_2P_1$ vector is generated: (up) discharging; (down) charging.

For low modulation indexes, it is possible to regulate the voltage of the $N-1$ dc-bus capacitors without modification of the conventional multilevel SVM or SM strategies, because the top of the reference voltage vector $V_m{}^*$ always lies inside the hexagon encompassing R_1, R_2, R_3, R_4, R_5, and R_6. According to the basic multi-level SVM strategy, the voltage–time Equation (5) referred to a stationary q,d reference frame gives the switching times t_a, t_b, and t_c from the voltage reference $V_m{}^*$ and the switching period T_m. As shown in Figure 8a, for low modulation indexes, V_a coincides with the null state V_0, while V_b and V_c are those LVVs whose tops coincide with the vertices of the triangle in which the top of the reference voltage vector lies. However, as shown in Figure 8b, for medium and high modulation indexes, no LVVs are selected according to the conventional SVM strategy; thus, they must be purposely introduced in the inverter switching path. This can be obtained by substituting one of the vectors V_a, V_b, or V_c of Equation (5) with a switching sequence including the LVV that must be activated, and other two voltage vectors V_1 and V_2. If, as shown in Figure 8c, the LVV R_1 must be activated, the switching times t_R, t_1, and t_2 are given by Equation (6), being T_a obtained by solving Equation (5). A similar procedure can be adopted to modify the switching patterns generated according to the standard multi-level SM.

$$\begin{cases} T_m V_{m\alpha}* = t_a V_{a\alpha} + t_b V_{b\alpha} + t_c V_{c\alpha} \\ T_m V_{m\beta}* = t_a V_{a\beta} + t_b V_{b\beta} + t_c V_{c\beta} \\ T_m = t_a + t_b + t_c \end{cases} \tag{5}$$

$$\begin{cases} T_a V_{a\alpha} = t_1 V_{1\alpha} + t_2 V_{2\alpha} + t_R V_{R1\alpha} \\ T_a V_{a\beta} = t_1 V_{1\beta} + t_2 V_{2\beta} + t_R V_{R1\beta} \\ T_a = t_1 + t_2 + t_R \end{cases} \tag{6}$$

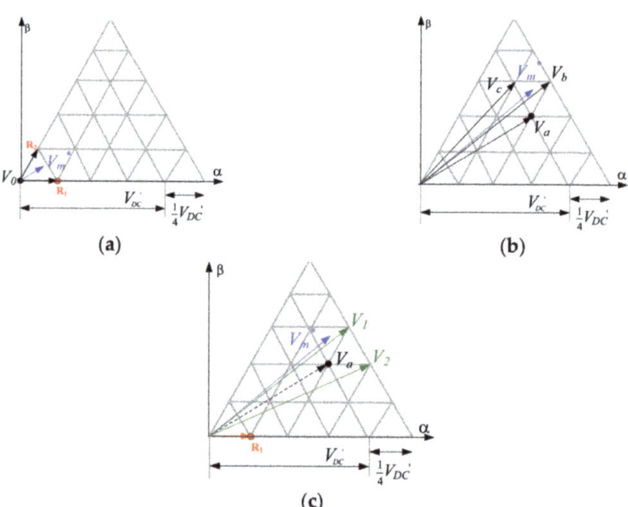

Figure 8. Five-level inverter (5LI) + TLI. (**a**) Space vector modulation (SVM) at a low modulation index. (**b**) SVM at a high modulation index. (**c**) Modified SVM.

3.2. TLI Operation and 3LI + TLI Control Algorithm

Transformer primary voltage harmonics generated by low-frequency modulation of the main inverter are then cancelled by the TLI. In fact, the TLI working at a high switching frequency plays the role of an active power filter while also managing the power flows originating from the $N-1$ energy sources connected to the MLI dc-bus and compensating for possible imbalances caused by different dc-bus capacitor voltages. This makes it possible to avoid the introduction of additional dc/dc converters, which otherwise would be necessary to connect each energy source to the system. The MMC control system encompasses a synchronous qd current controller regulating the transformer's primary current and the $N-1$ voltage controllers to manage the energy sources connected to the MLI dc-bus [28]. A schematic of the control system developed for the proposed MMC is shown in Figure 9 for a 3LI + TLI configuration. It consists of three main blocks, namely TLI dc-bus current (battery) control, NPC dc-bus voltage (energy sources) control, and regulation of active and reactive power at the primary side of the transformer. Moreover, a $N-1$ channels maximum power point tracking function can be provided in order to cope with PV arrays connected as energy sources. In this case, according to Figure 5, the MPPT sets the reference voltages V_{c1} and V_{c2}, whose sum $V_{DC}*'$ constitutes the reference for the q-axis 3LI current regulator. The reactive power is instead controlled by acting on the d-axis current. An independent control on V_{c1} and V_{c2} is obtained by two voltage controllers, whose outputs are processed in order to obtain the components V_{kja} of the TLI reference voltages as shown in Figure 5c. Whenever one LVV is active, k is set to 1 and the reference coefficient I is computed to charge or discharge the considered capacitor. At the same time, the correct MLI voltage vector path is selected according to Equations (5) and (6) in order to connect the specific energy source to the TLI, thus establishing a power stream from the energy source to the TLI d-bus. Further components of the TLI reference voltage V_{hj} and V_{battj} are computed dealing, respectively, with the compensation of harmonics generated by the low-frequency operation of the MLI (Equation (7)) and control of the battery current $i_{DC}{''}$. Hence, the j-phase TLI reference voltage is given by Equation (8) while V_{hj} is written in (7).

$$V_{hj} = V_{NPCj} - V_{1NPCj} \tag{7}$$

$$V_{TLIj}* = V_{hj} + V_{Battj} + V_{kj} \tag{8}$$

being V_{1NPCj} the fundamental component of the NPC output phase voltage V_{NPCj}. The TLI current control is also able to compensate for imbalanced voltages on the 3LI dc-bus by adapting the duty cycle of the TLI at each half-cycle. This capability depends on the value of $V_{DC}{''}$, which is the maximum capacitor voltage deviation value that can be compensated for. Hence, the TLI ensures a sinusoidal grid-current during $V_c = V_{cn} \pm V_{DC}{''}$ for all NPC dc-bus capacitors.

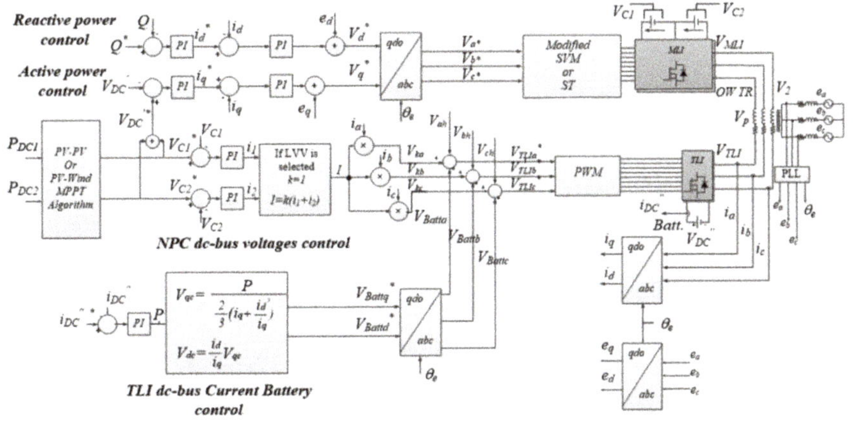

Figure 9. Six-level MMC (3LI + TLI).

4. Simulation Results

The effectiveness of the proposed topology was first evaluated through a simulation, taking into account a scaled model of a hybrid renewable energy generator tailored around an open-end winding 5 kVA-230/400 V three-phase transformer, a three-level NPC inverter exploiting a 1 kHz SVM strategy with a 400 V dc-bus voltage, a TLI PWM operating at 10 kHz, and a three-phase grid. The parameters of the system components are listed in Tables 4–11. These include the main data of the IGBT and diodes on the NPC MLI and those of the power MOSFET devices present on the TLI.

Table 4. Three-phase grid.

e_g (V)	400
f (Hz)	50
L_g (mH)	3

Table 5. Three-phase transformer.

A_n (kVA)	5
V_{n1} (V)	400
V_{n2} (V)	400
t	1

Table 6. PV modules (STC).

P_{nom} (W)	200
V_{mpp} (V)	40
I_{mpp} (A)	5
I_{cc} (A)	5.40
V_{open} (V)	47.8
string	5 modules

Table 7. Wind Turbine.

Pn (W)	1000
Vn (V)	220
ω_{wind} (m/s)	10
ω_{max} (m/s)	55
ω_0 (m/s)	2
Generator	Permanent Magnet Synchronous Generator PMSG

Table 8. Battery.

(Ah)	50
Vn (V)	400
Type	Lithium-Ion

Table 9. Diodes.

V_{DS} (V)	1000
I_n (A)	30
t_{rd} (ns)	67
Q_r (μC)	1.5

Table 10. MLI-IGBT (STGW40N120K).

Vce (V)	1200
$VenON$ (V)	2.7
In (A)	40
tr (ns)	48
tf (ns)	338

Table 11. TLI-MOSFET (IRFB5615PBF).

V_{DS} (V)	150
R_{DSON} (mΩ)	32
I_n (A)	35
t_r (ns)	17.2
t_f (ns)	35

Two cases are taken into consideration: a fully PV system with an ESS; and a hybrid PV–wind one with an ESS. The system model was developed using the MATLAB/Simulink environment and setting a 1 μs sample time. Output power characteristics of the PV modules are shown in Figure 10. Each PV string consists of five modules in order to achieve a peak voltage of 200 V on the NPC dc-bus capacitors C_1 and C_2. Operation of the first configuration is shown in Figure 11, dealing with NPC dc-bus voltages V_{c1} and V_{c2}, the output power of ES_1, ES_2, and the battery, the irradiances of PV strings G_{PV1} and G_{PV2}, the transformer's primary and secondary voltages, the NPC output voltages V_{NPCj}, the grid currents, the TLD bus voltage V_{DC}'' and current i_{DC}', the battery's state of charge (SOC), and the qd-axes current components. At $t = 0.5$ s, the solar insolation on the ES_2 PV string falls down from 1000 W/m^2 to 700 W/m^2. Then, the ES_2 output power decreases from 600 W to 560 W. According to Figure 10, the voltage V_{c2} is then reduced to track the maximum power point by acting on the TLI according to the voltage control scheme of Figure 5c. Once V_{c2} has reached the new optimal value, an imbalanced voltage condition ($V_{c1} = 200$ V, $V_{c2} = 165$ V) occurs. However, the three-phase grid currents are kept sinusoidal by the TLI current control system by drawing power from the ESS, as the SOC diagram confirms. In this case, in fact, the TLI dc-bus voltage $V_{DC}'' = 210$ V is sufficient to compensate for the $\Delta V_{c2} = (200 - 165)$ V = 35 V voltage deviation. Figure 12 deals instead with operation of the second configuration, when the rotor speed of the wind turbine ω_{rm} drops from 1500 to 0 rpm. Additionally, in this case, the TLI dc-bus voltage $V_{DC}'' = 210$ V is sufficient to compensate for the $\Delta V_{c2} = (200 - 200)$ V = 0 V voltage deviation. The proposed system is able to manage a bidirectional power stream towards/from the energy storage system as shown in Figure 13. The irradiances considered for PV$_1$ and PV$_2$ are not the same, being respectively 1000 W/m^2 and 700 W/m^2, while the battery current varies from 2.5 A to -2.5 A. The battery SOC trend demonstrates the bidirectional power capability of the proposed configuration.

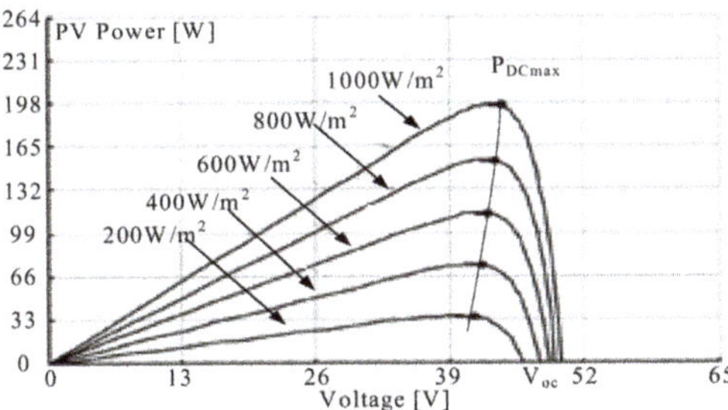

Figure 10. P-V diagrams of photovoltaic modules.

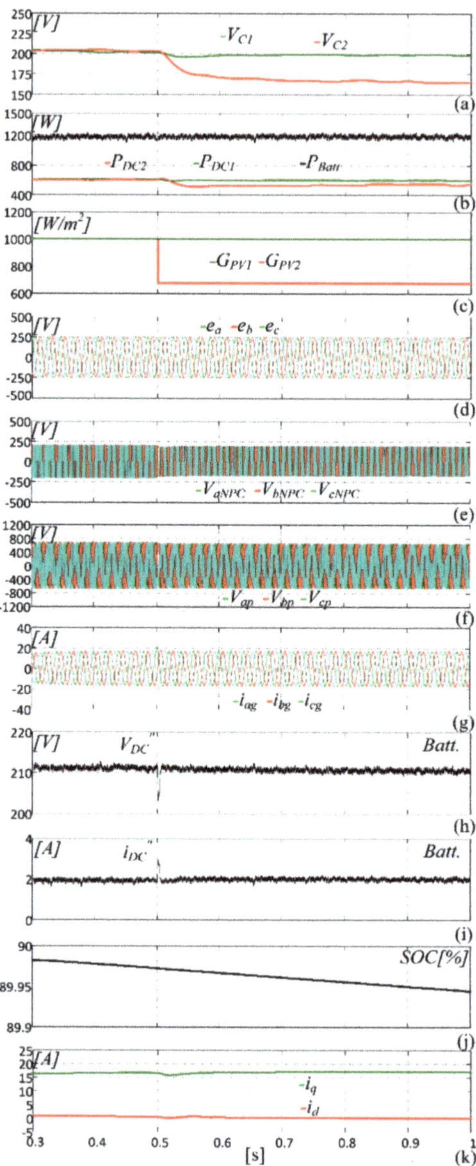

Figure 11. Six-level MMC-G_{PV2} drop from 1000 W/m^2 to 700 W/m^2. (**a**) String voltages V_{c1} and V_{c2}. (**b**) ES$_1$, ES$_2$, and energy storage system (ESS) output power. (**c**) G_{PV1} and G_{PV2}. (**d**) Grid voltage. (**e**) NPC output voltage. (**f**) Primary voltage. (**g**) Grid current. (**h**) Battery voltage. (**i**) Current. (**j**) State of charge (*SOC*). (**k**) Output *d,q* axes currents.

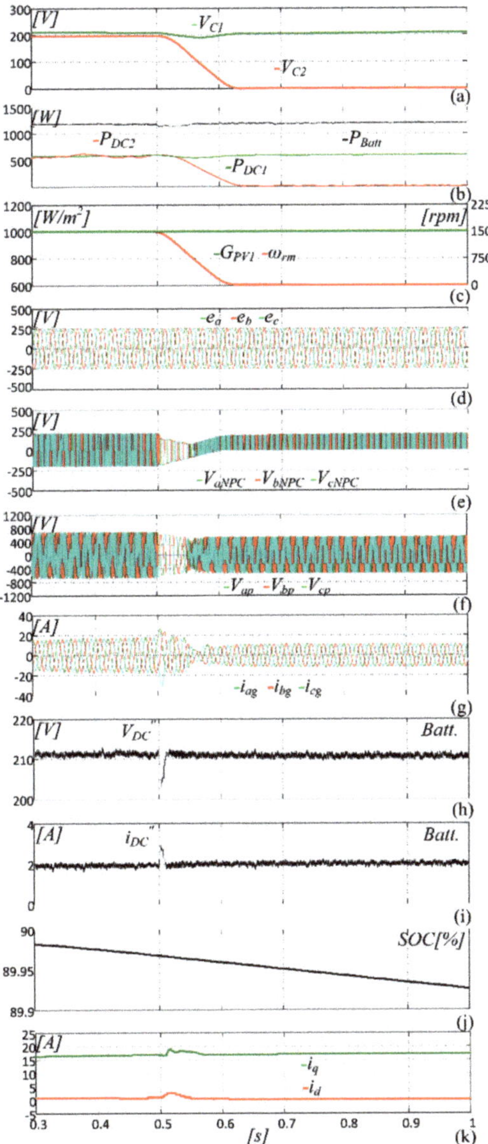

Figure 12. Six-level MMC-wind turbine (WT) shut-down $G_{PV1} = 1000$ W/ m^2. (**a**) V_{c1} and V_{c2}. (**b**) ES$_1$, ES$_2$, and ESS output power. (**c**) G_{PV1} and PMSG rotor speed ω_r. (**d**) Grid voltage. (**e**) NPC output voltage. (**f**) Primary voltage. (**g**) Grid current. (**h**) Battery voltage. (**i**) Current. (**j**) SOC. (**k**) Output d,q axes currents.

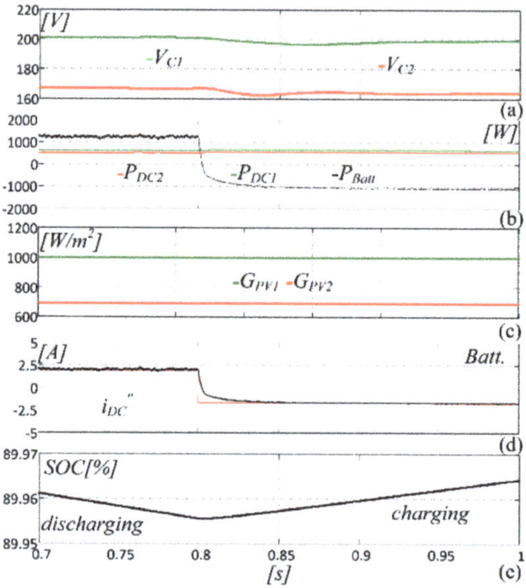

Figure 13. Si-level MMC-battery current transition from 2 A to -2 A with $G_{PV1} \neq G_{PV2}$. (**a**) String voltages V_{c1} and V_{c2}. (**b**) Output power of PV1, PV2, and the battery. (**c**) Solar power G_{PV1} and G_{PV2}. (**d**) Battery current. (**e**) Battery *SOC*.

5. Experimental Assessment

Experimental tests were accomplished on a six-level MMC encompassing an open-end winding 5 kVA–230 V/400 V three-phase transformer, a 3LI-NPC inverter with a 200 V dc-bus voltage exploiting an SM strategy, and a two-level inverter PWM operating at 10 kHz. A 100 V, 40 Ah Lithium-ion battery was connected to the dc-bus of the TLI. All system parameters are listed in Table 4, Table 5, Table 6, Table 7, Table 8, Table 9, Table 10, and Table 11. The control system was realized around a DSpace/1103 board running at 10 kHz. The primary winding of the transformer was connected, on one side, to the NPC MLI and, on the other side, to the TLI. Programmable PV module emulators played the role of two PV strings connected to the NPC dc-bus. Steady-state operation under balanced and imbalanced conditions is shown in Figure 14a,b, dealing with the 3LI + TLI output voltage, grid voltage, grid current, and 3LI output step voltage. In Figure 14a, a balanced condition is considered with $V_{c1} = 100$ V, $V_{c2} = 100$ V. Hence, the TLI acts only as an active power filter in order to compensate for the low-order harmonic generated by the NPC low switching frequency modulation. Figure 14b instead deals with the imbalanced condition with $V_{c1} = 100$ V, $V_{c2} = 70$ V, and $V_{DC}'' = 100$ V. Although the NPC dc-bus voltages are imbalanced, the grid current is sinusoidal with a THD as low as 1.5%. In this case, the TLI not only works as an active filter but also as a voltage imbalance compensator. A reduction in the dc voltage generated by PV_1 string is shown in Figure 15a, where V_{c1} is changed from 100 V to 70 V. Such an imbalance causes a variation in the peak voltage in the positive half-cycle of V_{jNPC}, but the current is kept sinusoidal by the TLI. Figure 15b deals with power generated by the two PV strings P_{PV1} and P_{PV2} provided by the battery P_{TLI} and the output one P_g. The reduction in V_{c2} causes a reduction in P_{PV2} from 320 W to 180 W according to the P–V profile of Figure 10. The same power variation is present in the grid power P_g because the active power produced by the TLI's battery is kept constant. Figure 16 shows a detailed view of the waveforms of V_{aNPC} and i_{ag} in the test of Figure 15. The TLI compensates for the unbalanced voltages and almost perfectly shapes the grid current.

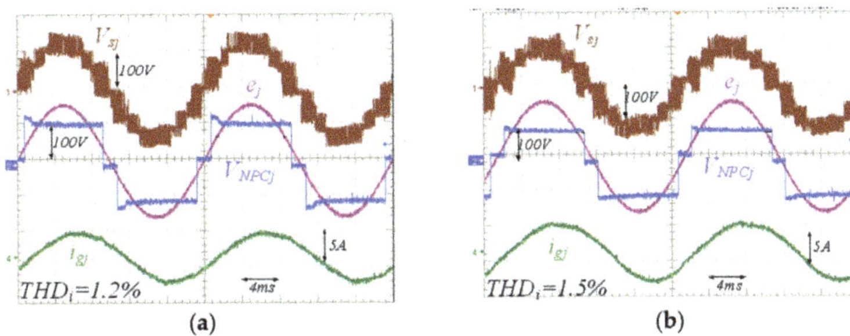

Figure 14. The (3LI + TLI) steady state. (**a**) Balanced voltage, $V_{DC}' = 200$ V, $V_{c1} = 100$ V, $V_{c2} = 100$ V, $V_{DC}'' = V_{DC}'/2 = 100$ V. (**b**) Unbalanced voltage, $V_{DC}' = 170$ V, $V_{c1} = 70$ V, $V_{c2} = 100$ V, $V_{DC}'' = 100$ V. Secondary phase voltage V_{sj}, grid phase voltage e_j, grid phase current i_{gj}, and NPC output voltage V_{NPCj}, ($V_g = 150$ V, 50 Hz, TLI PWM 10 kHz).

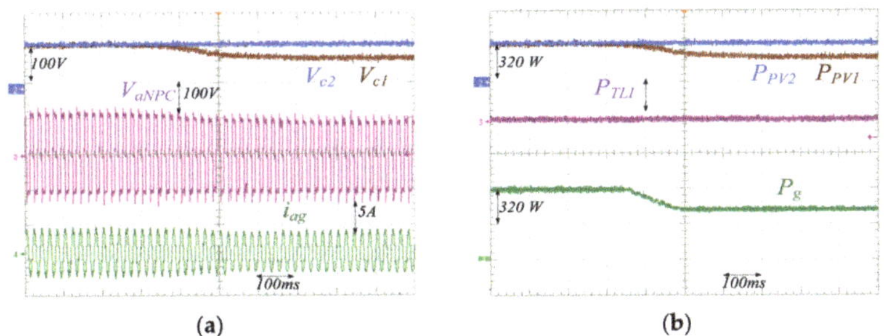

Figure 15. The (3LI + TLI) photovoltaic (PV) power variation. (**a**) PV voltages V_{c1} and V_{c2}, NPC output voltage V_{NPCj}, and grid current i_{gj}. (**b**) PV power P_{PV1} and P_{PV2}, TLI output active power P_{TLI}, and grid active power P_g. ($V_g = 150$ V, 50 Hz $V_{DC}'' = 100$ V, TLI 10 kHz PWM).

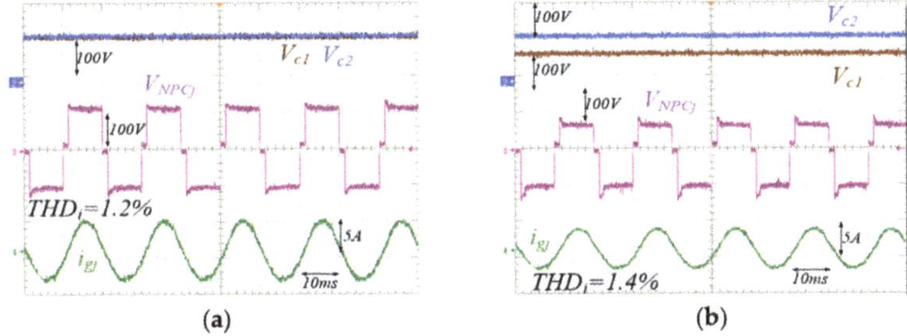

Figure 16. The (3LI + TLI) steady state. (**a**) Balanced voltage, $V_{DC}' = 200$ V, $V_{c1} = 100$ V, $V_{c2} = 100$ V, $V_{DC}'' = V_{DC}'/2 = 100$ V. (**b**) Unbalanced voltage, $V_{DC}' = 170$ V, $V_{c1} = 70$ V, $V_{c2} = 100$ V, $V_{DC}'' = 100$ V. Secondary voltage V_{sj}, grid voltage e_j, grid current i_{gj}, and NPC output voltage V_{NPCj} ($V_g = 150$ V, 50 Hz, TLI 10 kHz PWM).

A further test was performed dealing with battery current control, as shown in Figure 17, dealing with battery current, voltages V_{c1} and V_{c2}, and grid current. The voltages are kept balanced at $V_{c1} = 100$ V and $V_{c2} = 100$ V, while the battery current is changed from -3 A to 2 A. Hence, the battery is first discharged and then charged. A negative battery current means that the battery feeds power P_{batt} to the grid according to Figure 17. Vice-versa, a positive battery current means that the battery is charged from the grid. The harmonic spectrum of the grid current at a rated load is shown in Figure 18, fully complying with the IEC 61000-3-2 standard on power quality.

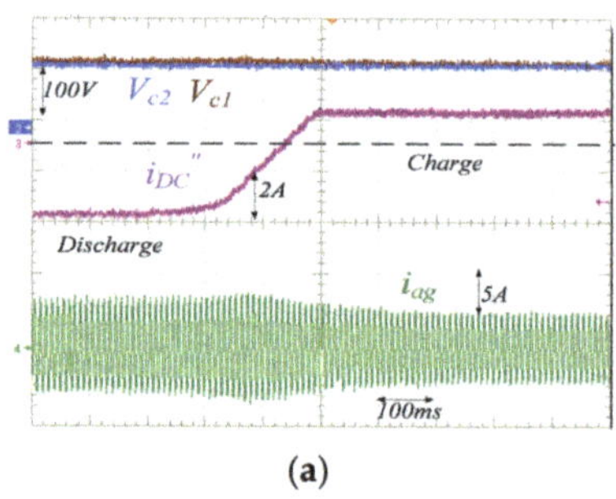

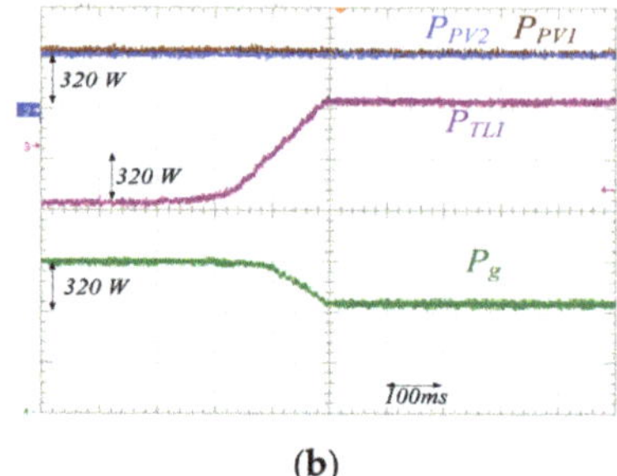

(a)   (b)

Figure 17. The (3LI + TLI) battery current variation from 2 A to −2A with $G_{PV1} \neq G_{PV2}$. (**a**) String voltages V_{c1} and V_{c2}. Battery current i_{DC}''. Grid current i_{gj} (**b**) String power P_{V1}, P_{V2}. Battery power P_{Batt}, grid active power P_g.

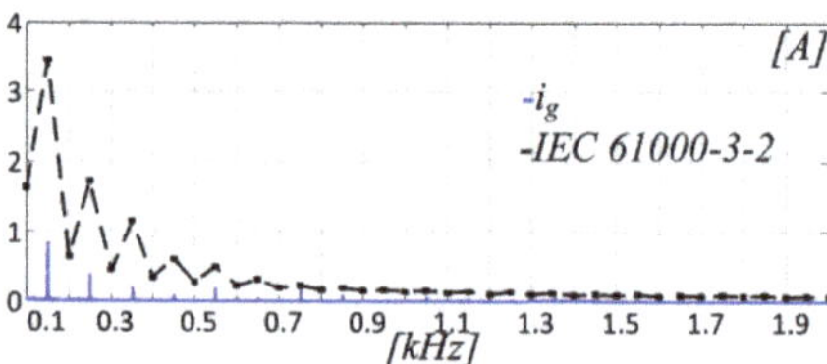

Figure 18. The (3LI + TLI) grid current spectrum vs. IEC 61000-3-2 l limits.

6. Power Losses Analysis

A power losses analysis was accomplished by considering the efficiency $\eta_{3LI + TLI}$ of the 3LI + TLI converter and that of the transformer η_{TR}. Hence, the total efficiency is obtained as Equation (9). The efficiency of the converter was estimated by computation of the power devices conduction P_c and the switching P_{sw} losses. According to [29], the 3LI is equipped with a high-voltage and low-frequency IGBT, while a low-voltage, high switching frequency MOSFET is used in the TLI. The main data on these power devices are listed in Tables 9–11. Conduction losses of the IGBT and MOSFET were computed according to Equations (10) and (11), respectively, while switching losses were evaluated by Equation (12) for both power switches. Furthermore, the diodes' reverse recovery power losses are also considered in Equation (13) and were included in the total power losses calculation.

$$\eta_{Tot} = \eta_{3LI+TLI}\eta_{TR} \tag{9}$$

$$P_{c\_MLI} = \delta V_{ce(on)}i_{RMS} \tag{10}$$

$$P_{c\_TLI} = R_{DS(on)}i_{RMS}^2 \tag{11}$$

$$P_{sw} = 0.5 V_{ce}i_{RMS}f_{sw}(t_{rise} + t_{fall}) \tag{12}$$

$$P_D = V_{DR}f_{sw}(t_{rd}i_{RMS} + Q_r) \tag{13}$$

$$\eta_{TR} = \frac{A_n\cos(\varphi)}{A_n\cos(\varphi) + iP_{fen} + P_{cun}/i} \tag{14}$$

where δ is the duty cycle, t_r and t_f are the rise and fall times of the power switches, i_{RMS} is the Root Mean Square (rms) value of the switch current, V_{ce} is the collector-to-emitter voltage, $V_{ce(on)}$ is the collector-to-emitter saturation voltage, $R_{DS(on)}$ is the static drain-to-source on resistance, f_{sw} is the switching frequency, V_{DR} is the diode's reverse voltage, t_{rd} is the diode's reverse recovery time, and Q_r is the reverse recovery charge. The efficiency of the three-phase transformer was computed as a function of iron losses P_{fen}, rated copper losses P_{cun}, power factor $\cos(\varphi)$, rated power A_n, and load coefficient $I = i_g/i_{gn}$, being i_{gn} the rated current of the transformer, Equation (14). Figure 19 shows the conduction and switching losses of the two inverters as a function of the load coefficient i. Specifically, P_{3LI_cond} and P_{TLI_cond} are the conduction losses of 3LI and TLI, respectively, while P_{3LI_sw} and P_{TLI_sw} are the switching losses, which increase with the load current from 2.6% at i = 0.1 to 8% at i = 1. The switching power losses of the 3LI are quite low due to the low switching frequency. The total efficiency of the two inverters $\eta_{3LI+TLI}$ and that of the transformer η_{TR} are shown in Figure 20 as function of the load ratio. The peak efficiency of the converter is 98.9% at i = 1, while the minimum is 95.2% at i = 0.2. The efficiency of the transformer, η_{TR}, reaches its peak value of 96.5% for i = 3/4. Finally, the total efficiency was obtained according to Equation (9) and is shown in Figure 20c. It reaches the maximum value of 95% at i = 0.7.

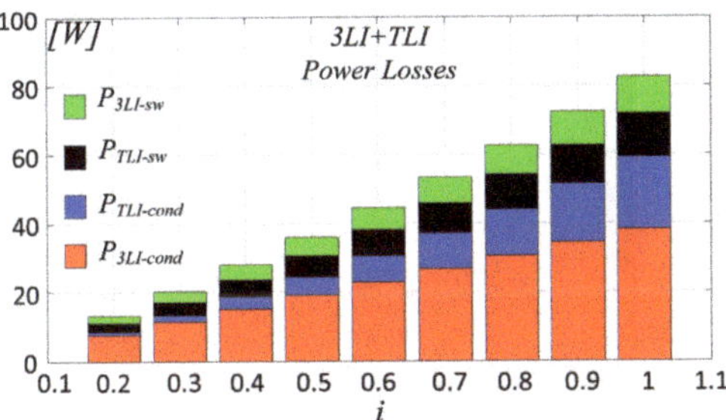

Figure 19. Power losses of 3LI + TLI vs. load current.

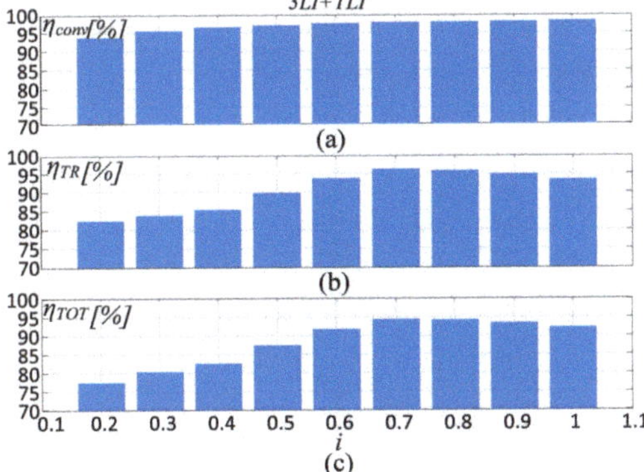

Figure 20. Efficiency vs. load current. (**a**) Three-level inverter (3LI) + TLI efficiency. (**b**) Transformer efficiency. (**c**) Total efficiency.

7. Discussion

Simulation and experimental results confirm that independent management of $N-1$ power sources and an ESS can be accomplished by using an AHMLI structure composed of a N-level inverter, an open primary winding transformer, and a two-level inverter. This makes unnecessary the introduction of additional dc–dc converters to connect the input ES and the ESS to the grid inverter. Such a structure is also able to compensate for a possible imbalance among the output voltages of energy sources, provided that it does not exceed the dc-bus voltage of the auxiliary TLI. Under this limit, the proposed configuration is also able to cope with a full shut-down of one of the input ESs. Moreover, an independent N-1-channel MPPT can be provided to manage multi-string PV arrays. As proved by experimental tests, the proposed configuration produces an almost perfectly sinusoidal grid current, despite the fact that the main inverter is operated at a low switching frequency in order to improve the efficiency. In fact, the current shaping is accomplished by the auxiliary TLI, which operates at a high switching frequency, but at a remarkably lower dc-bus voltage. This allows us to equip the TLI with fast and powerful MOSFET devices producing low switching and conduction power losses. The efficiency performance of the proposed structure is confirmed by a power losses analysis, which gives global efficiency levels similar to those obtainable with conventional conversion systems for HRESs, but with a more simple and less expensive structure.

8. Conclusions

The goal of this work was to prove that a multi-input conversion system can be constructed from an AHMLI topology exploiting an open-end primary winding transformer. The coherence of such a concept was confirmed first theoretically and then by simulation and experimental tests. Applied to hybrid renewable energy systems exploiting multiple energy sources and an energy storage system, the proposed approach allows us to largely reduce the complexity and cost of the power conversion systems, avoiding the introduction of additional dc–dc converters to interface each energy source with the grid-connected inverter. Further developments of the proposed concept will deal with applications in other sectors, such as electric vehicles and the aerospace industry.

Author Contributions: A.T. and S.F. conceived the original idea. A.T., S.F. and L.D.T. developed the theory and performed the computations. S.D.C., G.S. and M.C. verified the analytical methods. All authors discussed the results and contributed to the final manuscript. All authors have read and agreed to the published version of the manuscript.

Funding: This research received no external funding.

Institutional Review Board Statement: Not applicable.

Informed Consent Statement: Informed consent was obtained from all subjects involved in the study.

Data Availability Statement: Not applicable.

Conflicts of Interest: The authors declare no conflict of interest.

References

1. Sefa, I.; Altin, N. Grid interactive photovoltaic inverters—A review. *J. Fac. Eng. Archit. Gazi. Univ.* **2009**, *24*, 409–424.
2. Bhol, S.; Charansahu, N. Performance Analysis and Cost Calculation of Standalone PV-DG Generation System without Storag. In Proceedings of the CISPSSE, Keonjhar, Odisha, India, 29–31 July 2020; pp. 1–5.
3. Moniruzzaman, M.; Hasan, S. Cost analysis of PV/Wind/diesel/grid connected hybrid systems. In Proceedings of the ICIEV, Dhaka, Bangladesh, 18–19 May 2012; pp. 727–730.
4. Tyagi, A.; Dubey, M.; Gawre, S. Advance Monitoring of Electrical and Environmental Parameters of PV System: A Revie. In Proceedings of the SEEMS, Greater Noida, India, 26–27 October 2018; pp. 1–5.
5. Nehrir, M.H.; Wang, C.; Strunz, K.; Aki, H.; Ramakumar, R.; Bing, J.; Miao, Z.; Salameh, Z. A Review of Hybrid Renewable/Alternative Energy Systems for Electric Power Generation: Configurations, Control, and Application. *IEEE Trans. Sustain. Energy* **2011**, *2*, 392–403. [CrossRef]
6. Badin, R.; Huang, Y.; Peng, F.Z.; Kim, H.G. Grid interconnected Z-source PV system. In Proceedings of the IEEE PESC'07, Orlando, FL, USA, 17–21 June 2007; pp. 2328–2333.
7. Iannone, F.; Leva, S.; Zaninelli, D. Hybrid photovoltaic and hybrid photovoltaic-fuel cell system: Economic and environmental analysis. In Proceedings of the IEEE Power Engineering Society General Meeting, San Francisco, CA, USA, 16 June 2005; pp. 1503–1509.
8. Chakraborty, S.; Simões, M.G.; Kramer, W.E. *Power Electronics for Renewable and Distributed Energy Systems*; Springer: Berlin, Germany, 2015; pp. 1–620.
9. Núñez, J.V. Multilevel Topologies: Can New Inverters Improve Solar Farm Output. *Sol. Ind. J.* **2013**, *5*, 12.
10. Yalamanchili, K.P.; Ferdowsi, M. Review of multiple-input dc-dc converters for electric and hybrid vehicles. In Proceedings of the VPPC, Chicago, IL, USA, 7 September 2005; pp. 160–163.
11. Chen, Y.; Liu, Y.; Lin, S. Double-input PWM dc/dc converter for high-/low-voltage sources. *IEEE Trans. Ind. Electron.* **2006**, *53*, 1538–1545. [CrossRef]
12. Dobbs, B.G.; Chapman, P.L. A multiple-input DC-DC converter topology. *IEEE Power Electron. Lett.* **2003**, *99*, 6–9. [CrossRef]
13. Khaligh, A.; Cao, J.; Lee, Y. A Multiple-Input DC–DC Converter Topology. *IEEE Trans. Power Electron.* **2009**, *24*, 862–868. [CrossRef]
14. Wai, R.; Lin, C.; Liaw, J.; Chang, Y. Newly designed ZVS multiinput converter. *IEEE Trans. Ind. Electron.* **2011**, *58*, 555–566. [CrossRef]

15. Wai, R.; Lin, C.; Chen, B. High-efficiency dc–dc converter with two input power sources. *IEEE Trans. Power Electron.* **2012**, *27*, 1862–1875. [CrossRef]
16. Rosli, M.A.; Yahaya, N.Z.; Baharudin, Z. A multi-input converter for hybrid photovoltaic array/wind turbine/fuel cell and battery storage system connected AC grid network. In Proceedings of the ISGT ASIA, Kuala Lumpur, Malaysia, 20–23 May 2014; pp. 25–30.
17. Abdeen, E.; Gaafar, M.A.; Orabi, M.; Ahmed, E.M.; El Aroudi, A. Multi-Input Ćuk-Derived Buck-Boost Voltage Source Inverter for Photovoltaic Systems in Microgrid Applications. *Energies* **2019**, *12*, 2007. [CrossRef]
18. Zhoul, Z.; Wul, H.; Xingl, V. A non-isolated three-port converter for stand-alone renewable power system. In Proceedings of the IEEE Industrial Electronics Society Conference, Montreal, QC, Canada, 25–28 October 2012; pp. 3352–3357.
19. Taghvaee, M.H.; Radzi, M.A.M.; Moosavain, S.M.; Hizam, H.; Marhaban, M.H. A current and future study on nonisolated dc–dc converters for photovoltaic applications. *J. Renew. Sustain. Energy Rev.* **2013**, *17*, 216–227. [CrossRef]
20. Ding, S.; Wu, H.; Xing, Y.; Fang, Y.; Ma, X. Topology and control of a family of nonisolated three-port dc–dc converters with a bidirectional cell. In Proceedings of the IEEE Applied Power Electron Conference, Long Beach, CA, USA, 17–21 March 2013; pp. 1089–1094.
21. Wu, H.; Zhang, J.; Xing, Y. A family of multi-port buck-boost converters based on dc-link-inductors (DLIs). *IEEE Trans. Power Electron.* **2014**, *30*, 735–746. [CrossRef]
22. Foti, S.; Testa, A.; de Caro, S.; Scimone, T.; Scelba, G.; Scarcella, G. Multi-level Open End Windings Multi-Motor Drives. *Energies* **2019**, *12*, 861. [CrossRef]
23. De Caro, S.; Foti, S.; Sceiba, G.; Scimone, T.; Testa, A. A six-level asymmetrical Hybrid Photovoltaic Inverter with inner MPPT capability. In Proceedings of the ICCEP, Santa Margherita Ligure, Italy, 27–29 June 2017; pp. 631–635.
24. Foti, S.; Testa, A.; Scelba, G.; Cacciato, M.; Scarcella, G.; de Caro, S.; Scimone, T. Overvoltage mitigation in open-end winding AC motor drives. In Proceedings of the ICRERA, Palermo, Italy, 22–25 November 2015; pp. 238–245.
25. Foti, S.; Testa, A.; Scelba, G.; Sabatini, V.; Lidozzi, A.; Solero, L. Asymmetrical hybrid unidirectional T-type rectifier for high-speed gen-set applications. In Proceedings of the ECCE, Cincinnati, OH, USA, 1–5 October 2017; pp. 4887–4893.
26. Foti, S.; Scelba, G.; Testa, A.; Sciacca, A. An Averaged-Value Model of an Asymmetrical Hybrid Multi-Level Rectifier. *Energies* **2019**, *12*, 589. [CrossRef]
27. Foti, S.; de Caro, S.; Scelba, G.; Scimone, T.; Testa, A.; Cacciato, M.; Scarcella, G. An Optimal Current Control Strategy for Asymmetrical Hybrid Multilevel Inverters. *IEEE Transac. Ind. Appl.* **2018**, *54*, 4425–4436. [CrossRef]
28. Testa, A.; Foti, S.; de Caro, S.; Tornello, L.D.; Scelba, G.; Scarcella, G. Optimal Selection of the Voltage Modulation Strategy for an Open Winding Multilevel Inverter. In Proceedings of the ECCE, Detroit, MI, USA, 11–15 October 2020; pp. 2231–2237.
29. Rohner, S.; Bernet, S.; Hiller, M.; Sommer, R. Modulation, Losses, and Semiconductor Requirements of Modular Multilevel Converters. *IEEE Trans. Ind. Electron.* **2010**, *57*, 2633–2642. [CrossRef]

Article

Performance Improvement of Grid-Connected Induction Motors through an Auxiliary Winding Set

Luigi Danilo Tornello [1], Salvatore Foti [2], Mario Cacciato [1,*], Antonio Testa [2], Giacomo Scelba [1], Salvatore De Caro [2], Giuseppe Scarcella [1] and Santi Agatino Rizzo [1]

[1] Department of Eletrical Electronic and Computer Engineering, University of Catania, 95125 Catania, Italy; luigi.tornello@unict.it (L.D.T.); giacomo.scelba@unict.it (G.S.); giuseppe.scarcella@unict.it (G.S.); santi.rizzo@unict.it (S.A.R.)

[2] Department of Engineering, University of Messina, 98166 Messina, Italy; sfoti@unime.it (S.F.); atesta@unime.it (A.T.); sdecaro@unime.it (S.D.C.)

* Correspondence: mario.cacciato@unict.it

Abstract: A technique to improve the performance of grid-connected induction motors by exploiting an auxiliary winding set is proposed in this paper. This auxiliary winding features the same distribution of the main winding, although with a reduced number of turns and it is fed by an inverter a fraction of the power in comparison with the rated size of the induction motor. As shown in the paper, through the auxiliary winding, it is possible to set the machine power factor, increasing the efficiency of the power conversion system and mitigating speed oscillations due to torque disturbances. A mitigation of the grid current peaks due to motor start-up is obtainable. First, the proposed technique is theoretically introduced, then a feasibility assessment is accomplished by simulations.

Keywords: power factor correction; efficiency; induction machines; pulse width modulation inverters; dual stator winding machines; harmonic compensation

Citation: Tornello, L.D.; Foti, S.; Cacciato, M.; Testa, A.; Scelba, G.; De Caro, S.; Scarcella, G.; Rizzo, S.A. Performance Improvement of Grid-Connected Induction Motors through an Auxiliary Winding Set. *Energies* **2021**, *14*, 2178. https://doi.org/10.3390/en14082178

Academic Editor: Emilio Gomez-Lazaro

Received: 1 March 2021
Accepted: 6 April 2021
Published: 13 April 2021

Publisher's Note: MDPI stays neutral with regard to jurisdictional claims in published maps and institutional affiliations.

1. Introduction

The replacement of obsolete electric motors with new, more efficient ones could provide considerable advantages in terms of polluting emissions, energy resources exploitation, and manufacturing costs. For this reason, the mandatory compliance of industrial electromechanical systems with gradually increasing efficiency requisites has been enforced all over the world in the last decades. This can be accomplished either by increasing the efficiency of motors, for example, according to European Union (EU) IE2 and IE3 efficiency levels [1], or by replacing direct online electric motors with electric drives.

Despite the wide variety of electric motors available on the market, low- and medium-power three-phase direct online induction motors (DOL-IM) are by far the most dominant in the industrial sector, providing a wide variety of constant speeds and variable load applications where dynamic response requirements are not critical, such as pumps, fans, and compressors. Affordability, durability, easy functioning, and easy maintenance are some of the major factors driving the ever-increasing demand for induction motors. High-efficiency DOL-IMs [1] in industrial applications provide a significant reduction in energy consumption and environmental impact; however, they are still burdened by a low power factor (PF) at partial loads, which can be only mitigated by adding power factor correction capacitors. Moreover, large in-rush current is very commonly generated at start-up, leading to voltage dip, speed loss, torque pulsations, and possible activation of protection devices. A wide diffusion of variable speed drives (VSDs), featuring either torque or speed control, has been experienced in the past thirty years due to the advances in power electronics technology and control strategies, which have allowed an increase in the efficiency and reliability of electric drives [1–4]. Suitable Pulse Width Modulation (PWM) strategies have been developed that are able to cope with adverse effects related to shaft voltages, bearing currents, motor insulation breakdown, and electromagnetic interference (EMI) issues [5–11]. However, despite clear advantages achieved in terms of efficiency, induction motor drives are much more expensive than DOL-IMs, leading to the development of some intermediate or hybrid solutions. One of these solutions is represented by direct online double-winding IMs (DOL DWIM) [12–14]. They were proposed to overcome some typical drawbacks of conventional induction motors, mainly poor efficiency and power factor, by exploiting a partial power converter, which means an inverter rated at only a fraction of the motor rated power, cutting the cost. According to the DOL-DWIM approach, a special induction machine is used, encompassing two full sets of stator windings: one being directly connected to the grid and a second one being fed by an inverter supplied through a floating capacitor [12–17]. According to the basic DOL-DWIM configuration, the two winding sets feature the same number of turns; thus, the inverter supplying the second set is rated at the grid voltage, although at remarkably lower power. DOL-DWIMs are generally managed by simple control algorithms, mainly developed in order to control the PF under sinusoidal steady state operation [12–17].

An advanced DOL-DWIM configuration is presented in this paper, aimed at achieving new valuable functions in a cost-effective way, such as dynamic, active, and reactive power control; a reduction in in-rush current; and the mitigation of torsional vibrations generated by periodical torque disturbances or distorted grid voltages [18], which lead to acoustic noise and additional mechanical stress.

As shown in Figure 1, the proposed DOL-DWIM configuration features a grid-connected main winding and an auxiliary three-phase winding set featuring a lower number of turns, located in the same stator slots of the main one. The auxiliary winding is supplied by a voltage inverter featuring a floating DC-bus capacitor. As the auxiliary winding has less turns than the main one, the inverter power and voltage ratings are much lower than those of the motor.

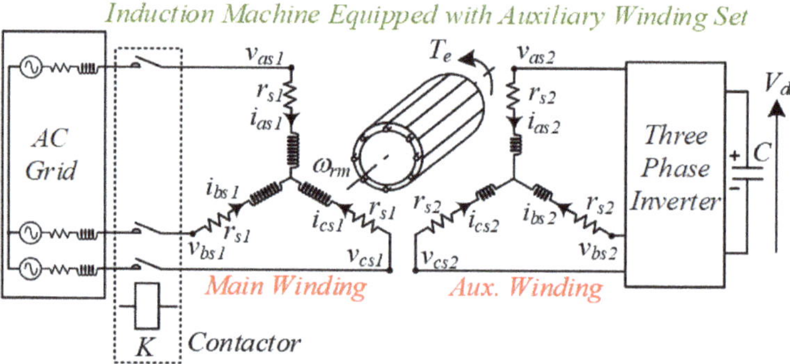

Figure 1. Double-winding induction motor (DWIM) featuring an auxiliary winding set, fed by a three-phase low power, low voltage, Voltage Source Inverter (VSI).

The outline of this paper is as follows: First, the modeling of the electromechanical system is presented in Section 2, and the control strategy implemented in the auxiliary winding set is described in detail in Section 3. Thereafter, the capability of the combined multi-winding sets motor, floating capacitor inverter, and control strategy to efficiently control the power factor PF, mitigate the detrimental effects related to distorted main grid voltages and mechanical torque vibrations, and reduce the large inrush current driven by the induction motor during the starting period is thoroughly investigated in Section 4. Finally, the conclusions are stated in the last section.

2. DWIM Model

It is assumed that the two polyphase stator windings [19–23] are distributed in order to produce sinusoidal magneto-motive forces; moreover, the effects of the saturation of the magnetic core are neglected, while iron losses are considered according to [24]. The voltage and flux equations of the DOL-DWIM, written in complex form $f_{qd} = f_q - jf_d$, according to an orthogonal qd reference frame rotating at the angular frequency ω_e of the input stator voltage vector, are given by [23]:

$$v_{qds1} = r_{s1}i_{qds1} + \frac{d\lambda_{qds1}}{dt} + j\omega_e\lambda_{qds1} \tag{1}$$

$$v'_{qds2} = r'_{s2}i'_{qds2} + \frac{d\lambda'_{qds2}}{dt} + j\omega_e\lambda'_{qds1} \tag{2}$$

$$v'_{qdr} = r'_r i'_{qdr} + \frac{d\lambda'_{qdr}}{dt} + j(\omega_e - \omega_{re})\lambda'_{qdr} \tag{3}$$

$$\lambda_{qds1} = L_{ls1}i_{qds1} + L_M\left[i_{qds1} + i'_{qds2} + i'_{qdsr}\right] \tag{4}$$

$$\lambda'_{qds2} = L'_{ls2}i'_{qds2} + L_M\left[i_{qds1} + i'_{qds2} + i'_{qdsr}\right] \tag{5}$$

$$\lambda'_{qdsr} = L'_{lsr}i'_{qdsr} + L_M\left[i_{qds1} + i'_{qds2} + i'_{qdsr}\right] \tag{6}$$

$$R_{Fe}i_{qdFe} = \frac{d\lambda'_{qdr}}{dt} + j\omega_e\lambda_{qdm} \quad \lambda_{qdm} = L_M i_{qdm} \tag{7}$$

$$i_{qds1} + i'_{qds2} + i'_{qdsr} = i_{qdm} + i_{qdFe} \tag{8}$$

where v_{qds1}, λ_{qds1} and v'_{qds2}, λ'_{qds2} are the stator voltages and fluxes of the two winding sets, separately; v'_{qdsr} and λ'_{qdr} are the rotor voltages and fluxes, respectively; r'_r, r_{s1} and r'_{s2} are the rotor and stator resistances, respectively; L_M is the mutual inductance; and L'_{lr}, L_{ls1}, and L'_{ls2} are the rotor and stator leakage inductances. All electrical quantities of the auxiliary winding set are referred to in the main stator winding. Moreover, $\omega_{re} = pp\omega_{rm}$, where ω_{rm} is the angular rotor speed and pp is the number of pole pairs. The equivalent circuit of the DWIM is depicted in Figure 2.

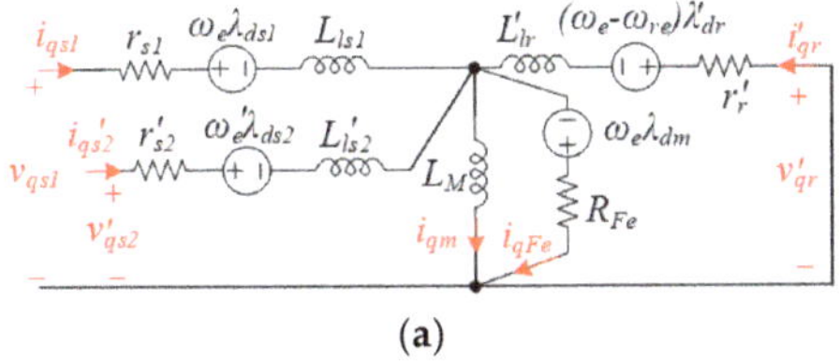
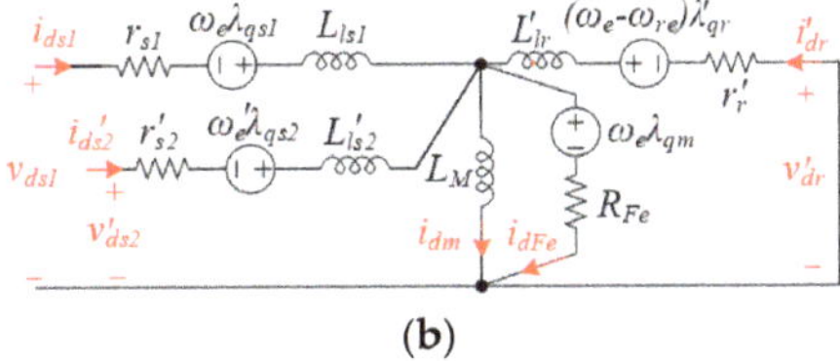

(a)   **(b)**

Figure 2. Equivalent circuits of the DWIM in the synchronous reference frame: (**a**) q-axis and (**b**) d-axis.

The electromagnetic torque is expressed in Equation (9), [25,26], and the relationships between the total electromagnetic torque T_e applied to the motor shaft and the torque related to the mechanical inertia J, the load torque T_L, and the frictional torque F are described in Equations (10) and (11). θ_{rm} is the rotor position.

$$T_e = 3/2 pp L_M[(i_{qs1} + i'_{qs2} - i_{qFe})i'_{dr} - (i_{ds1} + i'_{ds2} - i_{dFe})i'_{qr}] \tag{9}$$

$$T_e = T_L + J\frac{d\omega_{rm}}{dt} + F\omega_{rm} \tag{10}$$

$$\omega_{rm} = \frac{d\theta_{rm}}{dt} \tag{11}$$

3. Control System

The main stator winding set of the DOL DWIM is directly connected to the grid and it is tasked with handling the main power flow toward the mechanical load. Differently, the additional stator winding set is supplied by a floating power inverter, which enables a current vector control according to Figure 3. Low-cost devices are used to measure the AC quantities in both windings. An analysis of the different blocks composing the control system is presented in the following.

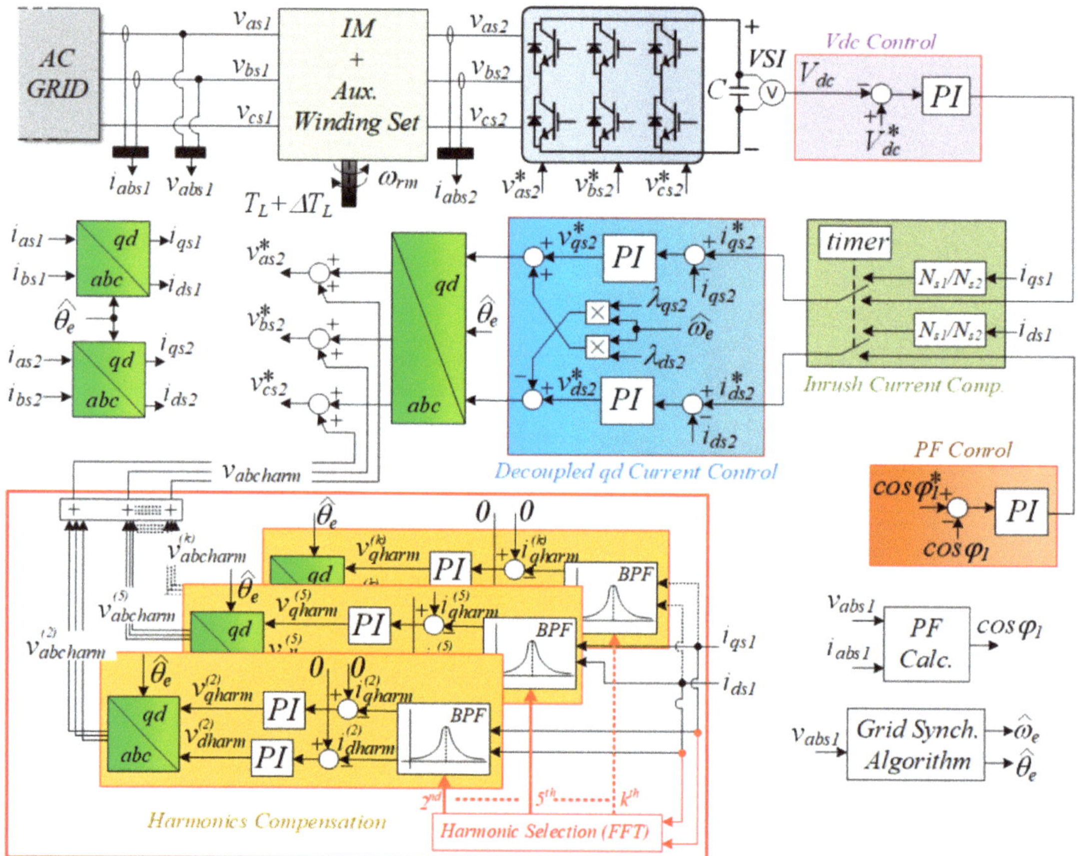

Figure 3. Block diagram of the proposed direct online (DOL) DWIM control solution.

3.1. Grid Synchronization

The control system driving the floating capacitor inverter is mainly based on a standard vector control approach approximating a stator flux orientation, which is achieved by transforming the three-phase utility voltages *abc* into a reference frame synchronous with the grid voltage vector phase θ_e. The last quantity can be determined with one of the well-known grid synchronization algorithms available in the literature [27,28]. In this work, a decoupling algorithm was exploited to remove all of the unwanted harmonic components from the measured voltages in order to keep the phase-locked loop (PLL) locked to fundamental harmonic, positive-sequence component at all times [29]. The block diagram of the grid synchronization algorithm based on a multi-harmonic synchronous reference frame (SRF) filtering (MSF) structure is shown in Figure 4. The basic idea is to estimate each voltage harmonic set by applying a low-pass filtering after a reference frame transformation synchronous with the corresponding harmonic frequency.

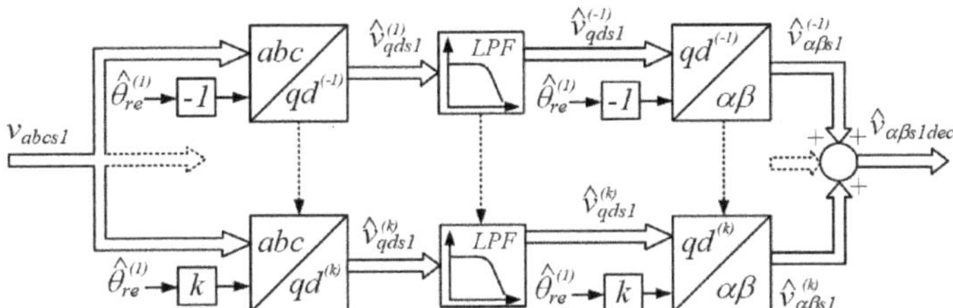

Figure 4. Block diagram of the multiharmonic SRF filtering (MSF) algorithm.

3.2. Active and Reactive Power Control

The machine active and reactive powers are functions of auxiliary winding currents i_{qs2} and i_{ds2}, which can be managed to control the PF under different load conditions. By neglecting the voltage drop due to stator resistances and leakage inductances, the active and reactive power generated by the auxiliary winding can be expressed as:

$$P_2 = 3/2(v_{qs2}i_{qs2} + v_{ds2}i_{ds2}) \quad Q_2 = 3/2(v_{qs2}i_{ds2} - v_{ds2}i_{qs2}) \tag{12}$$

If the q2-axis is aligned to the main grid voltage vector v_{qds1}, the active and reactive power can be approximated to:

$$P_2 = 3/2v_{qs2}i_{qs2} \quad Q_2 = 3/2v_{qs2}i_{ds2} \tag{13}$$

Hence, it is possible to control the machine reactive power by acting on the d-axis current of the auxiliary winding i_{ds2} in order to track a desired power factor at the main stator winding terminals. The q-axis current i_{qs2} can be managed to control the active power required to hold a stable voltage in the floating capacitor of the isolated two-level inverter supplying the auxiliary winding. A certain amount of active power is required by the floating capacitor inverter to compensate for power devices and capacitor losses. Hence, two external control loops were included in the control structure in order to achieve the desired PF and DC bus voltage V_{dc}.

3.3. Active Mitigation of Mechanical Vibrations

The auxiliary winding can be also exploited to mitigate vibrations generated by periodical torque disturbances or distorted grid voltages. Torque disturbances ΔT_L generate harmonic components in the stator currents i_{qs1} and i_{ds1}, yielding vibrations in the mechanical system. They can be mitigated by generating compensating components of the electromagnetic torque by means of the auxiliary winding set. In particular, in case of motor operation under a highly distorted grid voltage, some current harmonics set $i_{as1}^{(k)}$, $i_{bs1}^{(k)}$, $i_{cs1}^{(k)}$ of amplitude Ik and angular frequency $k\omega_e$ are superimposed to the fundamental current set i_{as1}^{I}, i_{bs1}^{I}, i_{cs1}^{I}, of magnitude I_1 and angular frequency ω_e. Therefore, it is possible to identify each couple of current harmonic components $i_{qs1}^{(k)}$ and $i_{ds1}^{(k)}$ in the qd current plane by applying a suitable reference frame transformation:

$$\begin{bmatrix} i_{qs1} \\ i_{ds1} \end{bmatrix} = K(\theta_e) \begin{bmatrix} i_{as1} \\ i_{bs1} \\ i_{cs1} \end{bmatrix} = \frac{2}{3} \begin{bmatrix} \cos(\omega_e t) & \cos(\omega_e t - \frac{2}{3}\pi) & \cos(\omega_e t + \frac{2}{3}\pi) \\ \sin(\omega_e t) & \sin(\omega_e t - \frac{2}{3}\pi) & \sin(\omega_e t + \frac{2}{3}\pi) \end{bmatrix} \begin{bmatrix} i_{as1} \\ i_{bs1} \\ i_{cs1} \end{bmatrix} \tag{14}$$

$$\begin{cases} i_{qs1} = i_{qs1}^{(1)} + \sum\limits_{k=1}^{+\infty} i_{qs1}^{(k)} = i_{qs1}^{(1)} + \sum\limits_{k=1}^{+\infty} I_{s1}^{(k)} \cos[(k \pm 1)\omega_e t] \\ i_{ds1} = i_{ds1}^{(1)} + \sum\limits_{k=1}^{+\infty} i_{ds1}^{(k)} = i_{ds1}^{(1)} + \sum\limits_{k=1}^{+\infty} I_{s1}^{(k)} \sin[(k \pm 1)\omega_e t] \end{cases} \begin{cases} +1 \to if \to k = 2,5,8,11,\ldots \\ -1 \to if \to k = 4,7,10,13,\ldots \\ 0 \to if \to k = 3,6,9,12,\ldots \end{cases} \tag{15}$$

where $i_{qs1}^{(1)}$ and $i_{ds1}^{(k)}$ are the DC components related to the fundamental current harmonic. These current harmonics are identified through a standard harmonic identification algorithm based on fast Fourier transform. Each undesired harmonic present on the main winding current is then reproduced with the opposite sign on the auxiliary winding through a specific closed-loop current control. A control structure composed of some paralleled current control modules is thus required to mitigate mechanical vibrations featuring different frequencies, each module being threshold-activated on the basis of the actual magnitude of the specific stator current harmonic to be attenuated. Similar considerations are valid in case of mechanical vibrations caused by shaft eccentricity, mechanical imbalance phenomena, or bearing aging.

3.4. Inrush Current Mitigation during Startup

The large line currents that occur on the grid at motor start-up can be mitigated by exploiting the charge stored in the floating capacitor by implementing a proper control strategy in the DOL DWIM. For achieving this additional feature, a suitable selection of the floating capacitor and of the inverter current rating is required. In particular, the inrush current in the main stator winding can be reduced by exploiting the energy stored in the floating capacitor to reduce the power drawn from the grid. This can be accomplished by generating a set of currents exactly in phase with the currents flowing in the main winding during the first few cycles of the grid voltage. The measured main winding currents i_{qs1} and i_{ds1} are firstly scaled by the turn ratio of the two stator windings; then, they are used as references of the closed loop current controllers managing the auxiliary windings voltages in order to generate three phase currents i_{abc2} in phase with i_{abc1}, thus allowing a sharing between the inrush currents of both windings.

4. Performance Assessment of the Grid-Connected DWIM

The control strategies described in the previous section were assessed via a simulation analysis realized in Simulink/MATLAB, as shown in Figure 5a. The model of the electromechanical system, shown in Figure 5b, was realized according to the data listed in Table 1. Implementation of the multiharmonic SRF filter and harmonic compensation system are presented, respectively, in Figure 6a,b. According to Table 1, the considered machine features a turns ratio between the main and auxiliary windings $N_{s1}/N_{s2} = 5$. The floating capacitor inverter was driven with a space vector modulation (SVM) at a switching frequency fs of 10 kHz and the DC bus voltage control held an average voltage V_{dc} of 150 V. The inverter was equipped with a 5 mF floating DC bus capacitor.

Table 1. Multi-winding induction motor data.

	Winding 1		Winding 2
Stator Voltage *Vs*	400 Vrms		60 Vrms
Stator Current *Is*	25 Arms		60 Arms
Stator Resistance *Rs*	0.2147 Ω		0.01 Ω
Leakage Inductance *Lls*	0.991 mH		40 μH
Rotor Resistance *r′r*		0.2205 Ω	
Magnetizing Inductance *LM*		64.2 mH	
Iron Losses Resistance *RFe*		700 Ω	
Nominal Rotor Speed *ωrm*		1460 rpm	
Mechanical Inertia *J*		0.102 kgm^2	

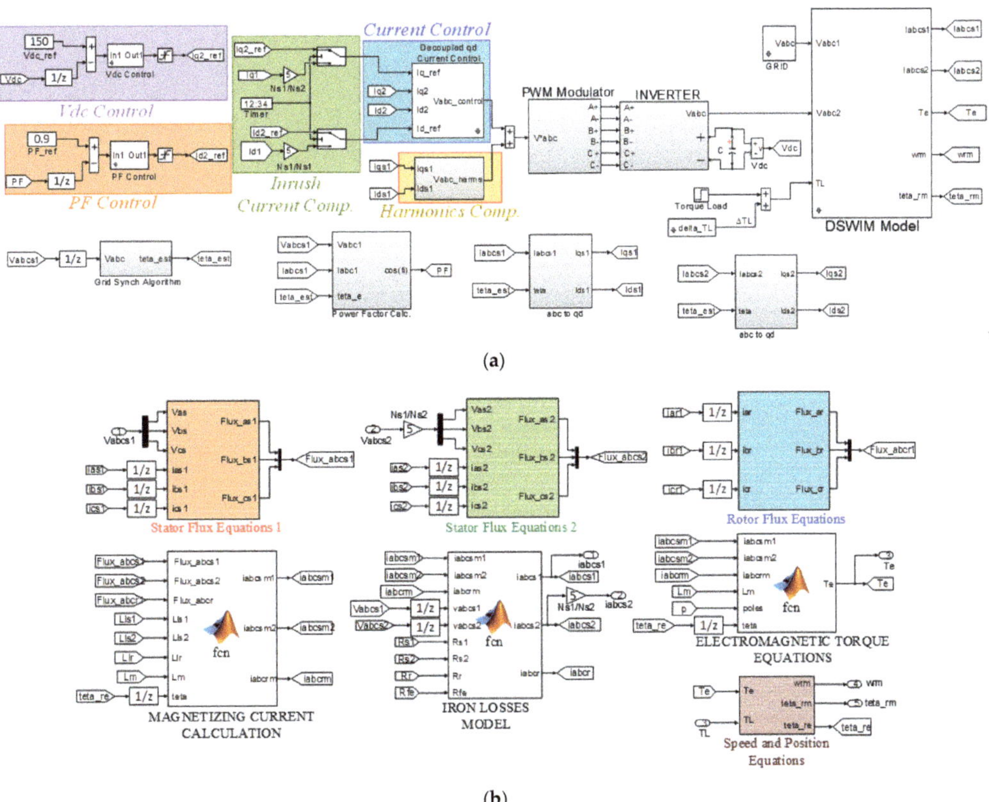

Figure 5. Simulink block diagram of the: (**a**) power conversion system and control strategies and (**b**) DOL-DWIM model.

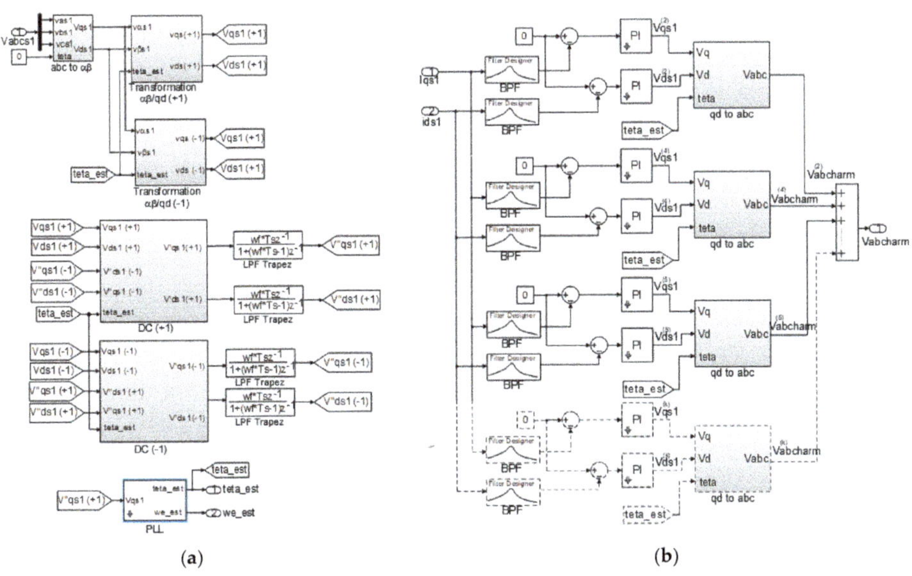

Figure 6. Simulink block diagram of the: (**a**) MSF algorithm and (**b**) harmonics compensation.

A first set of tests dealt with assessing the capability of the proposed solution to regulate the reactive power exchange between the power conversion system and the AC grid in a wide operating range. The performance of a standard IM featuring a single stator winding was compared with that of the DWIM under same load conditions. Active and reactive power flows associated to the fundamental harmonics in both main and auxiliary windings are shown in Figure 7. The simulation results are shown in Figure 8.

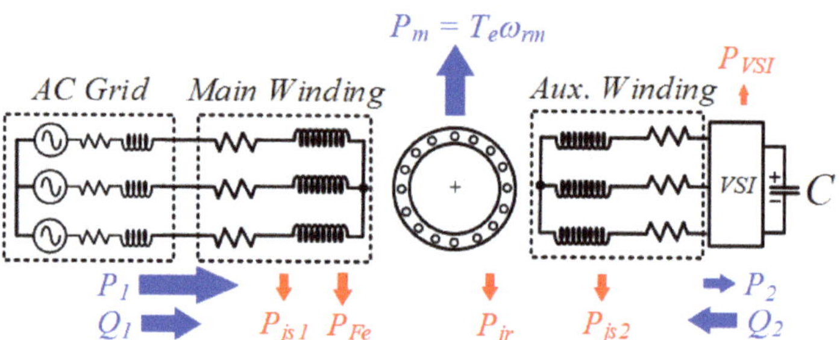

Figure 7. Power flow scheme of the DWIM.

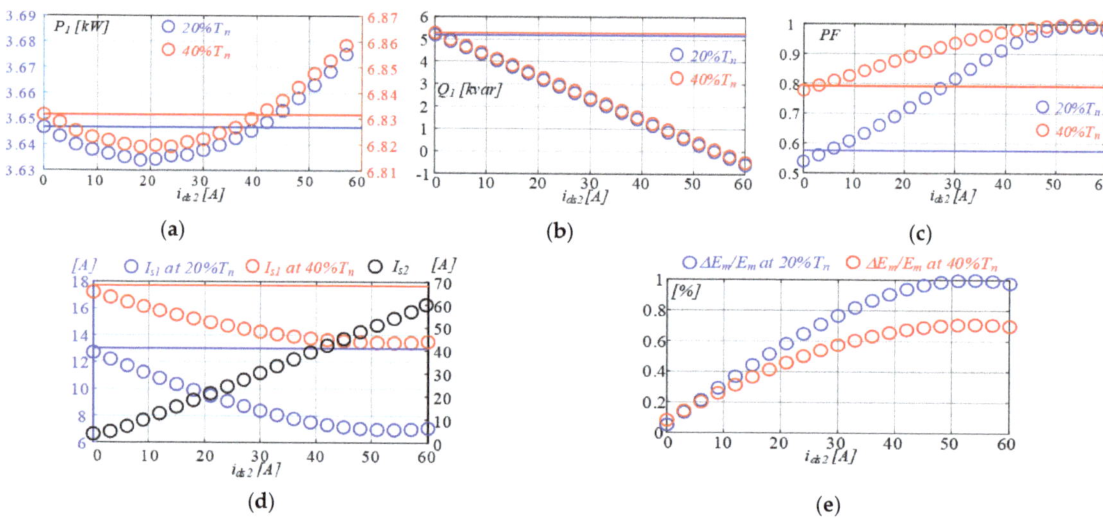

Figure 8. (a) Active and reactive power (P_1 and Q_1, respectively), (b) power factor *PF*, (c) main winding current magnitude I_{s1}, (d) auxiliary winding current magnitude I_{s2}, (e) p.u. back-electromotive force (EMF) variation ΔE_m, computed at 20% and 40% of rated torque T_n for different values of i_{ds2} (solid lines—standard IM; circles—DWIM).

The active power P_1 and reactive power Q_1 absorbed by the DWIM are reported in Figure 8a,b as a function of the current i_{ds2}. The magnitude of currents in both windings, PF, and the per-unit (p.u.) variation of the back-electromotive force (EMF) E_m are displayed in Figure 8c–e, respectively, considering two different load conditions: 20% and 40% of the rated torque. The same figures include the corresponding quantities computed for the single-winding IM configuration (the same electrical machine without auxiliary winding) operating under the same load conditions, whose values are indicated in the figures with horizontal solid lines. By varying i_{ds2}, a beneficial effect is achieved, consisting of a reduction in Q_1 and leading to an improved PF (Figure 8c). This causes a reduction in the current flowing through the main winding, as shown in Figure 8d. This last result is justified even by the increase in the back-emf E_m of the DWIM in correspondence with an increase in i_{ds2}, as clearly shown in Figure 8e; the last variation is in any case limited to a few percentages of the back-EMF of the DWIM with $i_{ds2} = 0$. The variation in E_m is associated to a variation in the magnetizing current.

The proposed approach can lead to a reduction in the active power P_1 absorbed by the DWIM at partial load, leading to an efficiency improvement. This result is confirmed by the analysis summarized in Figure 9, where the iron and total Joule losses are computed for the standard single-winding IM configuration and the considered DWIM as a function of the current i_{ds2}. In particular, Figure 9 displays the difference between the standard IM and DWIM total Joule losses ΔP_J (stator and rotor) and iron losses ΔP_{Fe}, with both motors operating under the same loads. The joule P_J and iron P_{Fe} losses referring to the single-winding IM are the points displayed in the three graphs at $i_{ds2} = 0$. Notably, in DWIM, by increasing i_{ds2}, iron losses rise and total joule losses initially drop and then rise. Thus, by suitably selecting i_{ds2}, the balance between iron losses increase and joule losses drop may lead to a neat reduction in total power losses and to a higher efficiency compared with the conventional IM. This beneficial result is strictly related to the motor design, load condition, and desired power factor PF. If we assume to operate the DWIM at PF = 0.9, thus requiring $i_{ds2} \approx 25A$ (Figure 8c) and a load equal to $40\% T_n$, a total loss reduction ΔP_{loss} equal to 3% can be achieved compared with the standard IM configuration. On the contrary, the same DWIM operated at PF = 0.9 and a load equal to $20\% T_n$, will require a $i_{ds2} = 40A$, yielding a $\Delta P_{loss}/P_{loss}$ less than 0.8%, significantly reducing the efficiency improvement of the DWIM. The implementation of a losses minimization algorithm is not the focus of the paper, and more details on how to approach this aspect can be found in [30].

Simulations were performed over a wide load range and the results shown in Figure 10 confirm the effectiveness of the proposed approach in increasing the power factor by acting on i_{ds2} with marginal variations in the additional power losses, although the improvement becomes progressively less relevant as the load increases.

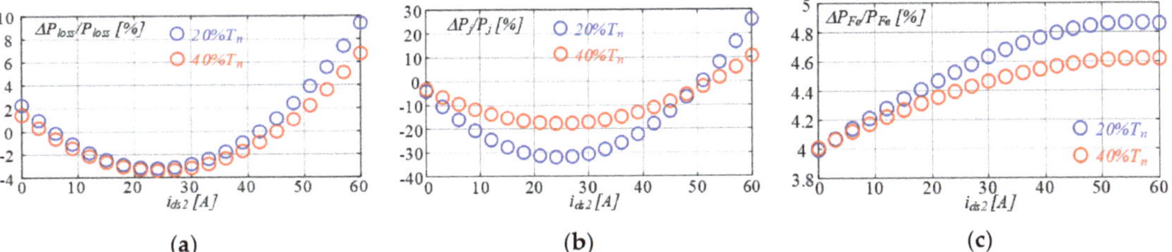

Figure 9. (a) Total loss variation in p.u. ΔP_{loss}, (b) joule losses variation in p.u. ΔP_J, and (c) iron losses variation in p.u. ΔP_{Fe}, vs. i_{ds2}, computed at 20% and 40% of rated torque T_n. Joule P_J and iron P_{Fe} losses referring to the single-winding IM are the points displayed in the three graphs at $i_{ds2} = 0$.

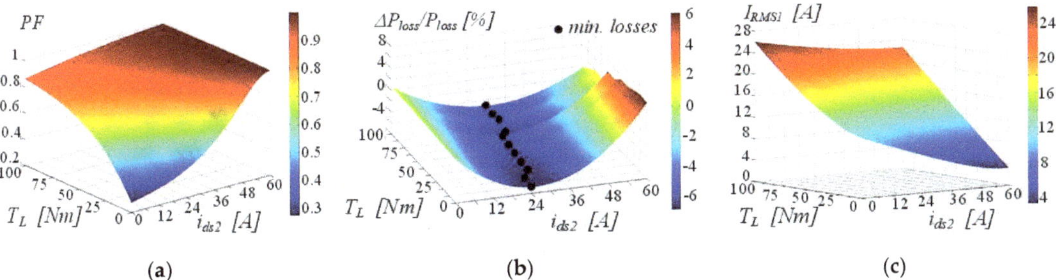

Figure 10. (a) PF, (b) $\Delta Ploss$, (c) root mean square (RMS) value of the main winding current $IRMS1$, determined under different loads TL.

According to Figure 10b, for a given load condition, negative losses ΔP_{loss} can be achieved by suitably selecting i_{ds2}, leading to higher efficiency compared with the standard IM. Figure 11 deals with some results obtained by holding the PF at 0.9 by means of the control system. As shown, under constant PF operation, i_{ds2} and the reactive power Q_2 decrease when decreasing the load torque and the absorbed active power. Moreover, only a small active power P_2 is transferred from the grid to the auxiliary winding. Finally, both the active power P_1 and reactive power Q_1 are changed by the controller in order to keep the power factor constant. The power handled by the auxiliary winding in the worst-case scenario, i.e., a very light load, is roughly one-fifth of the standard IM rated power.

The dynamic behavior of the PF controller was also evaluated and some results are shown in Figure 12, dealing with a sequence of step loads T_L using a standard DOL IM and the proposed DOL DWIM. A rather satisfying dynamic behavior was achieved with the DOL DWIM, which was able to hold the PF constant.

The test results shown in Figure 13a confirm the effectiveness of the proposed approach in mitigating the oscillations of the rotational speed caused by an externally applied sinusoidal shaft torque disturbance $\Delta T_L = 20$ Nm (20% of rated torque) at 100 Hz. We note the remarkable speed ripple reduction after the instant t_1, when the current control in the auxiliary winding set is activated.

Figure 13b depicts the mitigation of the speed ripple generated by a fifth harmonic component ($f_d = 250$ Hz) superimposed to the fundamental AC grid voltage. At t_1, the compensation algorithm is turned on, leading to the reduction in the speed ripple. A very small V_{dc} ripple was also observed in both cases.

The effects of the harmonic compensation on motor quantities are shown in Figure 14. The obtained results confirmed the theorical harmonic distribution given by Equation (15). Specifically, a second-order harmonic disturbance on the electromagnetic torque and on the mechanical speed produced a harmonic of the same order on the qd axis currents and therefore a third harmonic disturbance in the phase currents on both main and auxiliary windings. A different case is considered in Figure 15, where a fifth harmonic component is superimposed on the AC grid voltage. Th theoretical findings and the effectiveness of the harmonic compensation were also confirmed in this case.

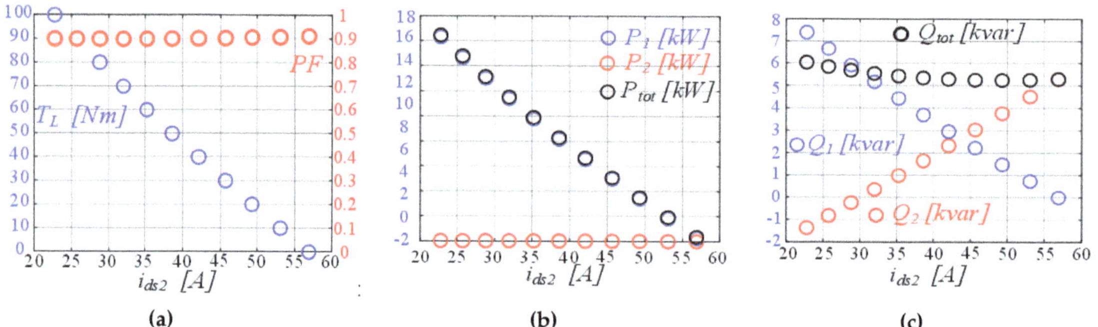

Figure 11. (**a**) Power factor control: *TL* and *PF* vs. *ids2*; (**b**) *P1*, *P2*, *Ptot* = *P1* + *P2*; (**c**) *Q1*, *Q2* and *Qtot* = *Q1* + *Q2*.

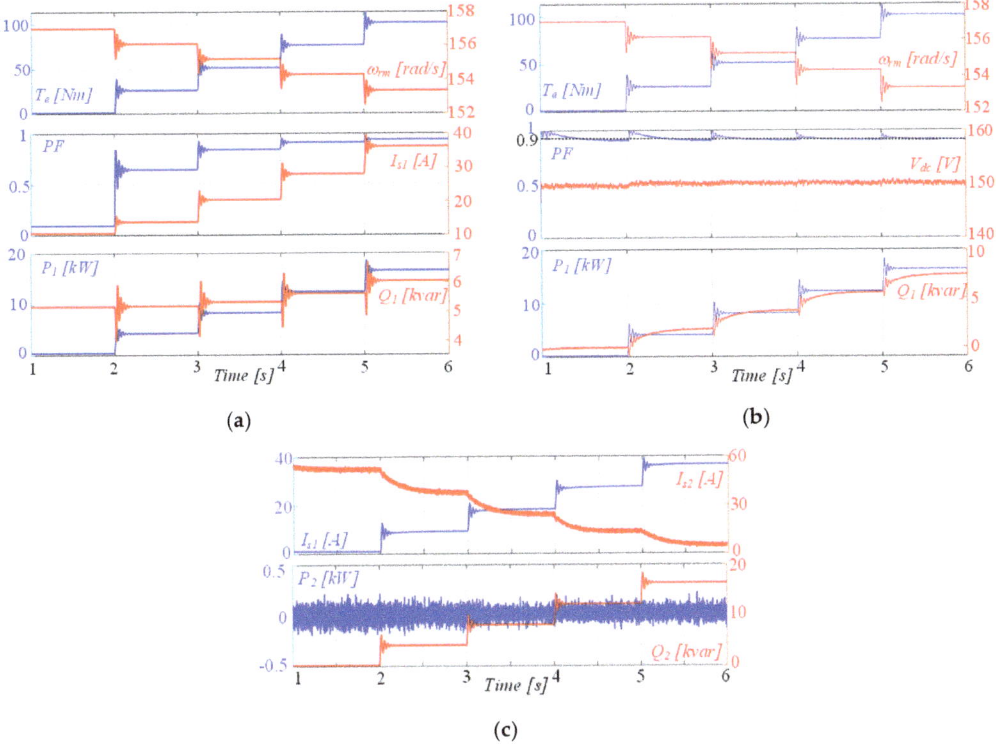

Figure 12. Dynamic transients under different torque load *TL* conditions: (**a**) in standard IM configuration and (**b,c**) in the proposed DWIM configuration.

Finally, the effectiveness of the proposed approach in mitigating inrush currents was evaluated by considering two motors rated at 3 and 15 kW. Figure 16 shows the currents in the two windings during the start-up without and with using the proposed technique for peak current reduction on the two motors. Figure 16a deals with the 3 kW IM using a 5 mF flying capacitor, whereas Figure 16b deals with the 15 kW machine using 120 mF capacitor. In both cases, at $t = t2$, the current control is disabled and the *PF* and V_{dc} controls are activated instead. A remarkable mitigation of the grid current was obtained, as well as an increase of the electromagnetic torque, although at the cost of rather high transitory currents on the auxiliary winding. The floating capacitor has to be charged before the start-up of the DOL DWIM. This goal can be reached in different ways by using low-power, low-voltage circuitries. A viable solution is presented in Figure 17, where a cascade connection of a small power transformer and diode rectifier was used to charge the floating capacitor. The performance in terms of inrush line current mitigation clearly improved by increasing the capacitance of the floating capacitor because this reduced the power drawn from the grid. However, a large flying capacitor may be impractical in some applications; hence, a smaller capacitor complemented by a parallel connected battery can be used instead. An alternative approach consists of operating the start-up at a higher DC-bus voltage in order to increase the energy stored in the floating capacitor. A cost-benefit analysis is also required to establish the inverter ratings and thus the overcurrent mitigation level during the start-up.

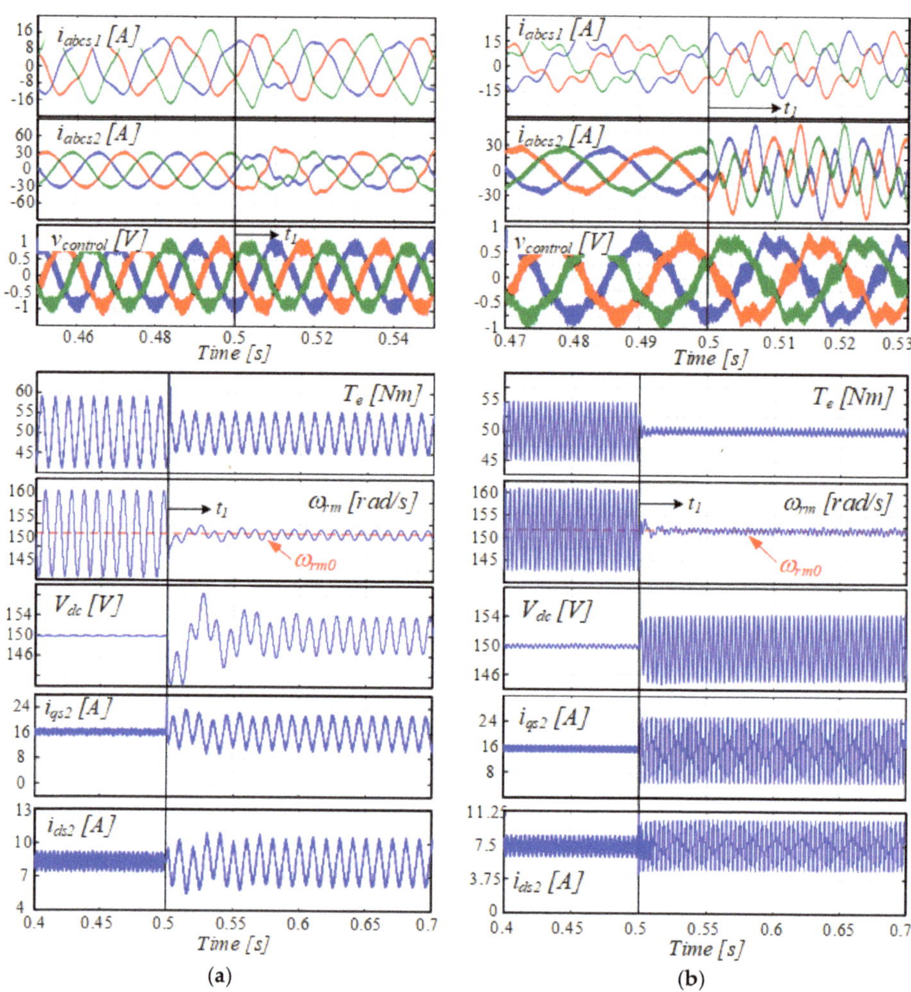

Figure 13. Simulation results: (**a**) compensation of *ωrm* with Δ*TL* = 20 Nm; (**b**) compensation of *ωrm* with *Vabc5th* = 30 V.

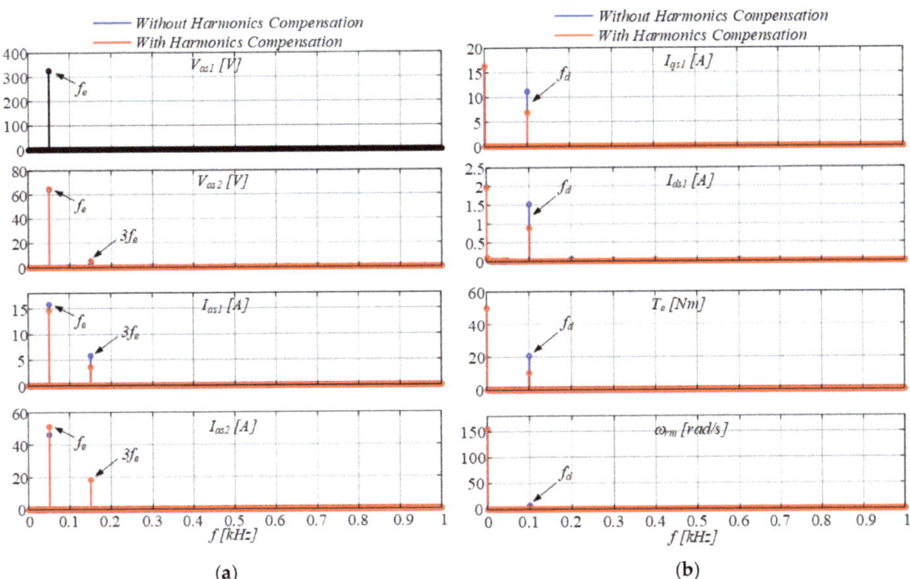

Figure 14. Harmonic content of motor quantities with and without the harmonic compensation when a sinusoidal torque disturbance is applied to the shaft: (**a**) main winding phase stator amplitude voltage *Vas1*, auxiliary winding phase stator voltage amplitude *Vas2*, main winding phase stator amplitude current *Ias1*, auxiliary winding phase stator amplitude current *Ias2*; (**b**) qd-axes main winding stator current *Iqs1* and *Ids1*, electromagnetic torque *Te*, mechanical speed *ωrm*.

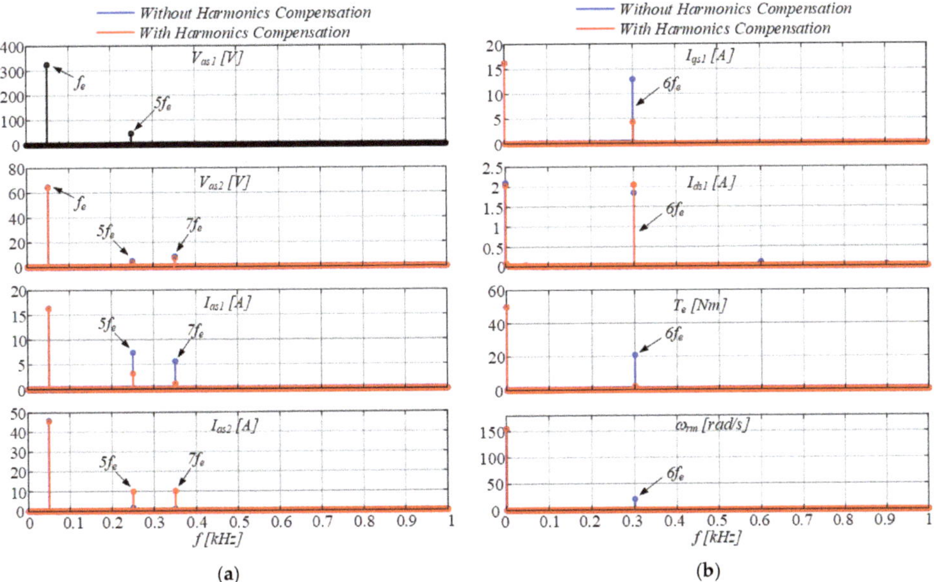

Figure 15. Harmonic content of motor quantities with and without harmonic compensation when a 5th harmonic component is superimposed on the AC grid voltage: (**a**) main winding phase stator amplitude voltage *Vas1*, auxiliary winding phase stator voltage amplitude *Vas2*, main winding phase stator amplitude current *Ias1*, auxiliary winding phase stator amplitude current *Ias2*; (**b**) qd-axes main winding stator current *Iqs1* and *Ids1*, electromagnetic torque *Te*, mechanical speed *ωrm*.

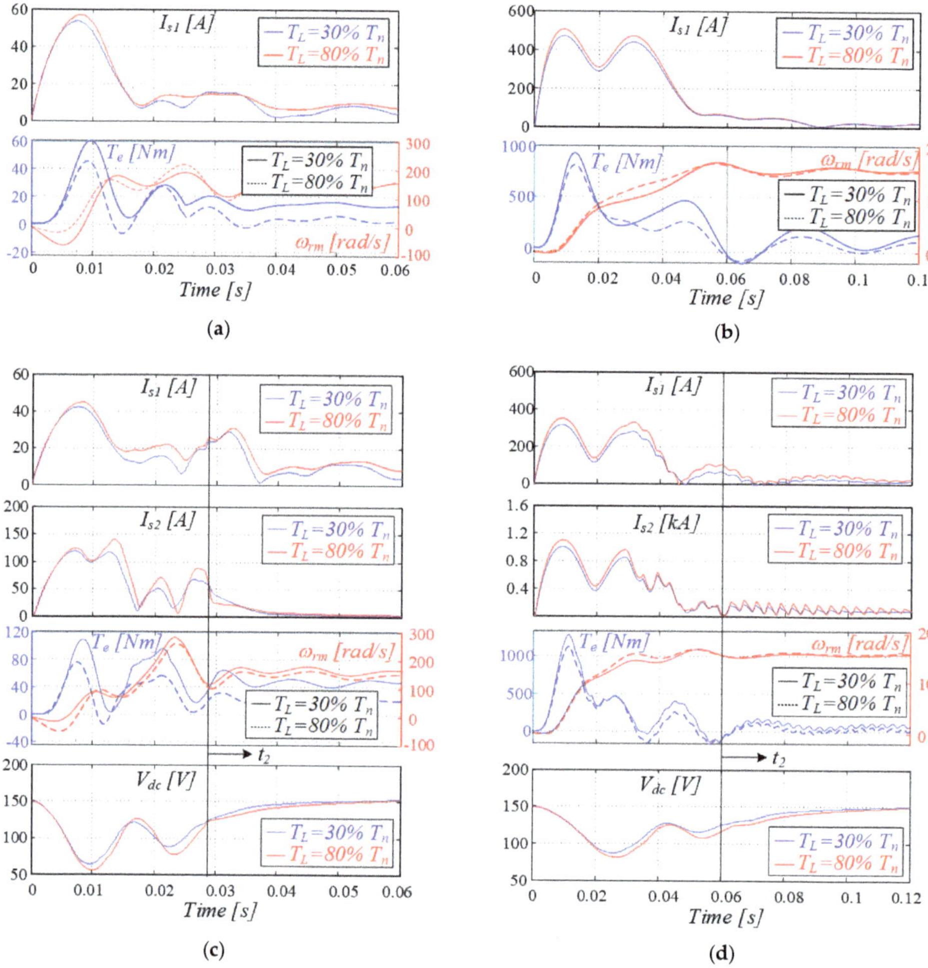

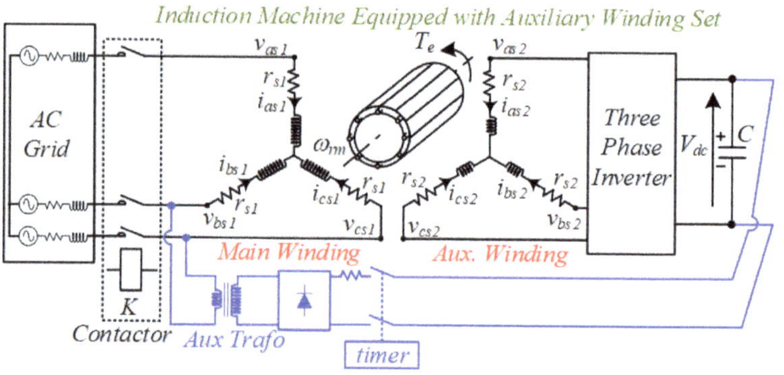

Figure 16. Induction motor start-up current: (**a**) standard IM (3 kW), (**b**) standard IM (15 kW), (**c**) 3 kW with the proposed technique, and (**d**) 15 kW with the proposed technique.

Figure 17. Floating capacitor low-power, low-voltage charging circuit.

5. Conclusions

The main aim of this paper was to demonstrate that the performance of direct line connected induction motors can be improved by introducing an auxiliary stator winding supplied by a partial power inverter. The auxiliary stator winding features fewer turns than the main one, and thus has lower voltage and power ratings. However, if suitably controlled, the additional winding allows for the achievement of an effective power factor control and a compensation of torque oscillations produced by the mechanical load or by distorted utility grid voltages. Moreover, under certain conditions, it is possible to mitigate the line overcurrent occurring during the start-up of the induction machine. The proposed solution is expected to be mainly applied for medium–high power range motors, where the cost of the system could be more affordable.

Author Contributions: G.S. (Giacomo Scelba) and L.D.T. conceived the original idea. M.C., S.F. and L.D.T. developed the theory and performed the computations. A.T., G.S. (Giuseppe Scarcella), S.A.R. and S.D.C. verified the analytical methods. All authors discussed the results and contributed to the final manuscript. All authors have read and agreed to the published version of the manuscript.

Funding: This work has been partially supported by the University of Catania under the interdepartmental project PIA.CE.RI. 2020–2022 Line 2-TMESPEMES.

Institutional Review Board Statement: Not applicable.

Informed Consent Statement: Informed consent was obtained from all subjects involved in the study.

Data Availability Statement: Not applicable.

Conflicts of Interest: The authors declare no conflict of interest.

References

1. G1 IEC 60034-30-1:2014. *Rotating Electrical Machines—Part 30-1: Efficiency Classes of Line Operated AC Motors*; IEC: Geneva, Switzerland, 2014.
2. Morimoto, S.; Tong, Y.; Takeda, Y.; Hirasa, T. Loss minimization control of permanent magnet synchronous motor drives. *IEEE Trans. Ind. Electron.* **1994**, *41*, 511–517. [CrossRef]
3. Kim, S.; Yoon, Y.; Sul, S.; Ide, K. Maximum Torque per Ampere (MTPA) Control of an IPM Machine Based on Signal Injection Considering Inductance Saturation. *IEEE Trans. Power Electron.* **2013**, *28*, 488–497. [CrossRef]
4. Uddin, M.N.; Nam, S.W. New Online Loss-Minimization-Based Control of an Induction Motor Drive. *IEEE Trans. Power Electron.* **2008**, *23*, 926–933. [CrossRef]
5. Cacciato, M.; Consoli, A.; Scarcella, G.; Scelba, G. Indirect maximum torque per ampere control of induction motor drives. In Proceedings of the EPE, Aalb, Den, 2–5 September 2007; pp. 1–10.
6. Attaianese, C.; Perfetto, A.; Tomasso, G. A space vector modulation algorithm for torque control of inverter fed induction motor drive. *IEEE Trans. Energy Convers.* **2002**, *17*, 222–228. [CrossRef]
7. Kim, H.-J.; Lee, H.-D.; Sul, S.-K. A new PWM strategy for common-mode voltage reduction in neutral-point-clamped inverter-fed AC motor drives. *IEEE Trans. Ind. Appl.* **2001**, *37*, 1840–1845.
8. Hava, A.M.; Kerkman, R.J.; Lipo, T.A. Carrier-based PWM-VSI overmodulation strategies: Analysis, comparison, and design. *IEEE Trans. Power Electron.* **1998**, *13*, 674–689. [CrossRef]
9. Zhong, E.; Lipo, T.A. Improvements in EMC performance of inverter-fed motor drives. *IEEE Trans. Ind. Appl.* **1995**, *31*, 1247–1256. [CrossRef]
10. Cacciato, M.; Caro, S.D.; Scarcella, G.; Scelba, G.; Testa, A. Improved space-vector modulation technique for common mode currents reduction. *IET Power Electron.* **2013**, *6*, 1248–1256. [CrossRef]
11. Cacciato, M.; Consoli, A.; Scarcella, G.; Scelba, G.; Testa, A. A novel space-vector modulation technique for common mode emissions reduction. In Proceedings of the International Aegean Conference on Electrical Machines and Power Electronics, Bodrum, Turkey, 2–4 September 2007; pp. 199–204.
12. Muljadi, E.; Lipo, T.A.; Novotny, D.W. Power Factor Enhancement of Induction Machines by Means of Solid-State Excitation. *IEEE Trans. Power Electron.* **1989**, *4*, 409–418. [CrossRef]
13. Spée, R.; Wallace, A.K. Comparative Evaluation of Power Factor Improvement Techniques for Squirrel Cage Induction Motors. *IEEE Trans. Ind. Appl.* **1992**, *28*, 381–386. [CrossRef]
14. Basak, S.; Chakraborty, C. Dual Stator Winding Induction Machine: Problems, Progress, and Future Scope. *IEEE Trans. Ind. Electron.* **2015**, *62*, 4641–4652. [CrossRef]

15. Muteba, M.C.; Jimoh, A.A.; Nicolae, D.V. Improving three-phase induction machines power factor using single phase auxiliary winding fed by an active power filter. In Proceedings of the IEEE Conference AFRICON, Windhoek, South Africa, 13–15 September 2007; pp. 1–7.
16. Malik, N.; Sadarangani, C.; Cosic, A.; Lindmark, M. Induction Machine at Uniy Power Factor with Rotating Power Electronic Converter. In Proceedings of the IEEE International Symposium SPEEDAM, Sorrento, Italy, 24–26 June 2012; pp. 401–408.
17. Knight, A.M.; Salomon, J.C.; Haque, R.; Perera, N.; Toulabi, M.S. A grid-connected induction machine capable of operation at unity and leading power factor. In Proceedings of the IEEE Energy Conversion Congress and Exposition, Denver, CO, USA, 10–14 October 2013; pp. 238–245.
18. Riley, C.M.; Lin, B.K.; Habetler, T.G.; Kliman, G.B. Stator current harmonics and their causal vibrations: A preliminary investigation of sensorless vibration monitoring applications. *IEEE Trans. Ind. Appl.* **1999**, *35*, 94–99. [CrossRef]
19. Foti, S.; De Caro, S.; Testa, A.; Tornello, L.D.; Scelba, G.; Scarcella, G. Grid-Connected Open-end Winding Induction Motor Drives. In Proceedings of the International Symposium SPEEDAM, Sorrento, Italy, 24–26 June 2020; pp. 30–35.
20. Basak, S.; Chakraborty, C.; Sinha, A.K. Dual Stator Induction Generator with Controllable Reactive Power Capability. In Proceedings of the IEEE Symposium ISIE, Instabul, Turkey, 1–4 June 2014; pp. 2584–2589.
21. Marouani, K.; Nounou, K.; Benbouzid, M.; Tabbache, B. Power factor correction of an electrical drive system based on multiphase machines. In Proceedings of the IEEE First International Conference on Green Energy ICGE, Sfax, Tunisia, 25–27 March 2014; pp. 152–157.
22. Ojo, O.; Davidson, I.E. PWM-VSI inverter-assisted stand-alone dual stator winding induction generator. *IEEE Trans. Ind. Appl.* **2000**, *36*, 1604–1611.
23. Li, Y.; Hu, Y.; Huang, W.; Liu, L.; Zhang, Y. The Capacity Optimization for the Static Excitation Controller of the Dual-Stator-Winding Induction Generator Operating in a Wide Speed Range. *IEEE Trans. Ind. Electron.* **2009**, *56*, 530–541.
24. Levi, E.; Sokola, M.; Boglietti, A.; Pastorelli, M. Iron loss in rotor-flux-oriented induction machines: Identification, assessment of detuning, and compensation. *IEEE Trans. Power Electron.* **1996**, *11*, 698–709. [CrossRef]
25. Scarcella, G.; Scelba, G.; Cacciato, M.; Spampinato, A.; Harbaugh, M.M. Vector Control Strategy for Multidirectional Power Flow in Integrated Multidrives Starter-Alternator Applications. *IEEE Trans. Ind. Appl.* **2016**, *52*, 4816–4826. [CrossRef]
26. Wee, S.-D.; Shin, M.-H.; Hyun, D.-S. Stator-flux-oriented control of induction motor considering iron loss. *IEEE Trans. Ind. Electron.* **2001**, *48*, 602–608.
27. Golestan, S.; Guerrero, J.M.; Vasquez, J.C. Single-Phase PLLs: A Review of Recent Advances. *IEEE Trans. Power Electron.* **2017**, *32*, 9013–9030. [CrossRef]
28. Zou, Z.; Liserre, M. Modeling Phase-Locked Loop-Based Synchronization in Grid-Interfaced Converters. *IEEE Trans. Energy Convers.* **2020**, *35*, 394–404. [CrossRef]
29. De Donato, G.; De Scelba, G.; Borocci, G.; Capponi, F.G.; Scarcella, G. Fault-Decoupled Instantaneous Frequency and Phase Angle Estimation for Three-Phase Grid-Connected Inverters. *IEEE Trans. Power Electron.* **2016**, *31*, 2880–2889. [CrossRef]
30. Nobile, G.; Scelba, G.; Cacciato, M.; Scarcella, G. Losses Minimization Control for an Integrated Multidrive Topology Devoted to Hybrid Electric Vehicles. *IEEE Trans. Ind. Electron.* **2019**, *66*, 8345–8360. [CrossRef]

Article

Modeling and Performance Assessment of the Split-Pi Used as a Storage Converter in All the Possible DC Microgrid Scenarios. Part I: Theoretical Analysis

Massimiliano Luna [1], **Antonino Sferlazza** [2,*], **Angelo Accetta** [1], **Maria Carmela Di Piazza** [1], **Giuseppe La Tona** [1] **and Marcello Pucci** [1]

[1] Istituto di Ingegneria del Mare (INM), Consiglio Nazionale delle Ricerche (CNR), Via Ugo La Malfa 153, 90146 Palermo, Italy; massimiliano.luna@cnr.it (M.L.); angelo.accetta@cnr.it (A.A.); mariacarmela.dipiazza@cnr.it (M.C.D.P.); giuseppe.latona@cnr.it (G.L.T.); marcello.pucci@cnr.it (M.P.)

[2] Dipartimento di Ingegneria (DI), Università degli Studi di Palermo, Viale delle Scienze ed.10, 90128 Palermo, Italy

[*] Correspondence: antonino.sferlazza@unipa.it

Abstract: The integration of an electrical storage system (ESS) into a DC microgrid using a bidirectional DC/DC converter provides substantial benefits but requires careful design. Among such converter topologies, the Split-pi converter presents several merits at the cost of non-isolated operation. However, the few works in the literature on the Split-pi presented only closed-loop control with a single control loop; furthermore, they neglected the reactive components' parasitic resistances and did not perform any experimental validation. This work aimed at investigating the use of the Split-pi converter as a power interface between an ESS and a DC microgrid. Five typical microgrid scenarios are presented, where each of which requires a specific state-space model and a suitable control scheme for the converter to obtain high performance. In this study, two different state-space models of the converter that consider the parasitic elements are presented, the control schemes are discussed, and criteria for designing the controllers are also given. Several simulations, as well as experimental tests on a prototype realized in the lab, were performed to validate the study. Both the simulation and experimental results will be presented in part II of this work. The proposed approach has general validity and can also be followed when other bidirectional DC/DC converter topologies are employed to interface an ESS with a DC microgrid.

Keywords: Split-pi; bidirectional converter; electrical storage system; DC microgrid; droop control; current control; feed-forward control

Citation: Luna, M.; Sferlazza, A.; Accetta, A.; Di Piazza, M.C.; La Tona, G.; Pucci, M. Modeling and Performance Assessment of the Split-Pi Used as a Storage Converter in All the Possible DC Microgrid Scenarios. Part I: Theoretical Analysis. *Energies* **2021**, *14*, 4902. https://doi.org/10.3390/en14164902

Academic Editor: Mohamed Benbouzid

Received: 12 July 2021
Accepted: 7 August 2021
Published: 11 August 2021

Publisher's Note: MDPI stays neutral with regard to jurisdictional claims in published maps and institutional affiliations.

1. Introduction

Over the last few years, DC distribution in terrestrial and marine power systems has attracted a growing interest in view of the implementation of the smart microgrid paradigm due to its advantages in terms of simpler and more efficient electrical architectures. Consequently, power electronic converters that interface distributed generation units, loads, and above all, electrical storage systems (ESSs) with a common DC bus are the subject of renewed interest [1–3]. ESSs have manifold beneficial impacts on DC microgrids: they allow for improving stability and resiliency, compensate for the intermittency of renewable generation, provide ramping support to generators, and act as backup power sources. Furthermore, ESSs ensure a power buffer that can be leveraged to apply suitable energy management strategies to microgrids. In particular, energy management systems (EMSs) can be used to compute the optimal values of power flows among the microgrid devices, which allow for pursuing chosen objectives, such as a minimum electricity bill, maximum efficiency, minimum fuel consumption, or minimum greenhouse gas emissions [1,4–6].

Depending on the designer's choice, the microgrid voltage can be controlled either stiffly using a single voltage generator or in a droop scheme using one or more voltage generators with a predefined power-sharing ratio, usually in proportion to their power rating (grid-forming generators). When an EMS is used, the other active devices of the microgrid must be controlled as current generators based on the optimal power flows that are computed by the EMS [1,5]. Therefore, depending on the specific configuration, the converter interfacing the ESS with the microgrid must be operated as a stiff voltage generator, a non-stiff voltage generator, or a current generator.

The scientific literature provides several contributions on bidirectional DC/DC converters (BDCs). Reference [7] presented an overview of BDCs, where, besides the review of both non-isolated and isolated configurations, the most relevant control schemes and switching strategies were analyzed. In some applications, galvanic isolation between the input and output side of the converter is required; in such cases, the most frequent choice is the dual active bridge converter (DAB) due to its many advantages [7–9]. However, isolation is mandatory only when very high voltage gain is needed. Non-isolated converters are thus more attractive when the goal is to improve the efficiency, size, weight, and cost of the system [7,10]. A review of non-isolated BDCs topologies was presented in [10]. The advantages and disadvantages of the considered converters were properly highlighted. For example, some converters provide an output voltage with opposite polarity than the input, some draw discontinuous current from the battery, while others require a tapped inductor or exhibit weak regulation capability or a high switch count. Overall, Tytelmaier et al. identified the half-bridge converter (HBC) and the related interleaving variants with coupled inductors as the most promising solutions from the efficiency and robustness standpoints [10]. Furthermore, Odo compared three non-isolated BDC topologies with the classical HBC in terms of their suitability for energy storage in a DC microgrid and identified the cascaded buck-boost HBC topology as the best compromise [11].

An interesting alternative is offered by the Split-pi converter, which is a non-isolated BDC that is based on two cascaded HBCs with a common bulk capacitor instead of a common inductor. As such, it is the dual topology of the cascaded buck-boost HBC that was analyzed in [7,10,11]. The Split-pi was initially developed for electric vehicles and patented in 2004 [12], and it is receiving increasing attention due to its distinct advantages [13–20]. It exhibits high efficiency, like the DAB, but with a reduced switch count (eight vs. four switches, of which, only two are actively commutated). Furthermore, LC filters at both ports of the Split-pi allow for small reactive components and reduce the switching noise. On the other hand, the additional phase delay introduced by such filters can slightly complicate this converter's control with respect to simpler converters [17,18]. Nonetheless, the control of the Split-pi is less complicated than that of the DAB because it requires conventional duty cycle control of the pulse width modulator (PWM) instead of phase shift control. An additional advantage of the Split-pi is its suitability for multiphase systems, where a significant reduction in component size and cost can be attained. These features make it attractive when high power density is required, such as in hybrid electric vehicles, renewable energy systems, and aerospace/marine/military applications [16,17].

Only three papers in the technical literature proposed applications in which the Split-pi converter was not controlled using an open-loop [16,17,19]. The aim of [17] was to study a system in which the Split-pi interfaced a flywheel with a 12 V, 120 W DC bus connecting a photovoltaic generator with a passive load. However, in the related control scheme, only one control loop was active at a time (either for the output voltage or the input current) using a shared controller, and the choice was made using a rule-based approach. Singhai et al. did not focus on a specific application and presented a state-space model of the Split-pi converter and a transfer function to design a closed-loop controller for its output voltage [16]. However, they considered the converter connected to a passive load and neglected the parasitic resistances. Finally, Monteiro et al. considered a Split-pi converter with a multilevel structure and controlled current or voltage in any of the two ports in an open-loop, whereas the common DC-link voltage was controlled in a closed-loop [19]. Only a few details about the control systems were given because the study mainly focused on comparing the multilevel Split-pi topology and the interleaved topology of [20].

Apart from [17], all these works only presented simulation results without experimental validation. Furthermore, they all studied closed-loop control with a single control loop, neglecting the reactive components' parasitic resistances. However, when the converter is used to interface an ESS with an active load, such as a DC microgrid, it is essential to control both the ESS current and the output voltage/current of the converter, as required by the microgrid designer. Furthermore, the parasitic elements cannot be neglected because they affect the losses and the maximum gain attainable using the converter.

To cover these aspects, in this work, the use of the Split-pi converter in such an application was investigated with particular attention to its model, the design of its control system, and the assessment of the expected performance in all the possible microgrid configurations. In this first part of the work, it was shown that the Split-pi must be modeled in two different ways depending on the microgrid scenario and that control schemes involving a different number of control loops are needed. In the case of the output voltage control, a feed-forward action was also required to obtain high performance. Furthermore, it was shown that conventional PI regulators alone were not sufficient to obtain the desired performance for output current control in stiff microgrids. The two state-space models considering the parasitic elements and the transfer functions of interest were given in the study, together with criteria to design the controllers. The chosen case study was a DC microgrid that was representative of both terrestrial and marine applications. In Part II of the work, a comprehensive performance assessment is presented based on simulations and experimental tests that validate the study. Finally, the proposed approach has general validity and can also be followed when other BDC topologies are used to interface a storage system with a DC microgrid.

2. Topology, Operation Modes, and Sizing of the Split-Pi Converter

The schematic of a symmetrical Split-pi converter is sketched in Figure 1, including its reactive components' parasitic resistances. Such a converter can be viewed as the cascaded connection of a first HBC at port 1, a bulk capacitor, and another HBC at port 2. Since the two HBCs are bidirectional, the whole Split-pi converter is also a BDC. If the two HBCs have equal reactive components, the Split-pi is said to be symmetrical.

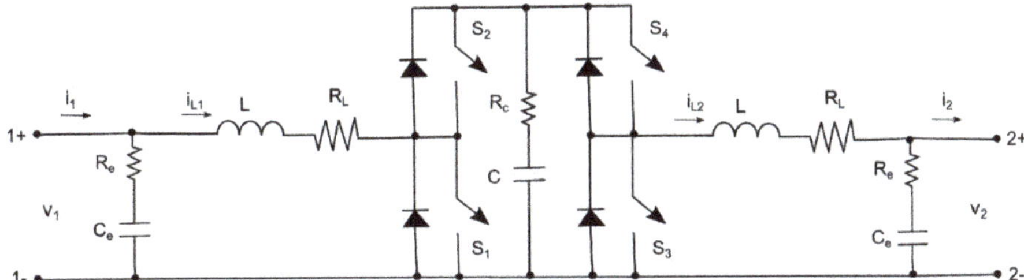

Figure 1. Schematic of the symmetrical Split-pi converter.

The four switches (S1–S4) were sketched in Figure 1 as ideal switches. In the original formulation of [12], they were implemented using MOSFETs to exploit the advantage of synchronous rectification. According to [12], the Split-pi has four operation modes, depending on the relationship between V_1 and V_2 and on the power direction, as summarized in the first four rows of Table 1, which is an extension of the table reported in [18]. In this work, the Split-pi converter was used to interface a storage system (connected to port 1) with a DC microgrid (connected to port 2). It is worth noting that, whenever possible, the storage system is chosen so that $V_1 \leq V_2$ for both technical and safety reasons. Hence, the present work considered the Split-pi converter operating in modes 1 and 2.

Table 1. Summary of the operating modes of the Split-pi converter.

Mode	$V_1 \leq V_2$	Power Direction	Split-Pi Operation	Duty Cycle of S1	Duty Cycle of S2	Duty Cycle of S3	Duty Cycle of S4	Gain of the Split-Pi in Direction $1 \rightarrow 2$
1	Yes	$1 \rightarrow 2$	Boost for $1 \rightarrow 2$	d	$1 - d$	0	1	$\frac{1}{1-d}$
2	Yes	$2 \rightarrow 1$	Buck for $2 \rightarrow 1$	d	$1 - d$	0	1	$\frac{1}{1-d}$
3	No	$1 \rightarrow 2$	Buck for $1 \rightarrow 2$	0	1	$1 - d$	d	d
4	No	$2 \rightarrow 1$	Boost for $2 \rightarrow 1$	0	1	$1 - d$	d	d

The Split-pi components can be sized according to the classical formulas used for buck and boost converters [17,21]. Specifically, the minimum inductance value is expressed by (1) for both boost and buck operations; on the other hand, the minimum capacitance of the input and output capacitors C_e can be computed using (2), which is valid for a buck converter, whereas the minimum value of the bulk capacitance C is expressed by (3), which is valid for a boost converter:

$$L_{min} = \frac{100\ V_2(1-d)d}{2F_{sw}r_{i\%}I_1}, \tag{1}$$

$$C_{e,\ min} = \frac{100(1-d)}{8F_{sw}^2 r_{ve\%}L}, \tag{2}$$

$$C_{min} = \frac{100\ I_2 d}{F_{sw}r_{v\%}V_2}. \tag{3}$$

In (1), (2), and (3), F_{sw} is the switching frequency; d is the duty cycle; V_2 and I_2 are the output voltage and current, respectively; I_1 is the input current; $r_{i\%}$ is the desired inductor current ripple; and $r_{v\%}$ and $r_{ve\%}$ are the desired voltage ripple values on the bulk and external capacitors, respectively.

3. Possible DC Microgrid Scenarios, Load Models, and Required Control Schemes for the ESS Converter

In the present study, it was supposed that the Split-pi converter was used to interface an ESS with a DC microgrid. Under this hypothesis, several scenarios could be considered depending on the control mode chosen for both the storage-side converter and the grid-side converters interfacing other microgrid generators, if any. Depending on the combinations of the control modes of such converters, five scenarios were identified and, for each of them, a different type of load model and control scheme must be used. The different DC microgrid scenarios and the required load models and control schemes are discussed in the following subsections.

3.1. Possible DC Microgrid Scenarios

In general terms, one of the following situations can occur as a control mode for the microgrid converters:

- All the generator converters are controlled in droop mode to regulate the microgrid voltage with a predefined power-sharing ratio. Thus, each generator behaves as an ideal voltage generator with a series-connected resistance. The droop scheme leads to a simple but very reliable microgrid and does not require a communication infrastructure. If a storage system is present, its droop characteristic is usually chosen so that no current is supplied at half the rated load of the microgrid, whereas charging/discharging occurs for load power below/above such a value. If there is only one active device in the microgrid (either a storage system or a generator), it could also be controlled with a null droop resistance (stiff microgrid) so that the microgrid voltage does not vary with load power

- The microgrid follows a master–slave architecture: some converters for generators or storage systems (i.e., the masters) are controlled in droop mode (a null droop resistance is possible only with a single master); the others (i.e., the slaves) behave as current generators and are managed by an EMS to pursue one or more predefined goals.

Therefore, three control modes are possible for both the storage-side and grid-side converters: stiff droop control, non-stiff droop control, and current control. The five scenarios resulting from the combinations of the control modes of such converters are described in the first three columns of Table 2. For the sake of clarity, such scenarios are referred to also by means of an abbreviation in the form Sx-Gy, where x and y specify the control mode for the storage-side and the grid-side converters, respectively. The range of options for x is:

- C for current control.
- D for non-stiff droop control.
- S for stiff droop control.

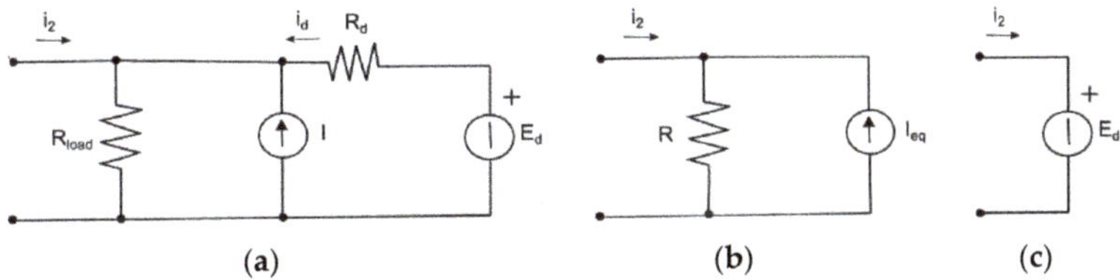

Figure 2. Equivalent active load transformations for non-stiff and stiff microgrids: (**a**) general load model for storage converter; (**b**) equivalent load model considered in scenarios #1–#4; (**c**) equivalent load model considered in scenario #5.

Table 2. Possible combinations of microgrid scenarios, converter and load models, and control schemes for the Split-pi.

Microgrid Scenario	Storage Converter	Other Generators	State-Space Model of the Split-Pi	Load for the State-Space Model of the Split-Pi	Control Scheme
#1 (SS-GN)	Droop mode with $R_d = 0$ (stiff microgrid)	No other generator present (passive load) or all current controlled by the EMS	Model A	R and I_{eq}, as in Figure 2b	2 loops + FF, as in Figure 3 with $R_{ds} = 0$
#2 (SD-GN)	Droop mode with $R_d \neq 0$	No other generator present (passive load) or all current controlled by the EMS	Model A	R and I_{eq}, as in Figure 2b	3 loops + FF, as in Figure 3
#3 (SD-GD)	Droop mode with $R_d \neq 0$	At least one is droop controlled and none has $R_d = 0$	Model A	R (low) and I_{eq}, as in Figure 2b	3 loops + FF, as in Figure 3
#4 (SC-GD)	Current mode	At least one is droop controlled and none has $R_d = 0$	Model A	R (low) and I_{eq}, as in Figure 2b	2 loops, as in Figure 4
#5 (SC-GS)	Current mode	One is droop controlled and has $R_d = 0$ (stiff microgrid); the others, if present, are current controlled by the EMS	Model B	E_d, as in Figure 2c	2 loops, as in Figure 4

As for the options for y:

- N if no grid-side generator is present or if none of the grid-side generators are operated with droop control.
- D if at least one generator is droop controlled but not in a stiff way.
- S if there is one generator operated with stiff droop control; in such a case, the other grid-side generators, if present, must be current controlled by the EMS.

3.2. Possible Load Models for the ESS Converter

The dynamic behavior of the storage converter is affected by the load that it must supply, which depends not only on the passive loads of the DC microgrid but also on the possible presence of grid-side generators. In the most general case, the load for the ESS converter can be modeled following these steps: (1) aggregating all the droop-controlled generators of the microgrid using Thevenin's theorem, obtaining parameters E_d and R_d; (2) combining all the current-controlled generators managed by the EMS to compute the overall current I; (3) including an aggregated passive load R_{load}. The resulting circuit model is shown in Figure 2a. However, it is convenient to reduce such a model to a suitable equivalent form comprising only one generator. Toward this aim, it is necessary to distinguish between the five scenarios. In scenarios #3 (SD-GD) and #4 (SC-GD), using Norton's theorem, the couple E_d, R_d can be substituted with an equivalent current generator E_d/R_d and a parallel-connected resistance R_d. Then, the load can be reduced to that of Figure 2b with the following assumptions: $I_{eq} = I + E_d/R_d$ and $R = R_{load}//R_d$, where $//$ denotes the parallel connection of circuit elements. In scenarios #1 (SS-GN) and #2 (SD-GN), the same scheme of Figure 2b can be used assuming $I_{eq} = I$ and $R = R_{load}$. Furthermore, in scenario #5 (SC-GS), the converter's output voltage is set equal to E_d because $R_d = 0$. Since both R_{load} and I are now parallel connected to an ideal voltage generator, they do not influence the converter's dynamics. Thus, the load can be modeled as shown in Figure 2c.

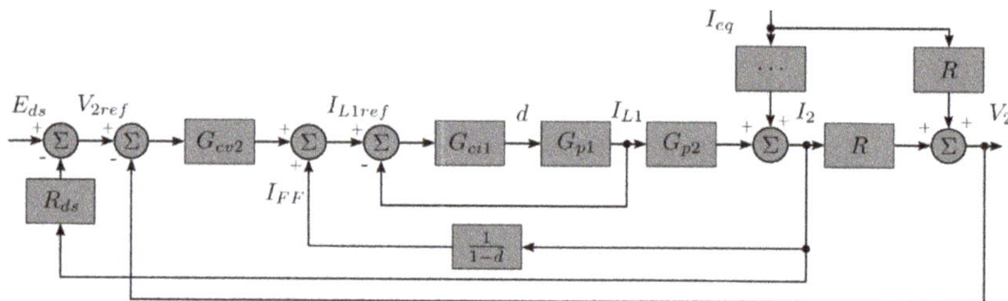

Figure 3. Control scheme that is used for the voltage control of the ESS converter in scenarios #1 (SS-GN), #2 (SD-GN), and #3 (SD-GD).

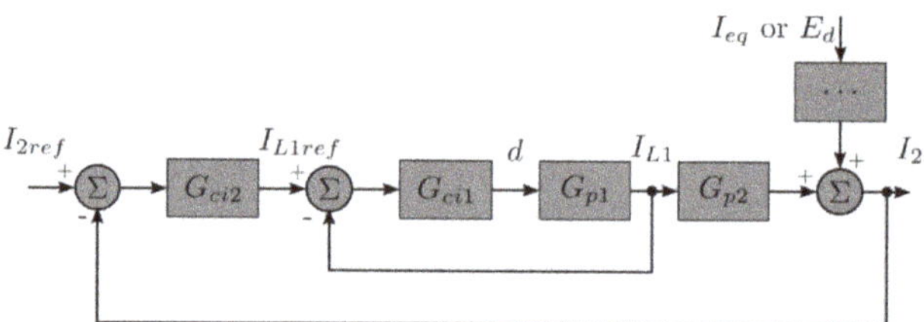

Figure 4. Control scheme that is used for the current control of the ESS converter in scenarios #4 (SC-GD) and #5 (SC-GS).

In the following, the state-space model of the storage converter connected to the load of Figure 2b is denoted as model A, whereas model B refers to the converter connected to the load of Figure 2c. The derivation of such models is given in Section 4. The relationship between the possible DC microgrid scenarios and the storage converter and load models is described in the first five columns of Table 2. It is worth noting that the value of the equivalent load resistance R considered in scenarios #3 (SD-GD) and #4 (SC-GD) is much lower than that of scenarios #1 (SS-GN) and #2 (SD-GN) because it results from the parallel connection of R_{load} and R_d. Finally, the last column of Table 2 reports the required control scheme for the ESS converter in each microgrid scenario according to the considerations given in the following section.

3.3. Required Control Schemes for the ESS Converter

In general, the closed-loop control scheme for the storage converter depends on the microgrid scenario. However, regardless of the scenario, an important goal is to control the current I_{L1} of the leftmost inductor to ensure that it is compatible with the storage system's current state. Thus, in the related control loop, its reference value should be dynamically saturated to avoid overcharging or overdischarging the storage system.

In the first three scenarios, i.e., SS-GN, SD-GN, and SD-GD, besides the current loop for I_{L1}, a voltage loop is needed to regulate the output voltage V_2 at the nominal voltage V_{2n}. A suitable controller is required in each control loop. Typically, PI or PID regulators with an anti-windup action are used, which must be designed to obtain a stable system with the desired dynamics. In scenarios #2 (SD-GN) and #3 (SD-GD), a third loop is also required to implement the droop characteristic of the storage converter by computing the voltage reference V_{2ref} based on the output current I_2 according to the equation $V_{2,ref} = E_{ds} - R_{ds} \cdot I_2$, where E_{ds} and R_{ds} are the parameters of the droop characteristic of such a converter. The control scheme used in scenarios #1 (SS-GN), #2 (SD-GN), and #3 (SD-GD) is depicted in Figure 3. In particular, the external loop is opened in scenario #1 (SS-GN) because $R_d = 0$. The presence of the current generator I_{eq} in parallel to the load resistance is considered a disturbance that affects the system's output, which will be suitably compensated for by the control system, regardless of the related transfer function. As will be shown in Section 6, when the output voltage V_2 is controlled, a feed-forward (FF) action is also required in addition to the voltage loop to suitably reduce the overshoot. Since I_{eq} cannot be measured, the output current I_2 is chosen as the FF action input.

On the other hand, the converter is current-controlled in scenarios #4 (SC-GD) and #5 (SC-GS). In these cases, besides the inner loop for I_{L1}, another current loop is needed to regulate the output current I_2 based on a reference I_{2ref} that is computed by the EMS. The related control scheme is shown in Figure 4. Again, the external voltage or current generator is considered a disturbance, and the related transfer function is irrelevant.

The type of control scheme to be used in each microgrid scenario and the required number of control loops are reported in the last column of Table 2. It is worth noting that suitable saturators are required in the controllers of each loop (G_{ci1}, G_{cv2}, G_{ci2}). Specifically, the output of the controller G_{ci1} (i.e., the duty cycle d) is bounded by the interval [0; 0.9] to avoid overcurrents due to prolonged transients with $d = 1$. On the other hand, the reference value for I_{L1} (i.e., the output of G_{cv} plus the FF term or the output of G_{ci2}) is bounded by the interval $[-I_{cx}; I_{dx}]$, where I_{cx} and I_{dx} are the maximum charging/discharging currents of the storage system. Finally, the upper or lower bound of such an interval is dynamically replaced with zero if the battery SOC reaches 100% or goes below the minimum allowed SOC, respectively.

4. State-Space Models of the Split-Pi Converter

The two state-space models of a Split-pi converter that interfaces a storage system with a non-stiff (scenarios #1~#4) or stiff (scenario #5) microgrid and operates with $V_1 \leq V_2$ are presented in the following. They consider the parasitic elements and were determined according to the state-space averaging technique [22].

4.1. State-Space Model A: Split-Pi Converter Connected to a Non-Stiff Microgrid

The state-space model of a Split-pi converter connected to a non-stiff microgrid and operating with $V_1 \leq V_2$ can be expressed in matrix form as follows:

$$\begin{cases} \dot{x} = Ax + Bu \\ y = Cx + Du \end{cases} \tag{4}$$

$$x = [I_{L1}, \ I_{L2}, \ V_c, \ V_e]' \tag{5}$$

$$u = [V_1, I_{eq}]' \tag{6}$$

$$y = [I_{L1}, V_2, I_2]' \tag{7}$$

$$\begin{cases} A = dA_{on} + (1-d)A_{off} \\ B = dB_{on} + (1-d)B_{off} \\ C = dA_{on} + (1-d)C_{off} \\ D = dD_{on} + (1-d)D_{off} \end{cases} \tag{8}$$

$$A_{on} = \begin{bmatrix} -\frac{R_L}{L} & 0 & 0 & 0 \\ 0 & -\frac{R_{tot}}{L} & \frac{1}{L} & -\frac{R}{LR_{sum}} \\ 0 & -\frac{1}{C} & 0 & 0 \\ 0 & \frac{R}{R_{sum}C_e} & 0 & -\frac{1}{R_{sum}C_e} \end{bmatrix} \tag{9}$$

$$A_{off} = \begin{bmatrix} -\frac{R_L+R_c}{L} & \frac{R_c}{L} & -\frac{1}{L} & 0 \\ \frac{R_c}{L} & -\frac{R_{tot}}{L} & \frac{1}{L} & -\frac{R}{LR_{sum}} \\ \frac{1}{C} & -\frac{1}{C} & 0 & 0 \\ 0 & \frac{R}{R_{sum}C_e} & 0 & -\frac{1}{R_{sum}C_e} \end{bmatrix} \tag{10}$$

$$B_{on} = B_{off} = \begin{bmatrix} \frac{1}{L} & 0 \\ 0 & -\frac{R_p}{L} \\ 0 & 0 \\ 0 & \frac{R}{R_{sum}C_e} \end{bmatrix} \tag{11}$$

$$C_{on} = C_{off} = \begin{bmatrix} 1 & 0 & 0 & 0 \\ 0 & R_p & 0 & \frac{R}{R_{sum}} \\ 0 & \frac{R_e}{R_{sum}} & 0 & \frac{1}{R_{sum}} \end{bmatrix} \tag{12}$$

$$D_{on} = D_{off} = \begin{bmatrix} 0 & 0 \\ 0 & R_p \\ 0 & -\frac{R}{R_{sum}} \end{bmatrix} \tag{13}$$

where $R_p = R//R_e$, $R_{sum} = R + R_e$, and $R_{tot} = R_p + R_L + R_c$.

Since $B_{on} = B_{off}$ and $D_{on} = D_{off}$, the small-signal behavior of the converter does not depend on the input values. According to the method described in [21], the model can be linearized around a chosen operating point corresponding to the duty cycle $\bar{d}$ and the state $\bar{x} = [I_{L10},\ I_{L20},\ V_{c0},\ V_{e0}]'$, obtaining the following transfer functions:

$$G_{p1}(s) = \frac{\tilde{I}_{L1}}{\tilde{d}} = \frac{n_3 s^3 + n_2 s^2 + n_1 s + n_0}{d_4 s^4 + d_3 s^3 + d_2 s^2 + d_1 s + d_0} \tag{14}$$

$$G_{p2}(s) = \frac{\tilde{I}_2}{\tilde{I}_{L1}} = \frac{R_k I_{L10}}{L^2 CC_e R_{sum}} \cdot \frac{(1 + sR_c C)(1 + sR_e C_e)\left(1 - s\frac{L}{R_k}\right)}{n_3 s^3 + n_2 s^2 + n_1 s + n_0} \tag{15}$$

whose coefficients n_i, d_i, and R_k are given in Appendix A. $G_{p1}(s)$ and $G_{p2}(s)$ are small-signal transfer functions that express the effect of perturbations of a variable on another one (both denoted with a tilde). According to Figure 3, $G_{p1}(s)$ expresses the relationship between d and I_{L1}, whereas $G_{p2}(s)$ describes the dependence of I_2 on I_{L1}. Clearly, the relationship between I_{L1} and V_2 is expressed by $G_{p2}(s){\cdot}R$.

4.2. State-Space Model B: Split-Pi Converter Connected to a Stiff Microgrid

In the case of a Split-pi converter operating with $V_1 \leq V_2$ and connected to a stiff microgrid, (4) and (8) are still valid, but (5), (6), (7),(9), (10) and (11), (12), (13) are replaced by (16), (17), (18), (19), (20) and (21), (22), (23), respectively.

$$x = [I_{L1},\ I_{L2},\ V_c,\ V_e]' \tag{16}$$

$$u = [V_1, E_d]' \tag{17}$$

$$y = [I_{L1}, I_2]' \tag{18}$$

$$A_{on} = \begin{bmatrix} -\frac{R_L}{L} & 0 & 0 & 0 \\ 0 & -\frac{R_c+R_L}{L} & \frac{1}{L} & 0 \\ 0 & -\frac{1}{C} & 0 & 0 \\ 0 & 0 & 0 & -\frac{1}{R_e C_e} \end{bmatrix} \tag{19}$$

$$A_{off} = \begin{bmatrix} -\frac{R_c+R_L}{L} & \frac{R_c}{L} & -\frac{1}{L} & 0 \\ \frac{R_c}{L} & -\frac{R_c+R_L}{L} & \frac{1}{L} & 0 \\ \frac{1}{C} & -\frac{1}{C} & 0 & 0 \\ 0 & 0 & 0 & -\frac{1}{R_e C_e} \end{bmatrix} \tag{20}$$

$$B_{on} = B_{off} = \begin{bmatrix} \frac{1}{L} & 0 \\ 0 & -\frac{1}{L} \\ 0 & 0 \\ 0 & \frac{1}{R_e C_e} \end{bmatrix} \tag{21}$$

$$C_{on} = C_{off} = \begin{bmatrix} 1 & 0 & 0 & 0 \\ 0 & 1 & 0 & \frac{1}{R_e} \end{bmatrix} \tag{22}$$

$$D_{on} = D_{off} = \begin{bmatrix} 0 & 0 \\ 0 & -\frac{1}{R_e} \end{bmatrix} \tag{23}$$

Again, the small-signal behavior does not depend on input values. The model can be linearized around a chosen operating point corresponding to $\bar{d}$ and $\bar{x}$, obtaining the transfer functions (24) and (25), whose coefficients n_i, d_i, and R_k are given in Appendix B. According to Figure 4, $G_{p1}(s)$ expresses the relationship between d and I_{L1}, whereas $G_{p2}(s)$ describes the dependence of I_2 on I_{L1}. It is worth noting that a zero-pole cancellation occurs due to the stiff voltage imposed on the $R_e C_e$ branch at port 2; thus, the resulting system's order is three instead of four and the voltage and current on C_e cannot be controlled.

$$G_{p1}(s) = \frac{\tilde{I}_{L1}}{\tilde{d}} = \frac{n_2 s^2 + n_1 s + n_0}{d_3 s^3 + d_2 s^2 + d_1 s + d_0} \tag{24}$$

$$G_{p2}(s) = \frac{\tilde{I}_2}{\tilde{I}_{L1}} = \frac{R_k I_{L10}}{L^2 C} \cdot \frac{(1 + s R_c C)\left(1 - s\frac{L}{R_k}\right)}{n_2 s^2 + n_1 s + n_0} \tag{25}$$

5. Case Study and Converter Sizing

In this section, the chosen case study is described and the droop characteristics of both the storage converter and the voltage generator of the microgrid are discussed. Then, the Split-pi converter's reactive components are sized based on the chosen case study's parameters.

With no loss of generality, the proposed investigation was performed by referring to a 48 V, 750 W storage system that was interfaced with a 180 V DC microgrid using a Split-pi converter. The chosen case study can represent the onboard grid of an unmanned marine vehicle or a scaled prototype of a residential DC microgrid with a 120 V, 60 Hz, single-phase, grid-connected inverter, whose DC link voltage must be higher than 170 V DC. The rated values of the system under study are shown in Table 3.

As for the chosen droop characteristic of the storage converter, in scenario #1 (SS-GN) it was defined by $E_{ds} = 180$ V and $R_{ds} = 0$, whereas in scenarios #2 (SD-GN) and #3 (SD-GD), it was expressed by $E_{ds} = 180$ V and $R_{ds} = 2.2\ \Omega$, i.e., imposing a 5% voltage reduction at the nominal current. For the equivalent droop-controlled microgrid generator, the same parameters as those of the storage converter were chosen in scenario #4 (SC-GD). On the other hand, in scenario #3 (SD-GD), the following parameters were used for the droop-controlled microgrid generator: $E_d = 198$ V and $R_d = 9\ \Omega$; with this choice, the storage system did not supply any power for half the rated load of the microgrid, as desired.

Finally, a constant voltage generator $E_d = 180$ V was considered in scenario #5 (SC-GS) to model the microgrid's stiff voltage generator.

Table 3. Rated values of the system.

Parameter	Symbol	Value
Switching frequency	F_{sw}	20 kHz
Nominal output voltage	V_{2n}	180 V
Nominal input voltage	V_{1n}	50 V
Nominal power	P_n	750 W
Nominal load resistance	R_n	43.2 Ω
Nominal input current	I_{1n}	15 A
Max. charge/discharge current	I_{cx}, I_{dx}	18 A
Nominal output current	I_{2n}	4.167 A
Nominal duty cycle	$\overline{d}$	0.722

The Split-pi's reactive components were sized using (1), (2), (3), and the following parameters were set: $r_{i\%} = \pm 6.0\%$, $r_{ve\%} = \pm 0.2\%$, $r_{v\%} = \pm 0.2\%$. Consequently, the following minimum inductor and capacitor ratings were obtained: $L_{min} = 803$ µH, $C_{e,min} = 113$ µF, $C_{min} = 419$ µF. Aiming to build a converter prototype and due to component availability, slightly higher values were chosen for the reactive components; they are reported in Table 4, together with their parasitic resistances.

Table 4. Reactive components of the Split-pi.

Parameter	Symbol	Value
Inductance value of L	L	1000 µH
Parasitic resistance of L	R_L	65 mΩ
Capacitance value of C_e	C_e	200 µF
Parasitic resistance of C_e	R_e	260 mΩ
Capacitance value of C	C	540 µF
Parasitic resistance of C	R_c	125 mΩ

6. Control System Design

The control system of the Split-pi converter must be designed by considering the specific microgrid scenario in which it will be used. However, in any case, it must be assumed that the converter supplies its rated power without any contribution from the external current or voltage generators, which are seen as disturbances. If the imposed stability margins are sufficiently wide, the designed controllers will be effective also at lighter loads, i.e., with higher values of R. Thus, models A and B must be linearized around the operating point corresponding to the rated values of the duty cycle and state variables: $\overline{d} = 0.722$ and $\overline{x} = [I_{L10}, I_{L20}, V_{c0}, V_{e0}]' = [15, 4.167, 180, 180]'$. No other condition is needed for model B in scenario #5 (SC-GS). As for model A, instead, it is $R = R_n$ in scenarios #1 (SS-GN) and #2 (SD-GN) and $R = R_n // R_d$ in scenarios #3 (SD-GD) and #4 (SC-GD).

All the above values must be substituted into (A1)–(A6), given in Appendix A, to obtain the coefficients of the transfer functions of interest. Then, designing of the controllers G_{ci1}, G_{cv2}, and G_{ci2} can be performed with classic techniques that involve imposing suitable values of the crossover frequency ω_c and phase margin m_φ and ensuring a suitable gain margin m_g [23]. For each control loop and scenario, the imposed values and the obtained PI coefficients and gain margins are summarized in Tables 5 and 6. Furthermore, a baseline scenario employing the control scheme of Figure 3 with $R_d = 0$ and without the FF action was also considered for comparison purposes to show the usefulness of such an action.

Table 5. Coefficients of PI regulators in the case of model A.

Loop	Scenario	Value of R	Controller	Imposed ω_c and m_φ	PI Coefficients	Obtained m_g
Current I_{L1}	Baseline #1 (SS-GN) #2 (SD-GN)	R_n	G_{ci_1}	$\omega_c = 3000$ rad/s and $m_\varphi = 85°$	$K_{pi} = 0.0160$ $K_{ii} = 5.3703$	∞
Current I_{L1}	#3 (SD-GD) #4 (SC-GD)	$R_n//R_d$	G_{ci_1}	$\omega_c = 3000$ rad/s and $m_\varphi = 85°$	$K_{pi} = 0.0161$ $K_{ii} = 5.2481$	∞
Voltage V_2 without FF	Baseline	R_n	G_{cv_2}	$\omega_c = 100$ rad/s and $m_\varphi = 85°$	$K_{pv} = 0.2659$ $K_{iv} = 18.6209$	12.7 dB
Voltage V_2 with FF	#1 (SS-GN) #2 (SD-GN)	R_n	G_{cv_2}	$\omega_c = 100$ rad/s and $m_\varphi = 85°$	$K_{pv} = 0.2712$ $K_{iv} = 10.4112$	13.1 dB
Voltage V_2 with FF	#3 (SD-GD)	$R_n//R_d$	G_{cv_2}	$\omega_c = 100$ rad/s and $m_\varphi = 85°$	$K_{pv} = 0.3284$ $K_{iv} = 14.6218$	24.2 dB
Current I_2	#4 (SC-GD)	$R_n//R_d$	G_{ci_2}	$\omega_c = 100$ rad/s and $m_\varphi = 85°$	$K_{pi2} = 0.3746$ $K_{ii2} = 389.045$	23.7 dB

Table 6. Coefficients of PI regulators in the case of model B.

Loop	Scenario	Controller	Imposed ω_c and m_φ	PI Coefficients	Obtained m_g
Current I_{L1}	#5 (SC-GS)	G_{ci_1}	$\omega_c = 3000$ rad/s and $m_\varphi = 85°$	$K_{pi} = 0.0161$ $K_{ii} = 5.0699$	∞
Current I_2	#5 (SC-GS)	G_{ci_2}	$\omega_c = 100$ rad/s and $m_\varphi = 85°$	$K_{pi2} = 1.1953$ $K_{ii2} = 362.22$	13.6 dB with $G_{add}(s)$

It is worth noting that the design of the controllers must be very conservative. By design, the currents I_{L1} and I_{L2} have a significant switching ripple compared to the input/output currents and voltages. Thus, the desired crossover frequency for the I_{L1} loop must be suitably lower than F_{sw} to avoid the switching ripple being processed by the controller. The crossover frequency of the loop for V_2 or I_2 must be even lower for proper decoupling with respect to the inner loop. When the converter supplies a passive load (R_{load}), some ringing can be tolerated, and the usually adopted phase margin (50–60°) is satisfactory. Instead, in the case of an active microgrid, the combined variations of I and R_{load} could determine a significant excursion from the nominal operating point and pronounced under/overshoots; thus, a higher phase margin (i.e., $m_\varphi > 80°$) is required to ensure stability under all the operating conditions. As for the gain margin, the usually adopted criterion (i.e., $m_g > 12$ dB) is enough to ensure robustness against parameter variations.

As for the chosen case study, some noteworthy remarks can be made:

- Regardless of the value of R, models A and B exhibited nearly the same dynamics for I_{L1}; therefore, the corresponding PI regulators G_{ci1} had similar coefficients. On the other hand, the dynamics of I_2 were quite dissimilar and required different G_{ci2} controllers.
- In the case of model B, it could be possible to achieve the desired ω_c and m_φ with a PI or I regulator alone, but the gain margin would be around 6.6 dB due to a resonance peak; thus, the second-order transfer function (26) must be included after the PI regulator to attenuate such a peak and achieve a gain margin of 13.6 dB.
- The dynamics of I_2 were very sensitive to the value of R in model A; thus, an unstable system was obtained in scenario #4 (SC-GD) if the controllers were designed assuming $R = R_n$ instead of $R = R_n//R_d$.
- Without the FF action, the dynamics of V_2 were sensitive to the value of R: a significantly slower behavior was obtained if the controllers were designed assuming $R = R_n$ in scenario #3 (SD-GD) instead of $R = R_n//R_d$; on the other hand, a significant ringing was obtained if the controllers were designed considering $R = R_n//R_d$ in scenarios #1 (SS-GN) and #2 (SD-GN) instead of $R = R_n$.

- Using the FF action, the dynamics of V_2 were pretty insensitive to the value of R; thus, almost no variation was obtained in scenarios #1 (SS-GN), #2 (SD-GN), and #3 (SD-GD) if the controllers were designed when considering either $R = R_n//R_d$ or $R = R_n$.

Several simulations were performed to assess the performance of the controlled system in all the scenarios. Then, a prototype of the Split-pi converter was built, and experimental tests were performed in several conditions that covered the baseline scenario and all the other five scenarios, obtaining successful results. The simulation and experimental results validating the study are presented in part II of this work.

$$G_{add}(s) = \frac{1}{\left(1 + \frac{s}{533}\right)\left(1 + \frac{s}{606}\right)} \tag{26}$$

7. Conclusions

The Split-pi converter is a suitable choice to interface electrical storage systems with DC microgrids. It offers distinct advantages, such as high efficiency, reduced switch count and switching noise, and suitability for multiphase systems at the cost of non-isolated operation. However, to obtain high performance, its control system must be suitably designed according to the specific microgrid scenario in which it will be used.

In this study, five typical microgrid scenarios were identified and analyzed, where each of which required a specific state-space model and a suitable control scheme for the converter. Two different state-space models were presented for the Split-pi converter operating with the storage-side voltage being lower than the grid-side voltage. Both models considered the parasitic elements of the reactive components. As for the control scheme, the number of required control loops depended on the scenario. It was shown that feed-forward action is needed to obtain a high performance in the case of voltage control and that, sometimes, conventional PI regulators alone were not sufficient for stable current control. The most relevant transfer functions of the Split-pi converter were given, together with criteria to design the controllers. Several simulations, as well as experimental tests on a prototype realized in the lab, were performed to validate the study, whose results will be presented in part II of this work.

The approach followed in this study has general validity and can also be followed to devise the state-space model of a Split-pi operating with a storage-side voltage that is higher than the grid-side voltage or when other bidirectional DC/DC converter topologies are employed to interface an ESS with a DC microgrid. Furthermore, the presented study builds the premises for designing unconventional control systems for the Split-pi that are suitable for operating in more than one microgrid scenario.

Author Contributions: Conceptualization, M.L., A.S., M.P. and M.C.D.P.; methodology, M.L., A.S. and A.A.; software, M.L., A.S., A.A. and G.L.T.; validation, M.L., A.A., M.C.D.P. and M.P.; writing—original draft preparation, M.L., A.S. and M.P.; writing—review and editing, M.L., A.A., M.C.D.P. and G.L.T.; supervision, M.P. and M.L.; funding acquisition, M.P. and M.L. All authors have read and agreed to the published version of the manuscript.

Funding: This research was funded by the Italian Ministry of University and Research (MUR), program PON R&I 2014/2020—Avviso n. 1735 del 13/07/2017—PNR 2015/2020, project "NAUSICA—NAvi efficienti tramite l'Utilizzo di Soluzioni tecnologiche Innovative e low CArbon," CUP: B45F21000680005.

Institutional Review Board Statement: Not applicable.

Informed Consent Statement: Not applicable.

Data Availability Statement: Not applicable.

Conflicts of Interest: The authors declare no conflict of interest.

Nomenclature

d	Duty cycle
$\overline{d}$	Average duty cycle for state-space model linearization
d_i	Coefficients of the denominator of the transfer function $G_{p1}(s)$
m_φ	Phase margin
m_g	Gain margin
n_i	Coefficients of the numerator of the transfer function $G_{p1}(s)$ and the denominator of $G_{p2}(s)$
$r_{i\%}$	Current ripple in input/output inductor L
$r_{v\%}$	Voltage ripple in the bulk capacitor C
$r_{ve\%}$	Voltage ripple in the external capacitor C_e
u	Input vector of the state-space model of the Split-pi converter
x	State vector of the state-space model of the Split-pi converter
$\overline{x}$	Average state vector for state-space model linearization
y	Output vector of the state-space model of the Split-pi converter
ω_c	Crossover frequency
A,B,C,D	Matrices of the state-space model of the Split-pi converter
C	Bulk capacitor of the Split-pi converter
C_e	External input/output capacitor of the Split-pi converter
$C_{e,min}$	Minimum capacitance value to obtain the chosen ripple $r_{ve\%}$
C_{min}	Minimum capacitance value to obtain the chosen ripple $r_{v\%}$
E_d	Equivalent no-load voltage of the aggregated droop-controlled generators of the microgrid in scenario #5 after load transformation
E_{ds}	No-load voltage used to control the storage converter in droop mode
F_{sw}	Switching frequency
$G_{add}(s)$	Additional transfer function to be included after the PI regulator of current loop for I_2 in the case of stiff microgrid
$G_{ci1}(s)$	Transfer function of the controller for the internal current loop for I_{L1}
$G_{ci2}(s)$	Transfer function of the controller for the external current loop for I_2
$G_{cv2}(s)$	Transfer function of the controller for the voltage loop for V_2
$G_{p1}(s)$	Transfer function of the process that expresses the relationship between d and I_{L1}
$G_{p2}(s)$	Transfer function of the process that expresses the relationship between I_{L1} and I_2
I	Current supplied by the aggregated current generators of the microgrid managed by the EMS
I_1	Input current of the Split-pi converter (port 1, storage-side)
I_{1n}	Nominal input current (storage-side)
I_2	Output current of the Split-pi converter (port 2, grid-side)
I_{2n}	Nominal output current (grid-side)
I_{2ref}	Reference current computed using the EMS for the storage converter controlled in current mode
I_d	Current supplied by the aggregated droop-controlled generators of the microgrid
I_{cx}	Maximum charging current of the storage system
I_{dx}	Maximum discharging current of the storage system
I_{eq}	Equivalent current generator considered as active load in scenarios #1–#4 after load transformation
I_{L1}	Current of the leftmost inductor of the Split-pi converter (port 1, storage-side)
I_{L10}	Average current value of the leftmost inductor for model linearization
I_{L2}	Current of the rightmost inductor of the Split-pi converter (port 2, grid-side)
I_{L20}	Average current value of the rightmost inductor for model linearization
K_{ii}	Integral gain of the PI regulator of current loop for I_{L1}
K_{ii2}	Integral gain of the PI regulator of current loop for I_2
K_{iv}	Integral gain of the PI regulator of voltage loop for V_2
K_{pi}	Proportional gain of the PI regulator of current loop for I_{L1}

K_{pi2}	Proportional gain of the PI regulator of current loop for I_2
K_{pv}	Proportional gain of the PI regulator of voltage loop for V_2
L	Inductor at input/output ports of the Split-pi converter
L_{min}	Minimum inductance value to obtain the chosen ripple $r_{i\%}$
P_n	Nominal power of the storage converter
R	Equivalent load resistance considered in scenarios #1–#4 after load transformation
R_c	Parasitic resistance of bulk capacitor of the Split-pi converter
R_d	Equivalent droop resistance of the aggregated droop-controlled generators of the microgrid
R_{ds}	Droop resistance used to control the storage converter in droop mode
R_e	Parasitic resistance of external input/output capacitor of the Split-pi converter
R_L	Parasitic resistance of inductor at input/output ports of the Split-pi converter
R_{load}	Equivalent load resistance of the microgrid
R_k	Fictitious resistance term appearing in the transfer function $G_{p2}(s)$
R_n	Nominal load resistance
SOC	State of charge of the storage system
V_1	Input voltage of the Split-pi converter (port 1, storage-side)
V_{1n}	Nominal input voltage (storage-side)
V_2	Output voltage of the Split-pi converter (port 2, grid-side)
V_{2n}	Nominal output voltage (grid-side)
V_{2ref}	Reference output (microgrid) voltage for the storage converter controlled in droop mode
V_c	Voltage of the bulk capacitor of the Split-pi converter
V_{c0}	Average voltage value of the bulk capacitor for model linearization
V_e	Voltage of the external input/output capacitor of the Split-pi converter
V_{e0}	Average voltage value of the external capacitors for model linearization

Appendix A. Coefficients of the Transfer Functions of Model A

The coefficients of the transfer functions (14) and (15) of state-space model A are expressed using the following equations:

$$\begin{cases} n_3 = \dfrac{V_{c0} + R_c(I_{L10} - I_{L20})}{L} \\[2mm] n_2 = \dfrac{(1 - \bar{d})I_{L10}(L - CR_c^2) + n_3 LC\left(R_p + R_L + R_c + \frac{L}{C_e R_{sum}}\right)}{L^2 C} \\[2mm] n_1 = \dfrac{(1 - \bar{d})I_{L10}\left(R_p + R_L - R_c + R_c\frac{\frac{L}{R_c} - R_c C}{C_e R_{sum}}\right) + n_3 L\left(\frac{CR^2}{C_e R_{sum}^2} + \frac{CR_{tot}}{C_e R_{sum}} + 1\right)}{L^2 C} \\[2mm] n_0 = \dfrac{(1 - \bar{d})I_{L10}(R + R_L - R_c) + n_3 L}{L^2 C} \cdot \dfrac{1}{C_e R_{sum}} \end{cases} \tag{A1}$$

$$\begin{cases} d_4 = 1 \\[2mm] d_3 = \dfrac{R_p + 2R_L + (2 - \bar{d})R_c}{L} + \dfrac{1}{C_e R_{sum}} \\[2mm] d_2 = \dfrac{R + 2R_L + (2 - \bar{d})R_c}{L} \cdot \dfrac{1}{C_e R_{sum}} + \dfrac{R_L^2 + R_L R_p + R_c\big((2 - \bar{d})R_L + \bar{d}(1-\bar{d})R_c + (1 - \bar{d})R_p\big)}{L^2} + \dfrac{1 + (1 - \bar{d})^2}{LC} \\[2mm] d_1 = \dfrac{(2 - \bar{d})R_L + \bar{d}(1 - \bar{d})(R_c - R_L - R_p) + (1-\bar{d})R_p}{L^2 C} + \left(\dfrac{R_L^2 + R_L R + R_c\big((2 - \bar{d})R_L + \bar{d}(1 - \bar{d})R_c + (1 - \bar{d})R\big)}{L^2} + \dfrac{1 + (1 - \bar{d})^2}{LC}\right) \cdot \dfrac{1}{C_e R_{sum}} \\[2mm] d_0 = \dfrac{(2 - \bar{d})R_L + \bar{d}(1 - \bar{d})(R_c - R_L - R) + (1 - \bar{d})R}{L^2 C} \cdot \dfrac{1}{C_e R_{sum}} \end{cases} \tag{A2}$$

$$R_k = \dfrac{(1 - \bar{d})(V_{c0} - I_{L20}R_c)}{I_{L10}} - R_L \tag{A3}$$

Appendix B. Coefficients of the Transfer Functions of Model B

The coefficients of the transfer functions (24) and (25) of state-space model B are expressed using the following equations:

$$\begin{cases} n_2 = \dfrac{V_{c0} + R_c(I_{L10} - I_{L20})}{L} \\[2mm] n_1 = \dfrac{(1 - \overline{d})I_{L10}\left(L - CR_c^2\right) + n_2 LC(R_L + R_c)}{L^2 C} \\[2mm] n_0 = \dfrac{(1 - \overline{d})I_{L10}(R_L - R_c) + n_2 L}{L^2 C} \end{cases} \tag{A4}$$

$$\begin{cases} d_3 = 1 \\[2mm] d_2 = \dfrac{2R_L + (2 - \overline{d})R_c}{L} \\[2mm] d_1 = \dfrac{R_L^2 + R_c\big((2 - \overline{d})R_L + \overline{d}(1 - \overline{d})R_c\big)}{L^2} + \dfrac{1 + (1 - \overline{d})^2}{LC} \\[2mm] d_0 = \dfrac{(2 - \overline{d})R_L + \overline{d}(1 - \overline{d})(R_c - R_L)}{L^2 C} \end{cases} \tag{A5}$$

$$R_k = \dfrac{\left(1 - \overline{d}\right)\left(V_{c0} - I_{L20}R_c\right)}{I_{L10}} - R_L \tag{A6}$$

References

1. Di Piazza, M.C.; Luna, M.; Pucci, M.; La Tona, G.; Accetta, A. Electrical Storage Integration into a DC Nanogrid Testbed for Smart Home Applications. In Proceedings of the 2018 IEEE International Conference on Environment and Electrical Engineering and 2018 IEEE Industrial and Commercial Power Systems Europe (EEEIC / I&CPS Europe), Palermo, Italy, 12–15 June 2018; pp. 1–5.
2. Chub, A.; Vinnikov, D.; Kosenko, R.; Liivik, E.; Galkin, I. Bidirectional DC–DC Converter for Modular Residential Battery Energy Storage Systems. *IEEE Trans. Ind. Electron.* **2020**, *67*, 1944–1955. [CrossRef]
3. Sim, J.; Lee, J.; Choi, H.; Jung, J.H. High Power Density Bidirectional Three-Port DC-DC Converter for Battery Applications in DC Microgrids. In Proceedings of the 2019 10th International Conference on Power Electronics and ECCE Asia (ICPE 2019—ECCE Asia), Busan, Korea, 27–30 May 2019; Volume 3, pp. 843–848.
4. Byrne, R.H.; Nguyen, T.A.; Copp, D.A.; Chalamala, B.R.; Gyuk, I. Energy Management and Optimization Methods for Grid Energy Storage Systems. *IEEE Access* **2018**, *6*, 13231–13260. [CrossRef]
5. La Tona, G.; Luna, M.; Di Piazza, A.; Di Piazza, M.C. Towards the real-world deployment of a smart home EMS: A DP implementation on the Raspberry Pi. *Appl. Sci.* **2019**, *9*, 2120. [CrossRef]
6. Accetta, A.; Pucci, M. Energy Management System in DC Micro-Grids of Smart Ships: Main Gen-Set Fuel Consumption Minimization and Fault Compensation. *IEEE Trans. Ind. Appl.* **2019**, *55*, 3097–3113. [CrossRef]
7. Gorji, S.A.; Sahebi, H.G.; Ektesabi, M.; Rad, A.B. Topologies and Control Schemes of Bidirectional DC–DC Power Converters: An Overview. *IEEE Access* **2019**, *7*, 117997–118019. [CrossRef]
8. Beheshtaein, S.; Cuzner, R.M.; Forouzesh, M.; Savaghebi, M.; Guerrero, J.M. DC Microgrid Protection: A Comprehensive Review. *IEEE J. Emerg. Sel. Top. Power Electron.* **2019**, *6777*. [CrossRef]
9. Han, S.; Divan, D. Bi-directional DC/DC converters for plug-in hybrid electric vehicle (PHEV) applications. In Proceedings of the 2008 Twenty-Third Annual IEEE Applied Power Electronics Conference and Exposition, Austin, TX, USA, 24–28 February 2008; pp. 784–789.
10. Tytelmaier, K.; Husev, O.; Veligorskyi, O.; Yershov, R. A review of non-isolated bidirectional dc-dc converters for energy storage systems. In Proceedings of the 2016 II International Young Scientists Forum on Applied Physics and Engineering (YSF), Kharkiv, Ukraine, 10–14 October 2016; pp. 22–28.
11. Odo, P. A Comparative Study of Single-phase Non-isolated Bidirectional dc-dc Converters Suitability for Energy Storage Application in a dc Microgrid. In Proceedings of the 2020 IEEE 11th International Symposium on Power Electronics for Distributed Generation Systems (PEDG), Dubrovnik, Croatia, 28 September–1 October 2020; pp. 391–396.
12. Crocker, T.R. Power Converter and Method for Power Conversion. U.S. Patent No. 20040212357A1, 28 October 2004.
13. Khan, S.A.; Pilli, N.K.; Singh, S.K. Hybrid Split Pi converter. In Proceedings of the 2016 IEEE International Conference on Power Electronics, Drives and Energy Systems (PEDES), Trivandrum, India, 14–17 December 2016; pp. 1–6.
14. Sobhan, S.; Bashar, K.L. A novel Split-Pi converter with high step-up ratio. In Proceedings of the 2017 IEEE Region 10 Humanitarian Technology Conference (R10-HTC), Dhaka, Bangladesh, 21–23 December 2017; pp. 255–258.
15. Ahmad, T.; Sobhan, S. Performance analysis of bidirectional split-Pi converter integrated with passive ripple cancelling circuit. In Proceedings of the 2017 International Conference on Electrical, Computer and Communication Engineering (ECCE), Cox's Bazar, Bangladesh, 16–18 February 2017; pp. 433–437.
16. Singhai, M.; Pilli, N.K.; Singh, S.K. Modeling and analysis of split-Pi converter using State space averaging technique. In Proceedings of the 2014 IEEE International Conference on Power Electronics, Drives and Energy Systems (PEDES), Mumbai, India, 16–19 December 2014; pp. 1–6.

17. Maclaurin, A.; Okou, R.; Barendse, P.; Khan, M.A.; Pillay, P. Control of a flywheel energy storage system for rural applications using a Split-Pi DC-DC converter. In Proceedings of the 2011 IEEE International Electric Machines & Drives Conference (IEMDC), Niagara Falls, ON, Canada, 15–18 May 2011; pp. 265–270.

18. Sabatta, D.; Meyer, J. Super capacitor management using a Split-Pi symmetrical bi-directional DC-DC power converter with feed-forward gain control. In Proceedings of the 2018 International Conference on the Domestic Use of Energy (DUE), Cape Town, South Africa, 3–5 April 2018; pp. 1–5.

19. Monteiro, V.; Oliveira, C.; Rodrigues, A.; Sousa, T.J.C.; Pedrosa, D.; Machado, L.; Afonso, J.L. A Novel Topology of Multilevel Bidirectional and Symmetrical Split-Pi Converter. In Proceedings of the 2020 IEEE 14th International Conference on Compatibility, Power Electronics and Power Engineering (CPE-POWERENG), Setubal, Portugal, 8–10 July 2020; pp. 511–516.

20. Alzahrani, A.; Shamsi, P.; Ferdowsi, M. Single and interleaved split-pi DC-DC converter. In Proceedings of the 2017 IEEE 6th International Conference on Renewable Energy Research and Applications (ICRERA), San Diego, CA, USA, 5–8 November 2017; pp. 995–1000.

21. Mohan, N.; Undeland, T.M.; Robbins, W.P. *Power Electronics. Converters, Applications and Design*, 3rd ed.; John Wiley and Sons, Inc.: Hoboken, NJ, USA, 2003.

22. Middlebrook, R.D.; Cuk, S. A general unified approach to modelling switching-converter power stages. In Proceedings of the 1976 IEEE Power Electronics Specialists Conference, Cleveland, OH, USA, 8–10 June 1976; pp. 18–34.

23. Fung, H.-W.; Wang, Q.-G.; Lee, T.-H. PI Tuning in Terms of Gain and Phase Margins. *Automatica* **1998**, *34*, 1145–1149. [CrossRef]

Article

Modeling and Performance Assessment of the Split-Pi Used as a Storage Converter in All the Possible DC Microgrid Scenarios. Part II: Simulation and Experimental Results

Massimiliano Luna [1], Antonino Sferlazza [2], Angelo Accetta [1], Maria Carmela Di Piazza [1,*], Giuseppe La Tona [1] and Marcello Pucci [1]

[1] Istituto di Ingegneria del Mare (INM), Consiglio Nazionale delle Ricerche (CNR), Via Ugo La Malfa 153, 90146 Palermo, Italy; massimiliano.luna@cnr.it (M.L.); angelo.accetta@cnr.it (A.A.); giuseppe.latona@cnr.it (G.L.T.); marcello.pucci@cnr.it (M.P.)

[2] Dipartimento di Ingegneria (DI), Università degli Studi di Palermo, Viale delle Scienze Ed. 10, 90128 Palermo, Italy; antonino.sferlazza@unipa.it

*   Correspondence: mariacarmela.dipiazza@cnr.it

Abstract: Bidirectional DC/DC converters such as the Split-pi can be used to integrate an energy storage system (ESS) into a DC microgrid providing manifold benefits. However, this integration deserves careful design because the ESS converter must behave like a stiff voltage generator, a non-stiff voltage generator, or a current generator depending on the microgrid configuration. Part I of this work presented a comprehensive theoretical analysis of the Split-pi used as an ESS converter in all the possible DC microgrid scenarios. Five typical microgrid scenarios were identified. Each of them required a specific state-space model of the Split-pi and a suitable control scheme. The present paper completes the study validating the theoretical analysis based on simulations and experimental tests. The chosen case study encompassed a 48 V, 750 W storage system interfaced with a 180 V DC microgrid using a Split-pi converter. It can represent a reduced-power prototype of terrestrial and marine microgrids. A prototypal Split-pi converter was realized in the lab, and several experimental tests were performed to assess the performance in each scenario. The results obtained from the experimental tests were coherent with the simulations and validated the study.

Keywords: Split-pi; bidirectional converter; electrical storage system; DC microgrid; droop control; current control; feed-forward control

Citation: Luna, M.; Sferlazza, A.; Accetta, A.; Di Piazza, M.C.; La Tona, G.; Pucci, M. Modeling and Performance Assessment of the Split-Pi Used as a Storage Converter in All the Possible DC Microgrid Scenarios. Part II: Simulation and Experimental Results. *Energies* **2021**, *14*, 5616. https://doi.org/10.3390/en14185616

Academic Editor: Mohamed Benbouzid

Received: 12 July 2021
Accepted: 3 September 2021
Published: 7 September 2021

Publisher's Note: MDPI stays neutral with regard to jurisdictional claims in published maps and institutional affiliations.

1. Introduction

Low voltage DC microgrids present well-known advantages over AC microgrids, such as increased efficiency, lower cost, simplified power electronic converters and related control systems, as well as easier power flow management. In general, their voltage level depends on the rated power. In fact, for a given power level, a higher voltage allows reducing the circulating currents and, in turn, the cross-section of the conductors; thus, a reduction of volume, weight, and cost can be attained by increasing the voltage level.

DC microgrids in stationary applications are mainly proposed for residential and commercial buildings with load powers ranging from 1 kW to hundreds of kilowatts. As explained in [1–3], several possibilities exist for their rated voltage. Typically, the rated DC voltage is chosen coherently with the standard AC ratings, i.e., 120 V DC, 230 V DC, or 400 V DC. Alternatively, if the DC microgrid is supplied by a passive diode rectifier connected to the mains, its voltage equals the rectified mains (i.e., 170 V DC, 325 V DC, or 566 V DC). Finally, if an active rectifier supplies the DC microgrid and the latter encompasses AC loads supplied by an inverter, slightly higher values than the rectified mains must be chosen to provide a suitable margin for voltage regulation.

As for DC microgrids in mobile applications, whenever possible, 48 V DC is chosen to exploit the advantages of Safety Extra-Low Voltage (SELV) systems. However, the rated power of the electrical equipment used in mobile applications is constantly increasing, and so does the nominal DC voltage. Contrarily to the 270 V DC bus imposed to aircraft and unmanned aerial vehicles (UAVs) by the standard MIL-STD-704F [4], no specific voltage level has been established for marine applications, except for large vessels and cruise ships, where 1 kV or 1.5 kV DC is commonly chosen.

For example, a 120 V DC microgrid supplied the 5 kW load of the small underwater remotely operated vehicle (ROV) described in [5]. Additionally, 200 V DC microgrids were considered as case studies in [6] and [7]. In the first case, a 1 kW docking system with bidirectional wireless power transmission capability was proposed for an autonomous underwater vehicle (AUV). In the other case, a centralized stabilizer was designed for a marine DC microgrid used in offshore and ship applications. A 400 V marine DC microgrid was considered in [8] and [9]. In particular, the microgrid of [8] was supplied by a fuel cell and delivered up to 90 kW to the electrical loads of a sailing boat; instead, the shipboard microgrid of [9] encompassed a 400 kW storage system, and it was studied to devise a power management strategy. A 500 V DC shipboard microgrid with two buses and two synchronous generators with 35 kW of total installed power was considered in [10]. Finally, shipboard microgrids with rated voltage ranging from 700 to 750 V DC were considered as case studies in [11] and [12]. In the first case, the microgrid supplied a 12 kW load connected to a 700 V DC bus, and a distributed secondary control strategy was proposed. In the other case, the microgrid of a ferry vessel was modeled for fuel cell integration studies; three 800 kW fuel cells powered three interconnected DC busbars at 750 V DC.

In any case, DC distribution in terrestrial and marine power systems requires extensive use of power electronic converters to interface distributed generation units, loads, and electrical storage systems (ESSs) [13–15]. In particular, bidirectional DC/DC converters are exploited to integrate ESSs into DC microgrids providing manifold benefits such as improved stability and resiliency, compensation of renewable generator's fluctuations, ramping support to generators, and the possibility to employ suitable energy management systems (EMSs) to optimize microgrid power flows [13,16–18].

Among the several topologies of bidirectional DC-DC power converters, the Split-pi is receiving increasing attention due to its distinct advantages such as high efficiency, reduced switch count, small reactive components and switching noise, as well as suitability for multiphase systems, at the only cost of non-isolated operation [19–27]. However, none of the previous works on the Split-pi focused on the connection to DC microgrids. Furthermore, only three papers proposed applications in which such a converter was not controlled in an open-loop fashion [23,24,26]. Such papers, however, presented only closed-loop applications with a single control loop; furthermore, they neglected the reactive components' parasitic resistances and did not perform any experimental validation.

On the other hand, the integration of DC-DC bidirectional converters into a DC microgrid requires a careful design of the power converter's control system, depending on how the microgrid voltage is controlled (stiff voltage control by a single generator, multiple droop-controlled voltage generators respecting a predefined power-sharing ratio or master-slave control supervised by an EMS). To the best of the authors' knowledge, no paper in the technical literature presented a comprehensive analysis of all the scenarios in which a bidirectional DC/DC converter can be used to interface an ESS with a DC microgrid. The traced contributions fall into two categories. On the one hand, many authors presented a specific converter topology for microgrid applications but modeled the converter considering its output closed on either a load resistor or an ideal voltage source. Therefore, the respect of the chosen stability margin was not guaranteed in real applications where mixed loads stem from the combination of passive loads, droop-controlled voltage generators, and current sources controlled by EMSs. On the other hand, other authors modeled the whole microgrid based on power system or control approaches with different goals such as assessing stabilization and fault or power management techniques or minimizing circulating currents. Thus, such studies were focused on a higher operation level than the present work.

This paper is the second part of a work aimed at analyzing all the scenarios in which a bidirectional DC/DC converter is used to interface an ESS with a DC microgrid. Part I dealt with the theoretical aspects of the problem [28]. First, the five possible DC microgrid scenarios were identified. Then, it was shown that a specific state-space model and a suitable control scheme were required to obtain high performance from the Split-pi in each scenario. The proposed models (denoted as models A and B) considered the parasitic resistances. The control schemes required two to three control loops plus a feed-forward (FF) action depending on the scenario. Finally, a more complex compensator than a conventional PI regulator was needed for output current control in stiff microgrids.

The present paper completes the study proposed in part I, validating the theoretical analysis based on a comprehensive set of simulations and experimental tests. Due to constraints imposed by the available lab equipment, a 180 V DC, 750 W microgrid was chosen as a case study. It can represent a reduced-power prototype of terrestrial and marine microgrids, e.g., a residential DC microgrid encompassing 120 V AC loads supplied by an inverter or the DC microgrid onboard a small-power ROV or AUV. Such an approach is coherent with [29], in which the microgrid of an electric ferry powered by fuel cells was scaled down to 110 V DC, 500 W to validate a stability enhancement technique. In any case, the choice of the voltage rating of the microgrid does not affect the validity of the conclusions, which can be straightforwardly extended to the whole range of low voltage DC.

A prototype of the Split-pi converter was realized in the lab to interface an emulated 48 V ESS with the 180 V microgrid. Then, several experimental tests were performed to assess the performance in each scenario. The results obtained from the experimental tests were coherent with the simulations and validated the study. Finally, it is worth remarking on the general validity of the proposed approach, which can also be followed when other bidirectional DC/DC converter topologies are employed to interface an ESS with a DC microgrid.

2. Overview of the DC Microgrid Scenarios for the Operation of the Split-pi as an ESS Converter

The complete theoretical analysis of the DC microgrid scenarios, state-space models, and required control schemes for the Split-pi converter used as an ESS converter (i.e., operating with lower storage-side voltage than grid-side voltage) was presented in Part I of the present work. In the following, only the main concepts will be recalled to ease the interpretation of the simulations and the experimental results.

Five scenarios can be considered depending on the control mode chosen for the storage-side converter (interfacing the ESS) and the grid-side converters (interfacing other microgrid generators, if any). They are described in Table 1, which summarizes the corresponding table reported in Part I of the present work. In particular, the first three columns describe each scenario; the fourth and fifth columns, instead, associate to each scenario the state-space model and the control scheme required for the Split-pi, which were devised in Part I.

For the sake of clarity, the five scenarios are referred to also using an abbreviation in the form Sx-Gy, where x and y specify the control mode for the storage-side and the grid-side converters, respectively. The range of options for x is:

- C for current control.
- D for non-stiff droop control.
- S for stiff droop control.

Table 1. Possible combinations of microgrid scenarios, converter models, and control schemes for the Split-pi.

Microgrid Scenario	Storage Converter	Other Generators	State-Space Model of the Split-pi	Control Scheme
#1 (SS-GN)	Droop mode with $R_d = 0$ (stiff microgrid)	No other generator present (passive load) or all current controlled by the EMS	Model A	2 loops (I_{L1} and V_2) + FF
#2 (SD-GN)	Droop mode with $R_d \neq 0$	No other generator present (passive load) or all current controlled by the EMS	Model A	3 loops (I_{L1}, V_2, and droop) + FF
#3 (SD-GD)	Droop mode with $R_d \neq 0$	At least one is droop controlled and none has $R_d = 0$	Model A	3 loops (I_{L1}, V_2, and droop) + FF
#4 (SC-GD)	Current mode	At least one is droop controlled and none has $R_d = 0$	Model A	2 loops (I_{L1} and I_{L2})
#5 (SC-GS)	Current mode	One is droop controlled and has $R_d = 0$ (stiff microgrid); the others, if present, are current controlled by the EMS	Model B	2 loops (I_{L1} and I_{L2})

Furthermore, a baseline scenario derived from scenario #1 excluding the FF action was also considered for comparison purposes to show the usefulness of such action.

3. Overview of the Chosen Case Study and Droop Characteristics of the Storage-Side and Grid-Side Converters

The considered case study encompassed a 48 V, 750 W storage system interfaced with a 180 V DC microgrid using a Split-pi converter. As explained in the Introduction, it can represent a reduced-power prototype of terrestrial and marine microgrids. Figure 1 shows the schematic of the Split-pi converter, which was sized for the chosen case study in Part I of the present work. Its main parameters are summarized in Table 2. Furthermore, in Part I of the work, five control systems were designed, i.e., one for each specific scenario. It was assumed that the converter supplied its rated power without an external current or voltage generator. The latter was seen as a disturbance, which the control system had to compensate promptly. Furthermore, suitably wide stability margins were imposed, so the designed controllers were also effective at lighter loads.

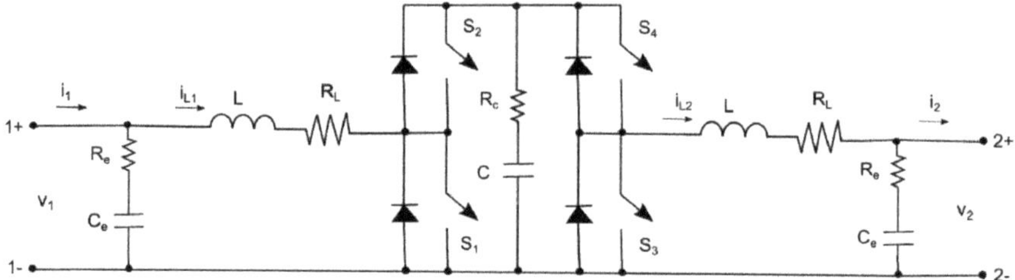

Figure 1. Schematic of the Split−pi converter.

For a proper interpretation of the simulation results, it is worth recalling the droop characteristics of both the storage converter and the voltage generator of the microgrid. As for the storage converter, its droop characteristic was defined by $E_{ds} = 180$ V and $R_{ds} = 0$ in scenario #1 (SS-GN); in scenarios #2 (SD-GN) and #3 (SD-GD), instead, it was expressed by $E_{ds} = 180$ V and $R_{ds} = 2.2\ \Omega$, i.e., imposing a 5% voltage reduction at the nominal current. As for the equivalent voltage generator of the microgrid, its droop parameters were equal to those of the storage converter in scenario #4 (SC-GD). Instead, in scenario #3 (SD-GD), they were expressed by $E_d = 198$ V and $R_d = 9\ \Omega$; thus, the storage system did not supply any power for half the rated load of the microgrid, as desired. Finally, in scenario #5 (SC-GS), a constant voltage generator $E_d = 180$ V was considered to model the microgrid's stiff voltage generator.

Table 2. Main parameters of the Split-pi converter.

Parameter	Symbol	Value
Switching frequency	F_{sw}	20 kHz
Nominal output voltage	V_{2n}	180 V
Nominal input voltage	V_{1n}	50 V
Nominal power	P_n	750 W
Nominal load resistance	R_n	43.2 Ω
Nominal input current	I_{1n}	15 A
Max. charge/discharge current	I_{cx}, I_{dx}	18 A
Nominal output current	I_{2n}	4.167 A
Nominal duty-cycle	$\bar{d}$	0.722
Inductance value of L	L	1000 μH
Parasitic resistance of L	R_L	65 mΩ
Capacitance value of C_e	C_e	200 μF
Parasitic resistance of C_e	R_e	260 mΩ
Capacitance value of C	C	540 μF
Parasitic resistance of C	R_c	125 mΩ
Max. ripple on input/output inductors	$r_{i\%}$	$\pm6.0\%$
Max. ripple on input/output capacitors	$r_{ve\%}$	$\pm0.2\%$
Max. ripple on bulk capacitor	$r_{v\%}$	$\pm0.2\%$

4. Simulation Results

Several simulations were performed to assess the performance of the controlled system in all the scenarios. The circuit model was implemented using PLECS based on the electrical parameters of Table 2, whereas the control system was realized in Simulink. The simulation parameters set in the Simulink environment were the following:

- solver type: variable-step
- solver: ode23tb (stiff/TR-BDF2)
- max step size: $1/(10 \cdot F_{sw})$
- solver reset method: robust.

The default values were confirmed for the other Simulink parameters, as well as for the PLECS parameters. The obtained simulation results are presented and commented on in the following.

4.1. Validation of FF Action and Performance Assessment in Scenario #1 (SS-GN)

As a starting point to prove the advantage of the FF action, the baseline scenario in which the storage converter supplied a microgrid stiffly at 180 V was simulated. Besides the passive load, a current generator was also present in the microgrid. Only the two innermost control loops for I_{L1} and V_2 were employed, the FF action was disabled, and the system was described by state-space model A.

The waveforms of the most meaningful electrical and control quantities are shown in Figures 2 and 3a. Before t = 0.2 s, the converter output was at a steady state with the following values of load resistance, load power, and external current reference: $R = R_n$, $P = P_n$, and $I = 0$. Then, a representative sequence of R and I values was applied, as shown in Table 3. From t = 0.2 s to t = 0.6 s, the load was passive and exhibited two stepwise variations from the rated power almost to a no-load condition and vice versa. At t = 0.6 s, the load resistance was kept constant, but the external current reference was increased to supply the load fully; hence, the converter's output current dropped to zero. Then, at t = 0.8 s, the converter operated again with a very low load, so the external current was almost entirely fed back to the converter to recharge the storage system at full power. At t = 1 s, the load power was increased to its rated value. Finally, the sequence applied during the remaining time intervals was such that the external current generator started and stopped recharging the storage system at full power while the load resistance was the rated one.

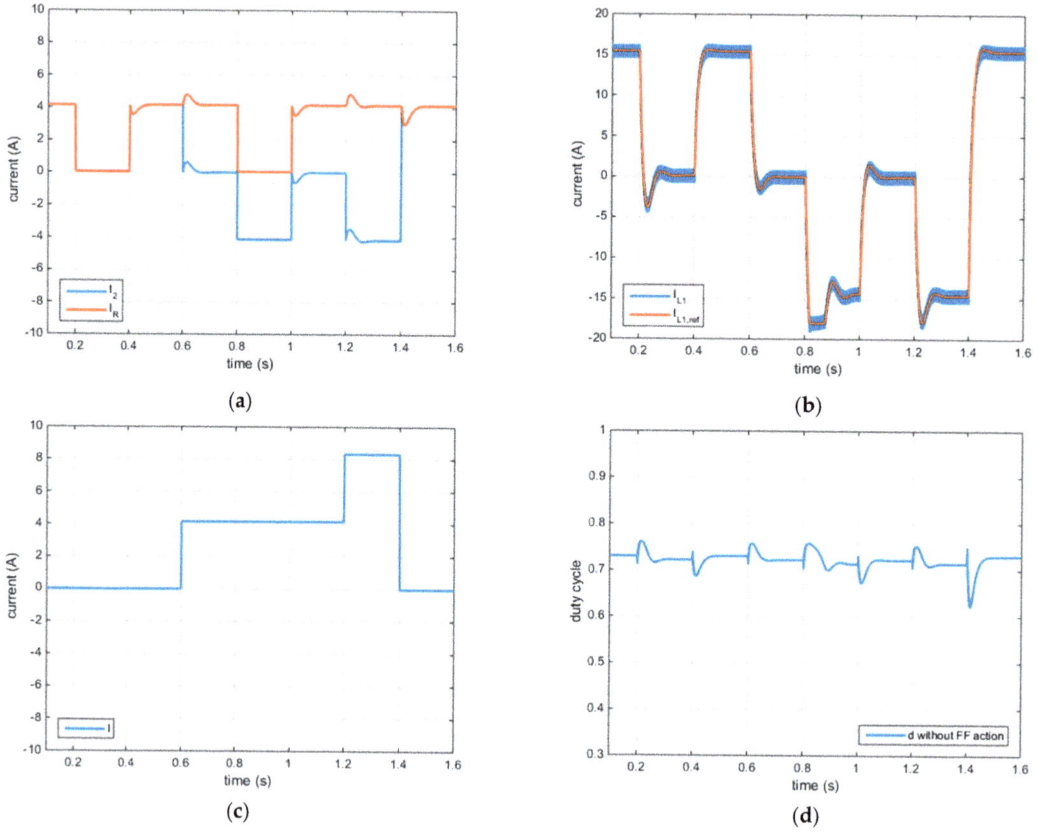

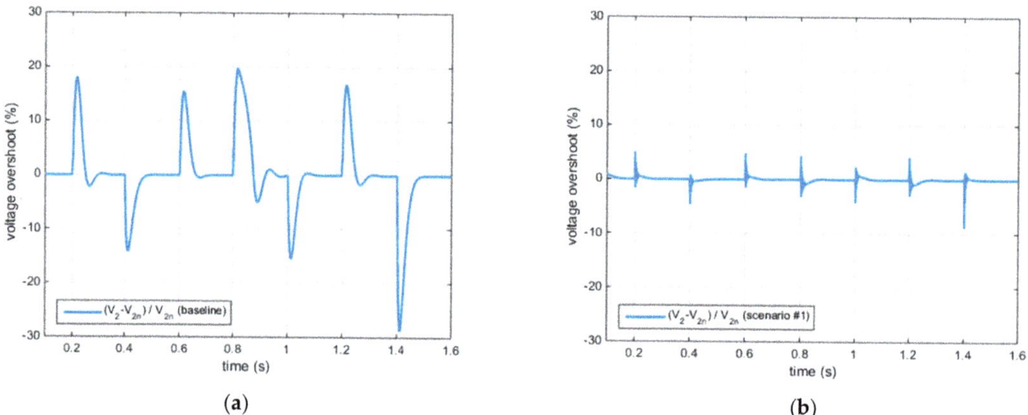

Figure 2. Baseline performance without FF action: (**a**) output current and load current; (**b**) input inductor current; (**c**) external generator's current; (**d**) duty cycle.

Figure 3. Grid voltage overshoot: (**a**) in baseline scenario; (**b**) in scenario #1 (SS−GN).

Table 3. Transitions considered in the baseline scenario and scenarios #1 (SS-GN) and #2 (SD-GN).

Time (s)	Load Resistance	Expected Load Power	External Current Reference
0.2	$100 \cdot R_n$	$P_n/100$	0
0.4	R_n	P_n	0
0.6	R_n	P_n	I_{2n}
0.8	$100 \cdot R_n$	$P_n/100$	I_{2n}
1.0	R_n	P_n	I_{2n}
1.2	R_n	P_n	$2 \cdot I_{2n}$
1.4	R_n	P_n	0

The normalized waveform of grid voltage is presented in Figure 3a and shows that the converter's output voltage was correctly regulated at the desired steady-state value as the passive/active load varied. As shown in Figure 2, the current loop dynamics were fast enough to track $I_{L1,ref}$ instantly. However, the voltage waveform exhibited significant under- and overshoots due to load variations, as shown in Figure 3a. In particular, the maximum overshoot and undershoot were 20% and -29%, respectively. On the other hand, the microgrid usually has an allowed transient voltage tolerance of ± 20% around its nominal voltage, so it would have been automatically de-energized by the protection devices, causing severe inconvenience.

A significant improvement of the system response can be achieved using the FF action; therefore, such action was enabled in the first three scenarios of Table 1, starting from scenario #1 (SS-GN). The difference between such a scenario and the baseline one is only the inclusion of the FF action. The related simulation results are shown in Figures 3b and 4. As shown in Figure 4d, the FF action varied the duty cycle instantaneously to compensate for the load variation. Consequently, the waveforms of I_{L1}, I_2, and I_R were pretty squared (Figure 4a,b), and the resulting output voltage exhibited acceptable over/undershoots (5% and -8.8%), as shown in Figure 3b. Thus, a correct operation of the microgrid was attained.

4.2. Performance Assessment in Scenario #2 (SD-GN)

Unlike the first scenario, in scenario #2 (SD-GN), the storage converter was controlled using a droop scheme with $R_d \neq 0$, i.e., with three control loops (I_{L1}, V_2, and droop) plus the FF action. All the other parameters, including the initial conditions and load sequence, were kept unchanged. The microgrid voltage was not constant at 180 V at steady-state due to the droop control; thus, the actual load power was slightly lower than the value reported in Table 3 depending on voltage reduction, as expected.

The simulation results in terms of grid voltage waveforms are shown in Figure 5. The converter's output voltage (V_2) quickly tracked the reference voltage (V_{2ref}) computed according to the droop characteristic during the entire timeframe. In particular, the voltage variation was ± 5% when the storage system was discharged/recharged at the rated current, as expected based on the droop characteristic. The over/undershoot superimposed to the steady-state voltage variation was less than ± 1.66%. Thus, the microgrid voltage stayed within ± 7% of its rated value even during transients. The input/output currents and duty cycle were not shown because their waveforms were like those obtained in scenario #1 (SS-GN).

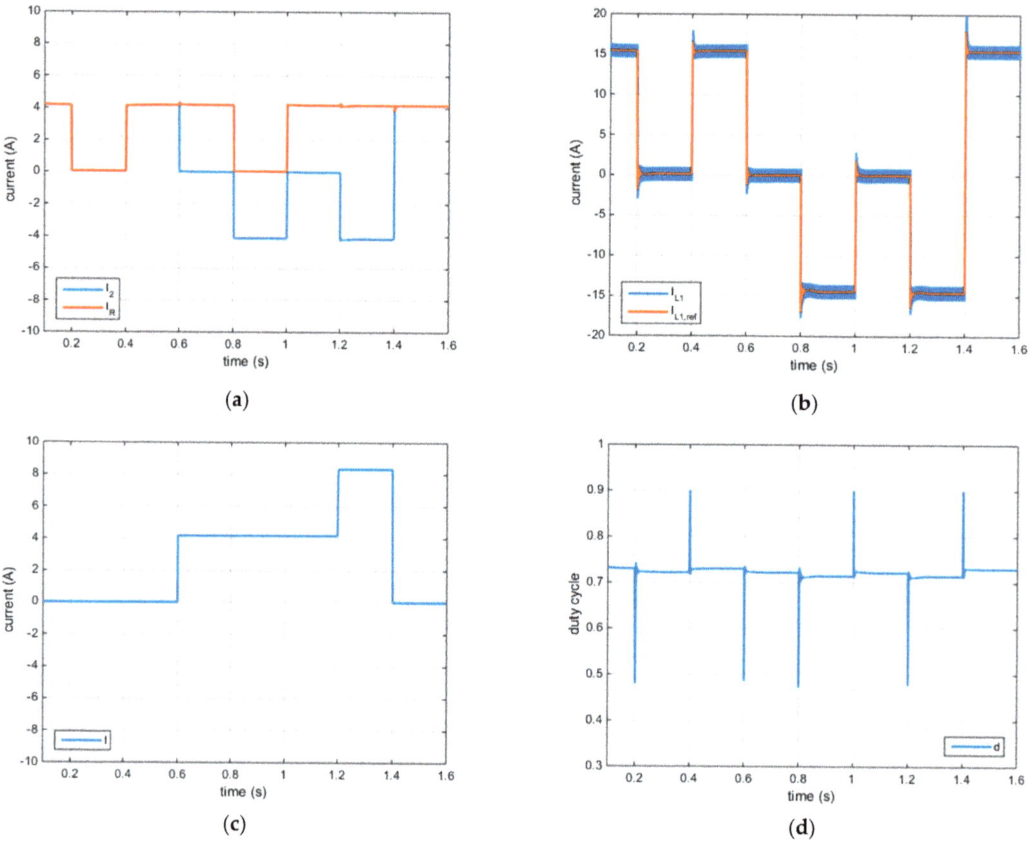

Figure 4. Performance obtained in scenario #1 (SS−GN): (**a**) output current and load current; (**b**) input inductor current; (**c**) external generator's current; (**d**) duty cycle.

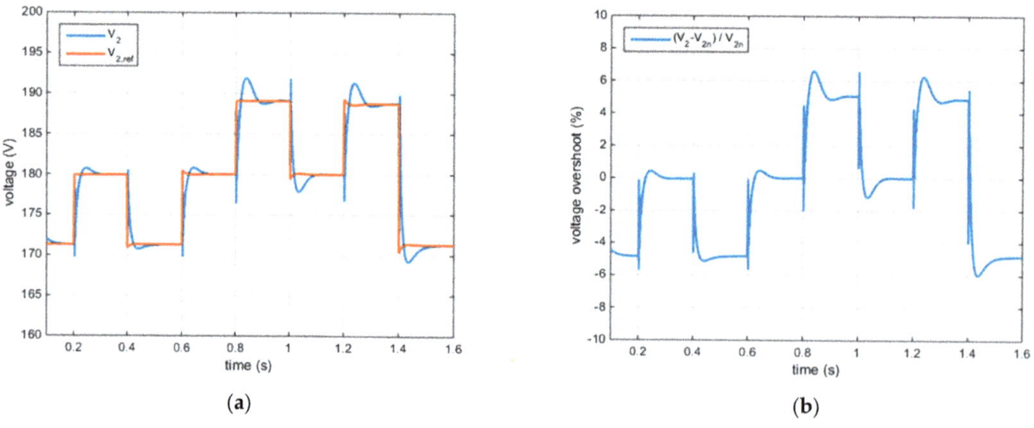

Figure 5. Performance obtained in scenario #2 (SD−GN): (**a**) actual and reference grid voltage; (**b**) grid voltage overshoot.

4.3. Performance Assessment in Scenario #3 (SD-GD)

In this scenario, both the storage converter and the equivalent voltage generator of the microgrid were controlled in droop mode with non-null droop resistances. Furthermore, the presence of an equivalent current generator in the microgrid was considered as well. The two droop characteristics were chosen so that the storage system did not supply any power for half the microgrid's rated load, as described in Section 3. Again, before t = 0.2 s, the converter output was at steady-state with the following values of load resistance, load power, and external current reference: $R = R_n$, $P = P_n$, and $I = 0$. However, the storage system delivered about half the rated power in this scenario, and the equivalent voltage generator supplied the remaining quota.

The load sequence that was applied to the converter after t = 0.2 s is shown in Table 4. It is like that of Table 3, with additional conditions in which the passive load drew half the rated power. Furthermore, the current reference values were chosen to inject the maximum allowed current that did not overload the droop-controlled generator and the storage system.

Table 4. Transitions considered in scenario #3 (SD-GD).

Time (s)	Load Resistance	Expected Load Power	External Current Reference
0.2	$2 \cdot R_n$	$P_n/2$	0
0.4	$100 \cdot R_n$	$P_n/100$	0
0.6	$100 \cdot R_n$	$P_n/100$	$0.75 \cdot I_{2n}$
0.8	$2 \cdot R_n$	$P_n/2$	$0.75 \cdot I_{2n}$
1.0	R_n	P_n	$0.75 \cdot I_{2n}$
1.2	R_n	P_n	$1.8 \cdot I_{2n}$

The simulation results for this scenario are shown in Figure 6. At t = 0.2 s, the external current reference was null while the load resistance doubled; hence, the converter's output current dropped to zero, as expected. At t = 0.4 s, the converter operated almost at no load, so the storage system was recharged at about half power. Then, at t = 0.6 s, the equivalent current generator injected current into the grid providing excess power; hence, the storage system was recharged at full power. From t = 0.8 s to t = 1.2 s, the load power increased from low to medium to high. The equivalent current generator injected constant current, so the charging power progressively decreased as the load power increased. Finally, at t = 1.2 s, the external current generator charged the storage system at the maximum power level besides supplying the rated load.

As in the previous scenario, the converter's output voltage quickly tracked the reference voltage computed according to its droop characteristic based on the output current; however, in scenario #3, the current also depended on the droop characteristic of the equivalent voltage generator. In particular, Figure 6d shows that the voltage variation of the microgrid was even lower compared to scenario #2 (SD-GN): from −2% to 5% at steady-state with minimal over/undershoots (i.e., less than ±1.05%). Finally, the waveforms of the input inductor current I_{L1} and the duty cycle d were not shown because they were like those obtained in scenario #1 (SS-GN).

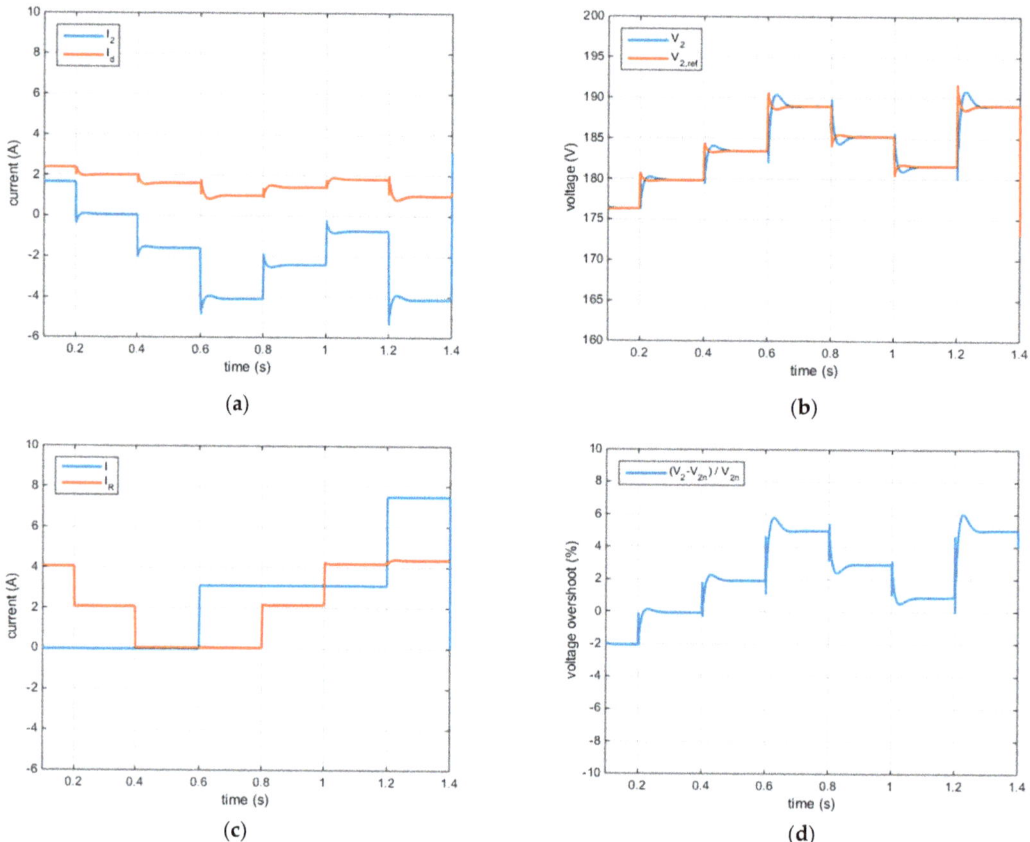

Figure 6. Performance obtained in scenario #3 (SD–GD): (**a**) output current and droop-controlled generator's current; (**b**) actual and reference grid voltage; (**c**) external generator's current and load current; (**d**) grid voltage overshoot.

4.4. Performance Assessment in Scenario #4 (SC-GD)

In scenario #4 (SC-GD), the voltage was regulated by the microgrid's equivalent voltage generator controlled in droop mode with non-null droop resistance. In contrast, the storage converter was operated as a current-controlled source based on a reference given by the EMS. Therefore, the control system of the Split-pi converter encompassed two loops for I_{L1} and I_{L2}, respectively. Furthermore, the presence of an equivalent external current generator in the microgrid was also considered.

The simulation results for this scenario are shown in Figure 7. Before t = 0.2 s, the converter output was at steady-state with the following values of load resistance, output current reference, and external current reference: $R = R_n$, $I_{2ref} = 0$, and $I = 0$. In other words, the storage system and the external current generator delivered no power while the load was supplied entirely by the equivalent voltage generator. Then, a representative sequence of R, I_{2ref}, and I values was applied, as shown in Table 5.

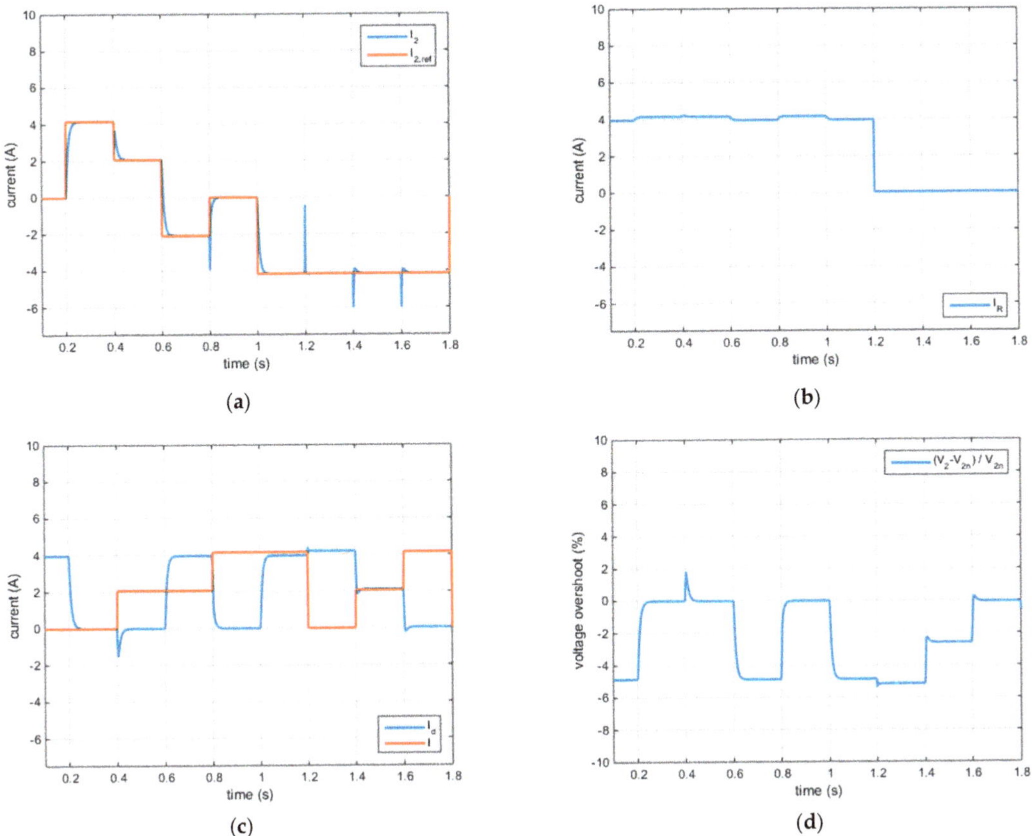

Figure 7. Performance obtained in scenario #4 (SC−GD): (**a**) actual and reference output current; (**b**) load current; (**c**) droop-controlled generator's current and external generator's current; (**d**) grid voltage overshoot.

Table 5. Transitions considered in scenario #4 (SC-GD).

Time (s)	Load Resistance	Expected Load Power	External Current Reference	Output Current Reference for the Converter
0.2	R_n	P_n	0	I_{2n}
0.4	R_n	P_n	$0.5 \cdot I_{2n}$	$0.5 \cdot I_{2n}$
0.6	R_n	P_n	$0.5 \cdot I_{2n}$	$-0.5 \cdot I_{2n}$
0.8	R_n	P_n	I_{2n}	0
1.0	R_n	P_n	I_{2n}	$-I_{2n}$
1.2	$100 \cdot R_n$	$P_n/100$	0	$-I_{2n}$
1.4	$100 \cdot R_n$	$P_n/100$	$0.5 \cdot I_{2n}$	$-I_{2n}$
1.6	$100 \cdot R_n$	$P_n/100$	I_{2n}	$-I_{2n}$

The load resistance kept the same value from t = 0.2 s to t = 1.2 s while the external current generator's reference progressively increased. Each time, the output current reference was set to the maximum value (either positive or negative) that respected the following conditions: it was compatible with the current combination of *R* and *I*, and it did not overload the droop-controlled generator and the storage system. From t = 1.2 s to t = 1.8 s, the load resistance was increased almost to a no-load condition, and the reference for the external current generator was progressively incremented. Again, the output current reference was set to the maximum value that did not overload the droop-controlled generator and the storage system; however, only negative values were allowed because of the no-load condition.

All the grid-side current transients were fast and clean even though challenging perturbations were applied, i.e., stepwise variations of circuit parameters. As for the microgrid voltage variation, Figure 7d shows that it went from −5.14% to 0% at steady-state with minimal over/undershoots, as expected. The FF action was not used with current-mode control; thus, the waveform of I_{L1} was like that obtained in the baseline scenario, and the waveform of the duty cycle was flat around its rated value.

4.5. Performance Assessment in Scenario #5 (SC-GS)

In this scenario, the storage converter was operated as a current-controlled source under the supervision of the EMS and connected to a microgrid in which one voltage generator was droop controlled with E_d = 180 V and R_d = 0 (stiff microgrid). Hence, model B was considered to design the control system. The initial condition was $I_{2ref} = 0$. Then, starting from t = 0.2 s, the same sequence of I_{2ref} as in scenario #4 (SC-GD) was applied, i.e., the one reported in the last column of Table 5. The simulation results for this scenario in terms of grid-side currents are shown in Figure 8 and confirmed the good system performance. The microgrid voltage V_2 was perfectly constant at 180 V because the voltage generator was controlled stiffly. Again, since the FF action was not used with current-mode control, the waveform of I_{L1} resembled that obtained in the baseline scenario. Furthermore, the duty cycle waveform was even flatter compared to the previous scenario because the converter had stiff voltages at both ports.

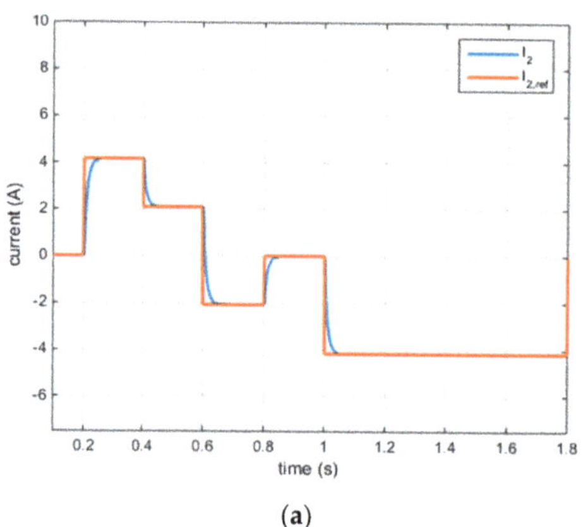

(a)

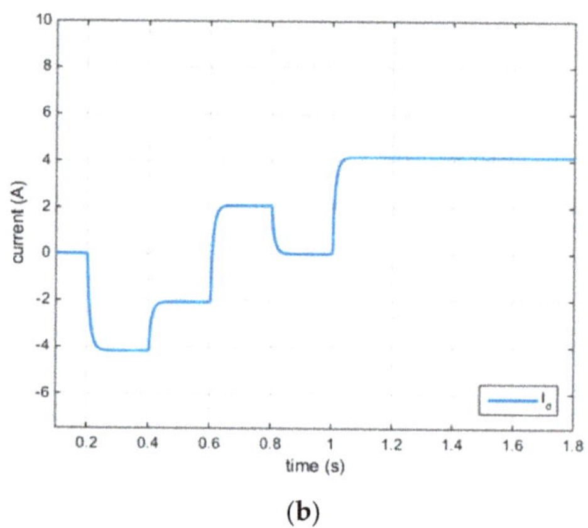

(b)

Figure 8. Performance obtained in scenario #5 (SC−GS): (**a**) actual and reference output current; (**b**) droop-controlled generator's current.

4.6. Comparison among the Five Scenarios

The results obtained in the five scenarios were summarized in Table 6 for a comparison. The chosen metrics were the maximum charging/discharging power, the percentage voltage variation against V_{2n}, and the output voltage and current settling time. The latter metric is defined as the time after which the output variable enters and remains within a specified error band after applying an instantaneous stepwise input. The simulation results showed that some applied stimuli determined no difference between the steady-state value after the transition and the initial value. Therefore, the error band was defined summing/subtracting a fixed quantity to/from the steady-state value, as if it was a measurement error. Such a quantity was $0.01 \cdot V_{2n}$ for the output voltage and $0.01 \cdot I_{2n}$ for the output current.

Table 6. Result comparison in the five scenarios.

Scenario	Max. Charging or Discharging Power	Max. Voltage Variation due to Droop Control	Max. Instantaneous Voltage Variation Including over/Undershoot	Max. Settling Time of V_2	Max. Settling Time of I_2
#1 (SS-GN)	P_N	0%	+5%, −8.8%	16.6 ms	14 ms
#2 (SD-GN)	Slightly lower than P_N	±5%	±6.66%	56.2 ms	53 ms
#3 (SD-GD)	Charging up to P_N. Discharging at about half P_N	+5%, −2%	+6.05%, −3.05%	8 ms	57 ms
#4 (SC-GD)	P_N	0%, −5.14%	+1.867%, −5.383%	14.5 ms	41.9 ms
#5 (SC-GS)	P_N	0%	0%	Not applicable	47.6 ms

It is worth remarking that, in general, the microgrid designer cannot freely choose from all five scenarios due to system constraints. For example, the first three scenarios are impracticable if the designer decides to put the ESS converter under the control of an EMS; on the other hand, only the first and last scenarios are practicable if a stiff microgrid is to be designed. Thus, the proposed comparison can provide helpful information when the designer is not obliged to choose a specific scenario.

According to Table 6, the fastest performance was achieved in scenario #1 at the expense of the voltage variation, which can reach −8.8%. This behavior descended from the simplicity of such a scenario. Comparing scenarios #1 and #2 shows that introducing a virtual droop resistance dampened the system response and slightly reduced the maximum charging/discharging power as a side effect. Thus, the microgrid's voltage variation decreased, and the settling time increased for both V_2 and I_2 (+ 238% and +278%, respectively).

The use of more than one droop-controlled generator in the microgrid has pros and cons. In fact, in scenario #3, the maximum discharging power of the ESS was halved by design; thus, the ESS and its converter were partly underutilized. On the other hand, a halved maximum voltage reduction was exhibited, and the maximum settling time of V_2 decreased significantly (i.e., −52% and −86% compared to scenarios #1 and #2, respectively). The settling time reduction was achieved because the voltage waveform's peak did not surpass the tolerance band. Instead, the maximum settling time of I_2 stayed almost constant.

Comparing scenarios #2 and #3 with the last two scenarios showed a slight reduction of the settling time of I_2. Instead, a comparison between scenarios #4~#5 and scenario #1 revealed an increase of such a metric equal to +200% and +240%, respectively. As for the percentage voltage variation and the settling time of V_2 in scenario #4, they both exhibited intermediate values compared to scenarios #2 and #3. Finally, scenario #5 presented no voltage variation thanks to the stiff control of the microgrid.

Summing up, the following guidelines can be given for the considered case study. If the ESS converter must be controlled by an EMS, the response time of I_2 is around 50 ms, whereas that of V_2 does not exceed 15 ms and depends on the stiffness of the microgrid. Otherwise, the ESS converter should be operated in scenario #1 to achieve the fastest performance, i.e., around 15 ms response time for both V_2 and I_2. In such a scenario, the ESS converter is the only voltage source of the microgrid and is controlled stiffly. If these two conditions cannot be met, the response time of I_2 worsens significantly (around 60 ms), and the waveform of V_2 depends on a compromise: lower voltage variation and better dynamic performance come with a partial underutilization of the ESS.

5. Experimental Validation

The designed Split-pi converter is shown in Figure 9. It was built as a prototype using the components available in the lab, among them an integrated power module STGIPS10K60A encompassing an IGBT-based three-phase H-bridge. In order to validate the converter and the related control laws experimentally, the storage system connected to port 1 was emulated using a TDK-Lambda GEN60-40 power supply (set to 50 V) and a parallel-connected 10 Ω, 300 W power resistor. The latter component was used to dissipate power during the recharge of the emulated storage system. On the other hand, a Sorensen SLH-500-6-1800 electronic load and a TDK-Lambda GEN600-5.5 power supply were connected to port 2 of the converter. The first device was used as the resistor R_{load} of the load model, whereas the other device was used as either the voltage generator E_d or the current generator I, depending on the specific test.

Figure 9. Picture of the realized Split-pi converter.

The control systems and the PWM modulator were implemented on a dSPACE DS1103 board connected to a desktop computer. Several electrical quantities were measured using LEM sensors, acquired and processed by the dSPACE board: I_{L1}, V_2, I_2, I_R. The current of the external generator was computed as $I_g = I_R - I_2$.

The prototypal Split-pi converter and the related control schemes were tested experimentally in several conditions covering the baseline and the other five scenarios. The obtained results showed good matching with the simulations presented in Section 4. Thus, they confirmed the good performance attainable in each scenario and, implicitly, the result comparison among the five scenarios. Furthermore, it is worth highlighting that a result comparison based on simulations rather than experiments inevitably gave a more accurate picture of the differences among the scenarios. In fact, the simulations allowed considering a much higher number of transients among operating conditions in each scenario compared to the experimental tests.

The waveforms of the most relevant signals acquired during the experimental tests were exported from dSPACE Control Desk to MATLAB and then plotted to analyze the results. The obtained plots are presented and commented on in the following.

5.1. Experimental Validation in the Baseline Scenario and Scenario #1 (SS-GN)

The Split-pi converter was used in these two tests to form a stiff microgrid supplying a passive load; no other generator was present in the microgrid. The converter was controlled using only the current and voltage loops in the baseline scenario, whereas the FF action was also included in scenario #1 (SS-GN). The electronic load was operated to reproduce a stepwise variation of load from 1300 Ω to 130 Ω; consequently, the load power instantly increased from 25 W to 250 W and produced a voltage undershoot.

The signals acquired during the tests are shown in Figure 10 and allow comparing the performance obtained with and without the FF action. All the waveforms were coherent with those obtained in the simulations. Thanks to the FF action, the reference for I_{L1} (and, in turn, the duty cycle) was instantly increased when the output current varied. Therefore, in scenario #1 (SS-GN), the current waveforms were pretty squared, the voltage undershoot on V_2 was reduced from -5.73% to -1.86%, and the dynamics of V_2 were faster compared with the baseline scenario. An even higher under/overshoot reduction would be expected for higher stepwise variations of output power.

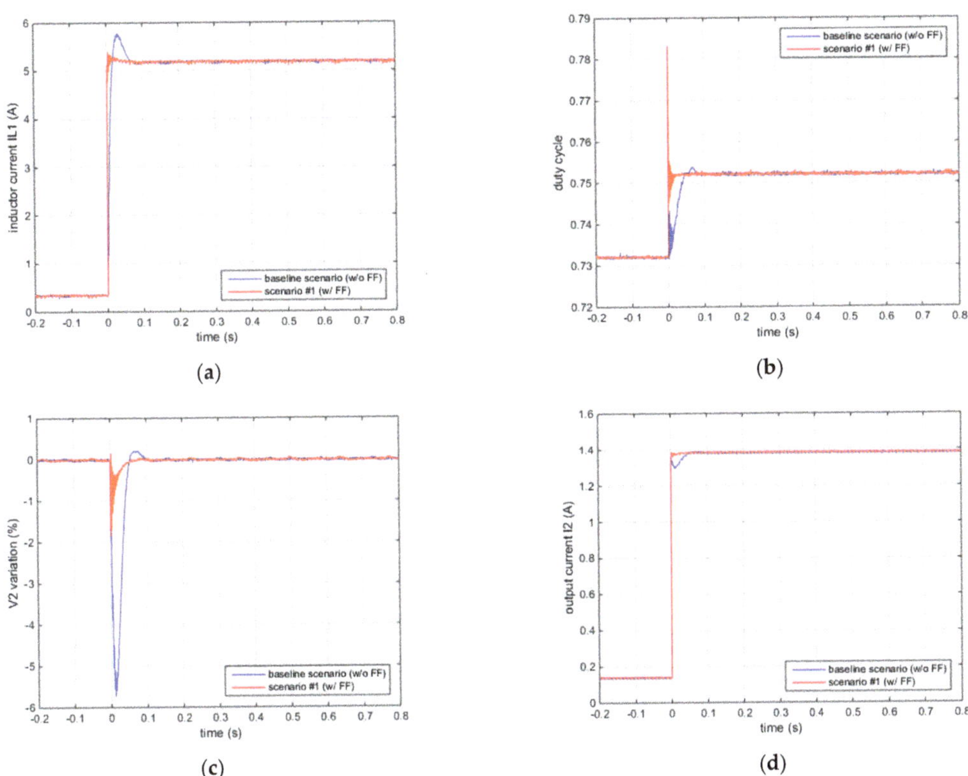

Figure 10. Experimental results in the baseline scenario and scenario #1 (SS−GN): (**a**) input inductor current; (**b**) duty cycle; (**c**) percentage variation of microgrid voltage; (**d**) grid−side current.

5.2. Experimental Validation in Scenario #2 (SD-GN)

In this test, the Split-pi converter was controlled using the FF action to form a non-stiff microgrid using a droop resistance of 1.33 Ω. The microgrid encompassed a passive load and an external power supply that was operated as a current generator. First, a current limit of 1.3 A and a voltage set point slightly higher than 180 V were set in the power supply; then, the device was turned on and off. The load resistance was 130 Ω, corresponding to a nominal current of about 1.385 A and a nominal power of 250 W.

The most relevant acquired signals are shown in Figure 11 and are coherent with those obtained in the simulations. Before t = 0 s, the load was entirely supplied by the converter. According to the droop resistance value, a 1.83 V voltage drop was exhibited, corresponding to a voltage variation of -1.02%; consequently, the load current was 1.375 A. At t = 0 s, the external current generator was turned on and supplied the load almost entirely, so the output current of the converter automatically dropped to almost zero. As a result, the voltage variation was almost zero, and the load current increased from 1.375 A to 1.382 A, reproducing the intended behavior. Finally, at about t = 1 s, the external current generator was turned off, and the converter took over the current again. In that condition, the load voltage and current again decreased slightly because of the droop control.

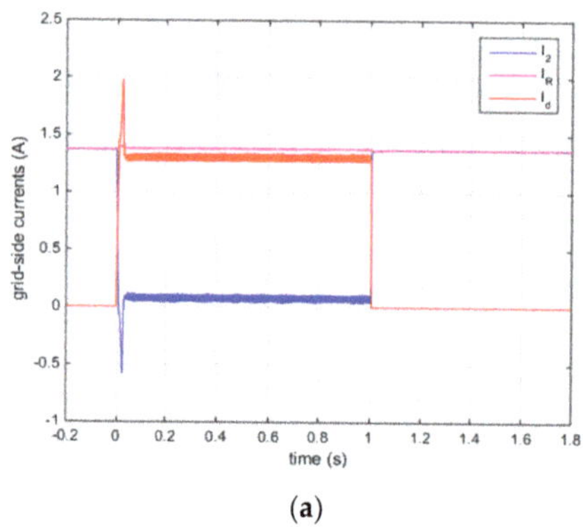 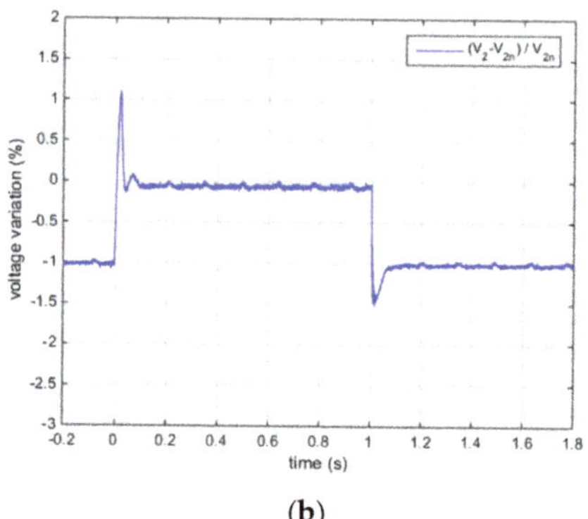

(**a**)   (**b**)

Figure 11. Experimental results in scenario #2 (SD−GN): (**a**) grid−side currents; (**b**) percentage variation of microgrid voltage.

It is worth noting that the external power supply did not offer a proper current-mode operation; actually, it tried to impose the preset voltage (185 V) and continuously adjusted its output voltage aiming to respect the current limit. This behavior explains the initial peak in the generator's current. Nevertheless, the converter's control system was fast enough to effectively compensate for the connection and disconnection of the external generator. Therefore, the over/undershoot of microgrid voltage was minimal (+1.1% and −0.5%).

5.3. Experimental Validation in Scenario #3 (SD-GD)

In this test, the Split-pi converter and the microgrid voltage generator were controlled in droop mode with $R_d \neq 0$, so the resulting microgrid was not stiff. For the sake of clarity, a droop resistance of 1.33 Ω and a no-load voltage of 180 V were chosen as droop parameters for both devices. In this way, such devices always contributed equally to supplying the load. A stepwise variation of load resistance from 130 Ω to 260 Ω and back again was applied using the electronic load; this sequence corresponded to load power transitions from 250 W to 125 W and back again. The waveforms obtained in the test are shown in Figure 12 and show that the system exhibited good dynamic behavior, as expected based on the simulation results.

The system response was aperiodic, thanks to the high phase margin that was imposed. Furthermore, the current sharing ratio was respected for both load power values. Finally, the voltage variation at each power level was precisely the one expected according to the droop resistance value and the delivered current (−0.26% and −0.52%).

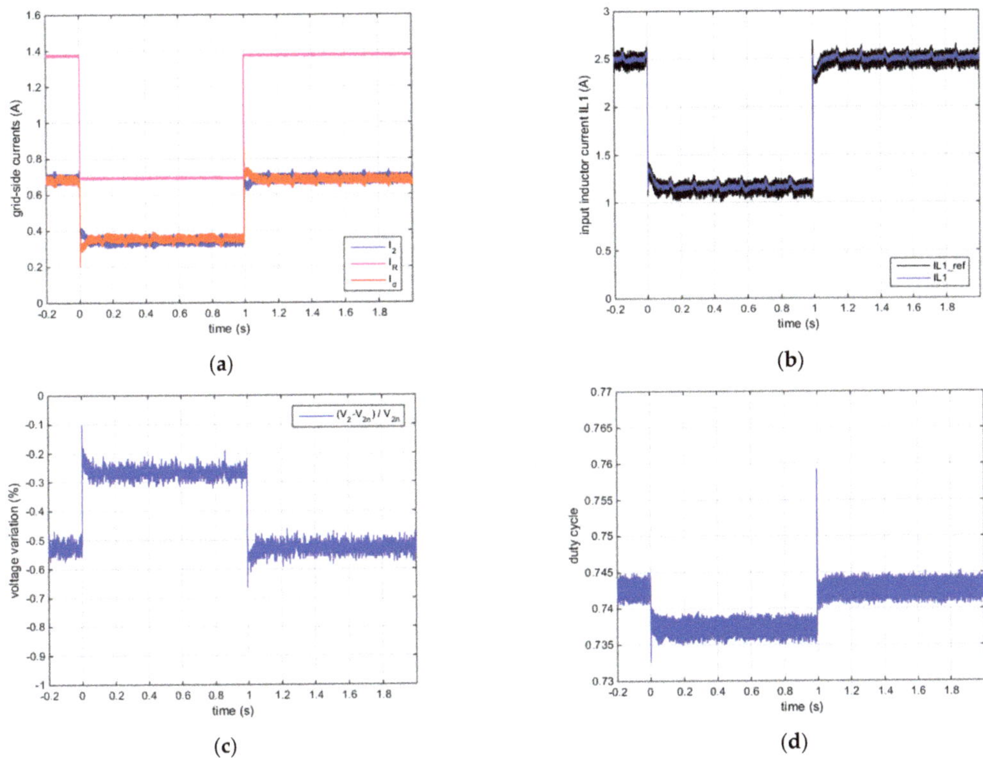

Figure 12. Experimental results in scenario #3 (SD−GD): (**a**) grid−side currents; (**b**) input inductor current; (**c**) percentage variation of microgrid voltage; (**d**) duty cycle.

5.4. Experimental Validation in Scenario #4 (SC-GD)

In this test, the Split-pi converter was operated as a current-controlled source exchanging power with a non-stiff microgrid. The external voltage generator was controlled in droop mode. The droop resistance was 1.33 Ω, and the no-load voltage was 180 V. The output current reference for the converter was changed at t = 0 s with a stepwise variation from −1.37 A to 1.37 A. Instead, the electronic load was operated to reproduce a stepwise variation of load resistance from 130 Ω to 65 Ω at about t = 0.7 s; these resistance values correspond to nominal load current (power) levels of 1.385 A (250 W) and 2.770 A (500 W), respectively.

The most relevant waveforms obtained in the test are presented in Figure 13 and are coherent with the simulation results. Before t = 0 s, the converter was controlled to draw −1.37 A from the DC microgrid recharging the emulated storage system. The external generator delivered 1.37 A to the converter and 1.345 A to the load for a total of 2.715 A. At t = 0 s, the current reference for the converter was increased to 1.37 A. Thus, the converter almost entirely supplied the load, and the current of the external generator automatically approached zero. Finally, at about t = 0.7 s, the load requested a current of 2.710 A. Since the converter's output was kept constant, the external generator automatically provided the additional current contribution of 1.34 A. In addition, the voltage variation was precisely the one expected according to the droop resistance value and the current supplied by the external voltage generator, i.e., −2%, 0%, and −1% for 2.715 A, 0 A, and 1.34 A, respectively. Furthermore, the system's dynamic behavior was good with fast aperiodic transients, coherently with the simulation results.

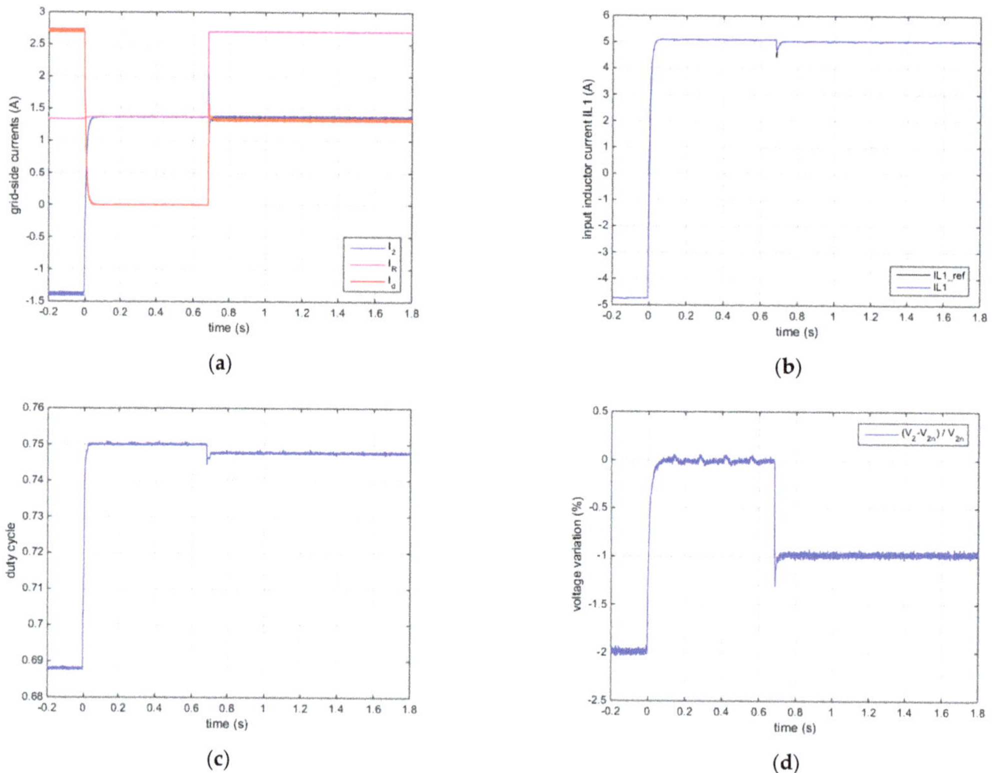

Figure 13. Experimental results in scenario #4 (SC−GD): (**a**) grid−side currents; (**b**) input inductor current; (**c**) duty cycle; (**d**) percentage variation of microgrid voltage.

5.5. Experimental Validation in Scenario #5 (SC-GS)

In this test, the Split-pi converter was operated as a current-controlled source exchanging power with a stiff microgrid. The external voltage generator still had a no-load voltage of 180 V but no droop resistance. The same combination of output current and load variations as in scenario #4 (SC-GD) was applied; however, the stepwise variation of load resistance occurred at about t = 1 s in scenario #5. Therefore, the output current reference was again changed with a stepwise variation from −1.37 A to 1.37 A at t = 0 s. Instead, the load resistance was varied from 130 Ω to 65 Ω at about t = 1 s, corresponding to nominal load current (power) levels of 1.385 A (250 W) and 2.770 A (500 W), respectively.

The most relevant among the acquired waveforms are presented in Figure 14. They are coherent with the simulation results and show that the converter's control system was also stable in this challenging condition. Furthermore, they are nearly identical to those of Figure 13; the main difference is the even smaller voltage variation, which determined slight differences in the load current and the external generator current. Before t = 0 s, the converter was controlled to draw −1.37 A from the DC microgrid recharging the emulated storage system. The external generator delivered 1.37 A to the converter and 1.385 A to the load for a total of 2.755 A. At t = 0 s, the current reference for the converter was increased to 1.37 A. Thus, the converter almost entirely supplied the load, and the current of the external generator automatically approached zero. Finally, at about t = 1 s, the load requested a current of 2.770 A. Since the converter's output was kept constant, the external generator automatically provided the additional current contribution of 1.385 A.

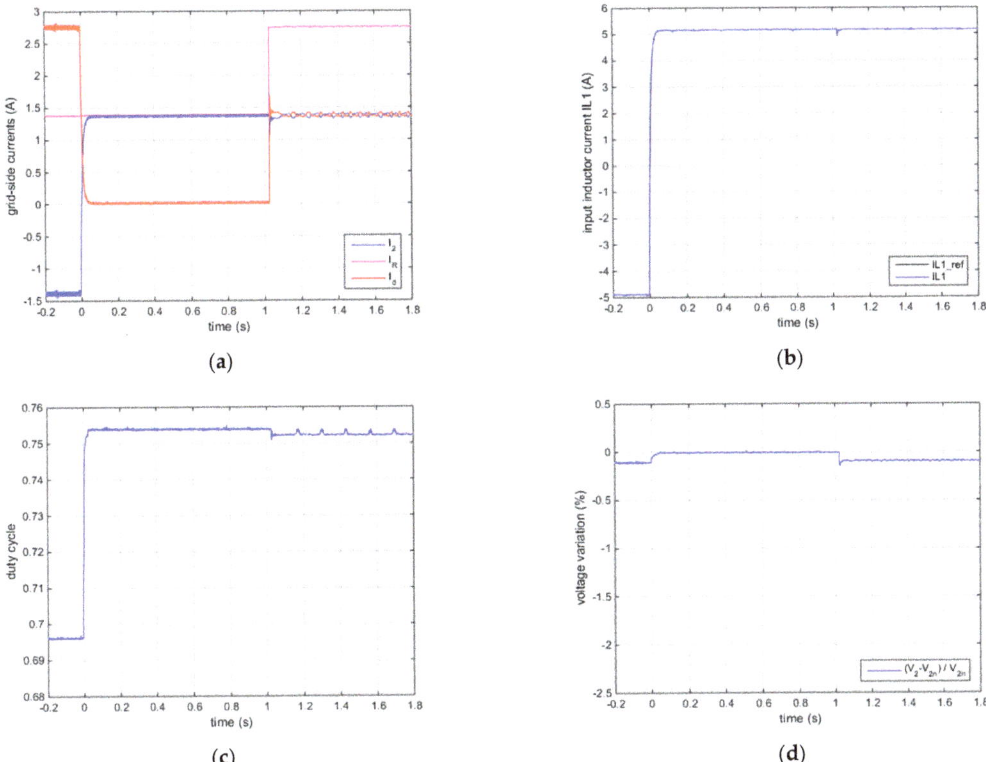

Figure 14. Experimental results in scenario #5 (SC−GS): (**a**) grid−side currents; (**b**) input inductor current; (**c**) duty cycle; (**d**) percentage variation of microgrid voltage.

In this scenario, the theoretical voltage variation was 0% because the DC microgrid was stiff. However, a small resistance was present due to cable connections. Nonetheless, the actual voltage variation was −0.1%, as shown in Figure 14d. Again, the system exhibited good dynamic behavior with fast aperiodic transients, as expected based on the simulation results.

6. Conclusions

The Split-pi converter is a good choice to integrate an energy storage system (ESS) into a DC microgrid. In part I of this work, five typical microgrid scenarios were identified. Each of them required a specific state-space model of the Split-pi and a suitable control scheme. Two different state-space models were devised for the Split-pi converter operating with lower storage-side voltage than grid-side voltage. Such models considered the parasitic elements of the converter. In addition, the most relevant transfer functions were given, together with control schemes with a different number of control loops and criteria to design the controllers.

The present paper completed the study validating the theoretical analysis based on comprehensive simulations and experimental tests. A 48 V, 750 W storage system interfaced with a 180 V DC microgrid using a Split-pi converter was chosen as a case study. It can represent a reduced-power prototype of terrestrial and marine microgrids. Several simulations were performed to assess the system performance in all the considered scenarios. Furthermore, a prototypal Split-pi converter was realized, and experimental tests were performed in conditions that covered all the scenarios. The obtained results were coherent with the simulations and validated the study.

Further activities will include theoretical studies and experiments with a twofold aim: (1) to devise the state-space model of the Split-pi operating with higher storage-side voltage than grid-side voltage; (2) to investigate the possibility of designing unconventional control systems for the Split-pi that are suitable for operating in more than one microgrid scenario.

Author Contributions: Conceptualization, M.L., A.S., M.P. and M.C.D.P.; methodology, M.L., A.S. and A.A.; software, M.L., A.S., A.A. and G.L.T.; validation, M.L., A.A., M.C.D.P. and M.P.; Writing—Original draft preparation, M.L., A.S. and M.P.; Writing—Review and Editing, M.L., A.A., M.C.D.P. and G.L.T.; supervision, M.P. and M.L.; funding acquisition, M.P. and M.L. All authors have read and agreed to the published version of the manuscript.

Funding: This research was funded by the Italian Ministry of University and Research (MUR), program PON R&I 2014/2020-Avviso n. 1735 del 13/07/2017-PNR 2015/2020, project "NAUSICA-NAvi efficienti tramite l'Utilizzo di Soluzioni tecnologiche Innovative e low CArbon," CUP: B45F21000680005

Institutional Review Board Statement: Not applicable.

Informed Consent Statement: Not applicable.

Data Availability Statement: Not applicable.

Conflicts of Interest: The authors declare no conflict of interest.

References

1. Anand, S.; Fernandes, B.G. Optimal voltage level for DC microgrids. In Proceedings of the IECON 2010—36th Annual Conference on IEEE Industrial Electronics Society, Glendale, AZ, USA, 7–10 November 2010; pp. 3034–3039.
2. Sharma, A.; Singh, S.N.; Srivastava, S.C. Optimal Selection of DC Microgrid Architecture. In Proceedings of the 8th IEEE Power India International Conference (PIICON 2018), Kurukshetra, India, 10–12 December 2018; pp. 1–5.
3. Wang, H.; Zhang, T.; Zhang, B. Selection of DC Voltage Level in household Hybrid AC/DC power supply system. In Proceedings of the 2019 IEEE 8th International Conference on Advanced Power System Automation and Protection (APAP), Xi'an China, 21–24 October 2019; pp. 1527–1530.
4. D.O.D. of United States of America. Military Std. MIL-STD-704F. In *Dep. Def. Interface Stand. Aircraft Electr. Power Charact*; D.O.D. of United States of America: Arlington County, VA, USA, 2004.
5. Ballou, P.J. A specialized ROV for inspection of salt dome caverns. In Proceedings of the IEEE Oceanic Engineering Society. OCEANS'98. Conference Proceedings (Cat. No.98CH36259), Nice, France, 28 September–1 October 1998; pp. 1584–1588.
6. Lin, M.; Lin, R.; Chen, L.; Yang, C.; Li, D. Underwater Docking System with Bidirectional Wireless Power Transmission Capability. In Proceedings of the Global Oceans 2020: Singapore—U.S. Gulf Coast, Biloxi, MS, USA, 5–30 October 2020; pp. 1–5.
7. Zadeh, M.K.; Zahedi, B.; Molinas, M.; Norum, L.E. Centralized stabilizer for marine DC microgrid. In Proceedings of the IECON 2013—39th Annual Conference of the IEEE Industrial Electronics Society, Vienna, Austria, 10–13 November 2013; pp. 3359–3363.
8. Wang, L. Dynamic analysis of a Microgrid system for supplying electrical loads in a sailing boat. In Proceedings of the IEEE Power Energy Society General Meeting, San Diego, CA, USA, 22–26 July 2012; pp. 1–7.
9. Xue, L.; Baxter, B.; Torrano, K.; Yao, G. A power management strategy of an onboard DC microgrid. In Proceedings of the IECON 2017—43rd Annual Conference of the IEEE Industrial Electronics Society, Beijing, China, 29 October–1 November 2017; pp. 6759–6764.
10. Kim, S.; Kucka, J.; Ulissi, G.; Kim, S.-N.; Dujic, D. Solid-State Technologies for Flexible and Efficient Marine DC Microgrids. *IEEE Trans. Smart Grid* **2021**, *12*, 2860–2868. [CrossRef]
11. Zeng, Y.; Zhang, Q.; Liu, Y.; Zhuang, X.; Lv, X.; Wang, H. Distributed Secondary Control Strategy for Battery Storage System in DC Microgrid. In Proceedings of the 2021 IEEE 4th International Electrical and Energy Conference (CIEEC), Wuhan, China, 28–30 May 2021; pp. 1–7.
12. D'Agostino, F.; Schiapparelli, G.P.; Silvestro, F.; Grillo, S. DC Shipboard Microgrid Modeling for Fuel Cell Integration Study. In Proceedings of the 2019 IEEE Power & Energy Society General Meeting (PESGM), Atlanta, GA, USA, 4–8 August 2019; pp. 1–5.
13. Di Piazza, M.C.; Luna, M.; Pucci, M.; La Tona, G.; Accetta, A. Electrical Storage Integration into a DC Nanogrid Testbed for Smart Home Applications. In Proceedings of the 2018 IEEE International Conference on Environment and Electrical Engineering and 2018 IEEE Industrial and Commercial Power Systems Europe (EEEIC / I&CPS Europe), Palermo, Italy, 12–15 June 2018; pp. 1–5.
14. Chub, A.; Vinnikov, D.; Kosenko, R.; Liivik, E.; Galkin, I. Bidirectional DC–DC Converter for Modular Residential Battery Energy Storage Systems. *IEEE Trans. Ind. Electron.* **2020**, *67*, 1944–1955. [CrossRef]
15. Sim, J.; Lee, J.; Choi, H.; Jung, J.H. High Power Density Bidirectional Three-Port DC-DC Converter for Battery Applications in DC Microgrids. In Proceedings of the 2019 10th International Conference on Power Electronics and ECCE Asia (ICPE 2019 - ECCE Asia), Busan, Korea, 27–30 May 2019; pp. 843–848.
16. Byrne, R.H.; Nguyen, T.A.; Copp, D.A.; Chalamala, B.R.; Gyuk, I. Energy Management and Optimization Methods for Grid Energy Storage Systems. *IEEE Access* **2018**, *6*, 13231–13260. [CrossRef]

17. La Tona, G.; Luna, M.; Di Piazza, A.; Di Piazza, M.C. Towards the real-world deployment of a smart home EMS: A DP implementation on the Raspberry Pi. *Appl. Sci.* **2019**, *9*, 2120. [CrossRef]
18. Accetta, A.; Pucci, M. Energy Management System in DC Micro-Grids of Smart Ships: Main Gen-Set Fuel Consumption Minimization and Fault Compensation. *IEEE Trans. Ind. Appl.* **2019**, *55*, 3097–3113. [CrossRef]
19. Crocker, T.R. Power Converter and Method for Power Conversion. U.S. Patent No. 20040212357A1, 28 October 2004.
20. Khan, S.A.; Pilli, N.K.; Singh, S.K. Hybrid Split Pi converter. In Proceedings of the 2016 IEEE International Conference on Power Electronics, Drives and Energy Systems (PEDES), Trivandrum, India, 14–17 December 2016; pp. 1–6.
21. Sobhan, S.; Bashar, K.L. A novel Split-Pi converter with high step-up ratio. In Proceedings of the 2017 IEEE Region 10 Humanitarian Technology Conference (R10-HTC), Dhaka, Bangladesh, 21–23 December 2017; pp. 255–258.
22. Ahmad, T.; Sobhan, S. Performance analysis of bidirectional split-Pi converter integrated with passive ripple cancelling circuit. In Proceedings of the 2017 International Conference on Electrical, Computer and Communication Engineering (ECCE), Cox's Bazar, Bangladesh, 16–18 February 2017; pp. 433–437.
23. Singhai, M.; Pilli, N.; Singh, S.K. Modeling and analysis of split-Pi converter using State space averaging technique. In Proceedings of the 2014 IEEE International Conference on Power Electronics, Drives and Energy Systems (PEDES), Mumbai, India, 16–19 December 2014; pp. 1–6.
24. Maclaurin, A.; Okou, R.; Barendse, P.; Khan, M.A.; Pillay, P. Control of a flywheel energy storage system for rural applications using a Split-Pi DC-DC converter. In Proceedings of the 2011 IEEE International Electric Machines & Drives Conference (IEMDC), Niagara Falls, ON, Canada, 15–18 May 2011; pp. 265–270.
25. Sabatta, D.; Meyer, J. Super capacitor management using a Split-Pi symmetrical bi-directional DC-DC power converter with feed-forward gain control. In Proceedings of the 2018 International Conference on the Domestic Use of Energy (DUE), Cape Town, South Africa, 3–5 April 2018; pp. 1–5.
26. Monteiro, V.; Oliveira, C.; Rodrigues, A.; Sousa, T.J.C.; Pedrosa, D.; Machado, L.; Afonso, J.L. A Novel Topology of Multilevel Bidirectional and Symmetrical Split-Pi Converter. In Proceedings of the 2020 IEEE 14th International Conference on Compatibility, Power Electronics and Power Engineering (CPE-POWERENG), Setubal, Portugal, 8–10 July 2020; pp. 511–516.
27. Alzahrani, A.; Shamsi, P.; Ferdowsi, M. Single and interleaved split-pi DC-DC converter. In Proceedings of the 2017 IEEE 6th International Conference on Renewable Energy Research and Applications (ICRERA), San Diego, CA, USA, 5–8 November 2017; pp. 995–1000.
28. Luna, M.; Sferlazza, A.; Accetta, A.; Di Piazza, M.C.; La Tona, G.; Pucci, M. Modeling and Performance Assessment of the Split-Pi Used as a Storage Converter in All the Possible DC Microgrid Scenarios. Part I: Theoretical Analysis. *Energies* **2021**, *14*, 4902. [CrossRef]
29. Khooban, M.H.; Gheisarnejad, M.; Farsizadeh, H.; Masoudian, A.; Boudjadar, J. A New Intelligent Hybrid Control Approach for DC–DC Converters in Zero-Emission Ferry Ships. *IEEE Trans. Power Electron.* **2020**, *35*, 5832–5841. [CrossRef]

Article

High-Performance Power Converter for Charging Electric Vehicles

Nikolay Madzharov [1] and Nikolay Hinov [2],*

[1] Department of Electronics, Faculty of Electrical Engineering and Electronics, Technical University of Gabrovo, 4 H. Dimitar, 5300 Gabrovo, Bulgaria; madjarov@tugab.bg
[2] Department of Power Electronics, Faculty Electronic Engineering and Technology, Technical University of Sofia, 8, Kliment Ohridski Blvd., 1000 Sofia, Bulgaria
* Correspondence: hinov@tu-sofia.bg; Tel.: +359-2965-2569

Abstract: This paper presents the analysis, modeling, simulation and practical studies of resonant inverters with a voltage limitation on the resonant capacitor. The power circuits obtained in this way are characterized by the fact that the power consumption does not depend on the load changes, but is a function of the operating frequency, the value of the resonant capacitor and the supply voltage—these are the so-called inverters with energy dosing. Analytical dependences, simulations and experimental results were determined, which described the behavior of the studied power electronic devices. The obtained expressions for the inverter current in the different stages of the converter operation were the basis for the creation of the engineering methodology for their design and prototyping. Based on the derived basic ratios and characteristics, the capabilities of these devices for self-adaptation to the needs and changes of the load were demonstrated. A comparison of the characteristics of classical resonant inverters and those with energy dosing was made, thus demonstrating their qualities and advantages. The presented results display the properties of this class of circuits and the challenges to their effective application to find the optimal solution for the implementation of charging stations for different specific needs. On the other hand, the limitations in the use of these circuits were that no power was consumed from the power supply during the whole period, the lack of limitation of the maximum current through the transistors and the need for sufficient time to dissipate energy in the resonant inductor when working with high-resistance and low-power loads.

Keywords: charging stations; energy dosing; resonant converters; electric vehicles

Citation: Madzharov, N.; Hinov, N. High-Performance Power Converter for Charging Electric Vehicles. *Energies* **2021**, *14*, 8569. https://doi.org/10.3390/en14248569

Academic Editor: Massimiliano Luna

Received: 3 November 2021
Accepted: 15 December 2021
Published: 19 December 2021

Publisher's Note: MDPI stays neutral with regard to jurisdictional claims in published maps and institutional affiliations.

1. Introduction

Mobility is one of the most important characteristics of a modern and smart society. The dynamics and nature of human relations in recent years are unthinkable without the constant movement of people, goods and capital. Environmental problems and challenges have necessitated a new concept for the provision of transport connections and systems through the implementation and growing dominance of electric vehicles. Regarding this aspect, there are several main obstacles to the distribution of electric vehicles: the provision of the necessary electricity, the underdeveloped infrastructure of the electricity transmission network with its insufficient capacity, minimizing the impact of the charging infrastructure on the energy transmission network and the development of a new class of power electronic devices and systems designed to charge energy storage elements [1–3]. Despite the diversity of these research problems, what they have in common is that their sustainable and effective solution is related to the development of power electronics. The aim of the present work was to present the capabilities of a class of power circuits of electronic energy converters, known as resonant converters with energy dosing, which, due to their unique properties and characteristics, are very suitable for the realization of charging stations [4,5].

2. Overview of Different Methods and Power Converters for the Realization of Charging Stations

The development of power electronics has led to a wide variety of applications and, accordingly, topologies and operating modes of power electronic converters and systems. In particular, the development of power electronic devices with applications for charging electric vehicles can be divided into two main groups: direct and contactless charging [6–12].

Characteristic of both large groups is the use of resonant transducers in the process of converting electricity. This is due to some of their main advantages over other types of converters, namely [13,14]:

- They work with electrical signals in a shape close to a sinusoid and correspondingly low content of harmonics;
- The realization of soft switching and hence of high efficiency of the devices in a natural way without the participation of additional elements;
- The use of parasitic elements as the main reason for the realization of resonant circuits and processes.

On the other hand, the use of resonant converters is associated with several difficulties and challenges, such as:

- The use of variable frequency control, which creates preconditions for difficult elimination of electromagnetic interference;
- The strong dependence of the operating modes on the tolerances of the building elements that participate in the resonant circuit;
- The incomplete use of power by semiconductor elements.

Addressing these challenges is usually done in two ways: either by improving the methods of and synthesis of power electronic systems control, including the use of artificial intelligence techniques [15–19] or by proposing power schemes that are weakly dependent on the operating modes of the load changes and have possibilities for self-adjustment of the converter to the requirements of the specific application [20–24].

This study proposed the use of resonant converters with improved characteristics that eliminate the most significant part of these shortcomings through the use of voltage-limiting circuits (fully or partially) on the resonant capacitor. These schemes have gained wide popularity in the specialized literature, such as energy dosing schemes. The present research involved the development of the long-term work of the authors in this field and was aimed at the implementation of this class of converters for the purpose of charging stations.

3. Basic Relations in the Analysis of Resonant Converters with Energy Dosing

The method of energy dosing (ED) has been used in several applications of AC and DC power supplies, where their use initially began in industrial technologies based on induction heating [4,5,24,25]. The loads of the converters in these applications are characterized by highly variable parameters, which can often vary from idle to short-circuiting. These properties of ED converters make them very suitable for the development of contactless charging stations for static and dynamic charging of electric vehicles. Their stable operation is obtained by fulfilling the following condition:

$$k_S = (dU_T/dI_T - dU/dI) > 0 \tag{1}$$

where k_S is the coefficient of stability of the system, while dU_T/dI_T and dU/dI are the dynamic resistances of the load and the DC-DC converter, respectively.

Figure 1 shows a power scheme of a half-bridge DC-DC converter with energy dosing. It consists of a half-bridge resonant inverter with energy dosing (RI with ED) without reverse diodes, a high-frequency matching transformer and an output rectifier with a capacitive filter and an equivalent load.

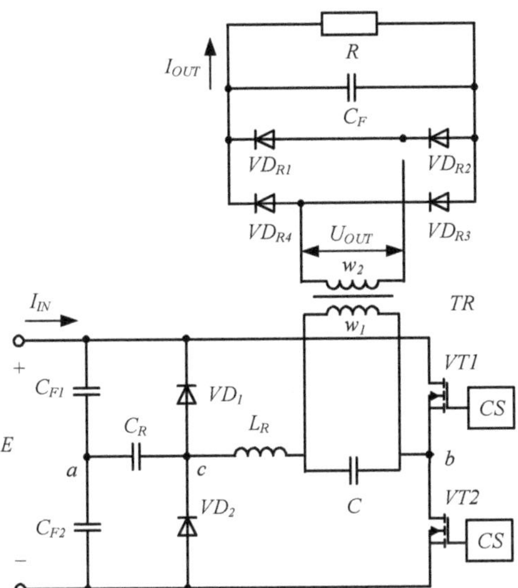

Figure 1. Half-bridge DC–DC converter with energy dosing.

Of the possible operating modes, the most suitable when using the circuit is the mode when the operating frequency is less than the resonant frequency of the alternating current (AC) circuit of RI with ED, i.e., $f < f_0$. Figure 2 shows the time diagrams of the current through the resonant inductor L_R, the voltage of the resonant capacitor C_R, the output current and the voltage of the transistor VT_1, which clarifies the principle of operation. It should be noted that the transistors operate with zero on and off current (ZCS).

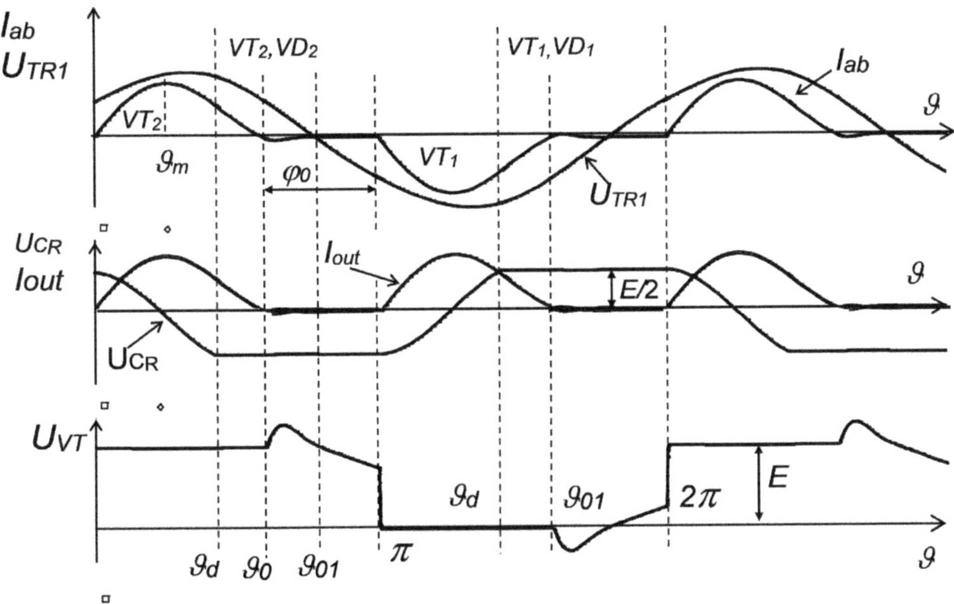

Figure 2. Basic timing diagrams of a half-bridge DC–DC converter with energy dosing.

The indicated time diagrams show that in the operation of the power circuit during each half-period the following intervals are distinguished: $0 \div \vartheta_d$, $\vartheta_d \div \vartheta_0$, $\vartheta_0 \div \vartheta_{01}$ and $\vartheta_{01} \div \pi$. On the other hand, for the analysis of the resonant inverter, it is important to distinguish between the intervals $(0 \div \vartheta_d)$ in which energy is consumed from the power source and those in which the energy accumulated in the resonant elements supplies the output of the circuit. The operation of the considered power scheme for one of the two half-periods of its operation $(0 \div \pi)$ is explained in detail using a modal diagram, which is shown in Figure 3. In order to achieve convenient ratios, the following assumptions are made: the active and passive elements are ideal, and the times for their commutation are neglected. In addition, the processes in the scheme are considered after reaching a set mode of operation when there is periodicity and repeatability of the state variables.

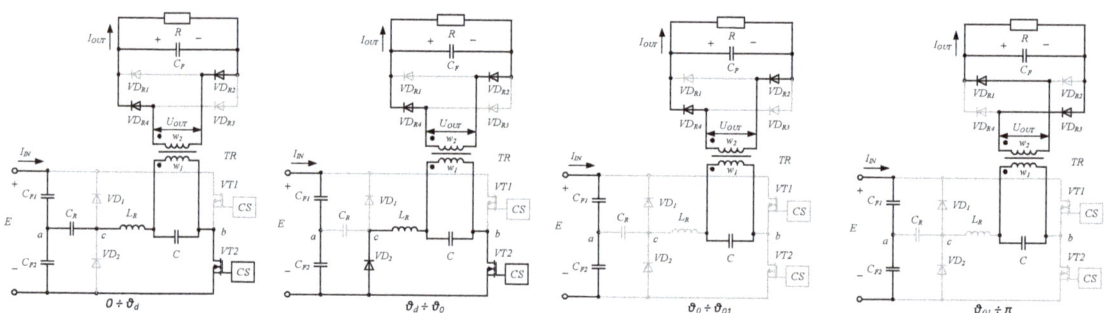

Figure 3. Modal diagram of a half-bridge DC–DC converter with energy dosing.

In the analysis of the circuit, the secondary circuit is brought to the primary circuit; as a result of which, the output capacitor, the load and the transformer are replaced by a voltage source nU_{OUT}, where n is the transformation coefficient ($n = W_1/W_2$). The behavior of the converter is determined mainly by the values of the elements L_R and C_R, and more generally, by the following parameters [5,24,25]:

- Characteristic impedance:

$$Z_0 = \sqrt{L_R/C_R} \tag{2}$$

- Natural resonant frequency of a resonant circuit composed of ideal elements:

$$\omega_0 = \sqrt{1/L_R C_R} \tag{3}$$

- Period of the resonant frequency:

$$T_0 = 2\pi/\omega_0 \tag{4}$$

- Control frequency period—T;
- Fill factor (duty cycle):

$$\gamma = T_0/2T \tag{5}$$

If the losses in the elements of the converter are ignored, then its input power will be equal to the output. Taking this assumption into account, the following electrical ratios for the input current and output voltage can be recorded as follows:

$$I_{IN} = I_{OUT}/2n \qquad U_{OUT} = E/2n \tag{6}$$

Diodes VD_1 and VD_2 start to conduct when the maximum and minimum values of the voltage on the capacitor C_R becomes equal to $+E/2$ or $-E/2$. If the load resistance or operating frequency has values that are significantly different from the nominal ones, the capacitor C_R will not be recharged from $-E/2$ to $+E/2$ and the diodes VD_1 and VD_2

will not turn on. In this case, the voltage to which the capacitor will be recharged can be determined using the expressions from Equations (2) to (6).

$$U_{CR} = E/4 + \pi E Z_0 / 8\gamma R n^2 \tag{7}$$

The latter expression has a very important practical meaning. It displays the ratio for the values of the load resistance at which the energy dosing mode is performed, i.e., limiting the voltage on the capacitor C_R to $\pm E/2$ by VD_1 and VD_2, namely,

$$R \leq \pi Z_0 / 2\gamma n^2 \tag{8}$$

This mode, which is considered optimal, is characterized by two main intervals: the consumption of energy from the power supply and the short circuit of the alternating current circuit when energy from the power supply is not consumed.

3.1. Electromagnetic Analysis of the Converter at the Time of Energy Consumption from the Power Source $0 \div \vartheta_d$

During the interval $0 \div \vartheta_d$ (Figure 2), only the transistor VT_2 is unlocked and the equivalent circuit consists of the power supply connected in series, the reduced load circuit and the resonant inductor and capacitor. The natural frequency of the AC circuit (a-b in Figure 2) must be higher than the control frequency, i.e., $\omega_0/\omega > 1$, in order to obtain a tendency for the current to drop to zero at the beginning of the interval φ_0. The following frequency ratio is valid for the considered operating mode:

$$\frac{\omega_0}{\omega} = \frac{\pi}{\pi - \varphi_0 - (0.65 \div 1.7)\varphi_0} > 1 \tag{9}$$

$$\text{i.e., } \omega_0/\omega = 1.2 \div 1.4 \tag{10}$$

Taking into account the expression for the current through the resonant inductor:

$$i_{L_R}(\vartheta) = \frac{E}{\omega_0 L_R} \exp^{-\vartheta/2Q} \sin \frac{\omega_0}{\omega}\vartheta \tag{11}$$

and the voltage across the resonant capacitor C_R:

$$u_{CR}(\vartheta_d) = \frac{1}{\omega_0 C_R} \int_0^{\vartheta_d} i(\vartheta) = \frac{E}{2} \tag{12}$$

as determined at the end of the interval of energy consumption from the power source, the moment that corresponds to the angle ϑ_d is

$$\vartheta_d = [\pi - arctg(2Q\omega_0/\omega)]/(\omega_0/\omega) \tag{13}$$

where $Q = \omega L_R / R_E$ is the quality factor of the resonant circuit, R_E is the equivalent resistance of the AC circuit between points b and c of Figure 1 and $\vartheta = \omega t$.

3.2. Operation of the Converter in the Short Circuit Interval of the Alternating Current Circuit $\vartheta_d \div \vartheta_0$

At the moment corresponding to the angle ϑ_d, the resonant capacitor C_R is charged to voltage $-E/2$ and the diode VD_2 is turned on. In essence, the electromagnetic processes in the second interval $\vartheta_d \div \vartheta_0$ are aperiodic. The expression for the current through the resonant inductor is of the form

$$i_{LR}(\vartheta) = i_{LR}(\vartheta_d) \exp^{-\frac{R}{\omega L_R}\vartheta} \tag{14}$$

where $R = R_E$.

It is very important from a practical point of view to determine the value of the current at the end of the interval (moment corresponding to an angle $\pi - \varphi_0$) because the level of switching losses in the transistors depends on it. It is calculated using the current expression $i_{LR}(\vartheta)$ at $\vartheta = \pi - \varphi_0$. In this regard, additional information about the processes in the circuit is given by the expressions for the voltages across the resonant inductor and the equivalent load:

$$u_{LR} = -L di_{LR}/dt \tag{15}$$

$$u_{OUT} = R\left[n i_{LR}(\vartheta_d)\exp^{-\vartheta/Q}\right] \tag{16}$$

From the presented relations, it is clear that if at the moment corresponding to an angle $\pi - \varphi_0$, the resonant current has zero value, then the voltages U_{LR} and U_{OUT} will also have zero values. At this point, the voltage on the transistors will also have a zero value, i.e., they will turn on at zero current and will turn off at zero current and zero voltage.

3.3. Stabilization and Regulation of the Output Power and Voltage of the Converter

The energy of the capacitor when recharging from $-E/2$ to $+E/2$ is equal to

$$W = C_R E^2/2 \tag{17}$$

When transistors VT_1 or VT_2 are turned on, this energy is transmitted to the load, i.e.,

$$C_R E^2/2 = U_{OUT} I_{OUT} T/2. \tag{18}$$

In this case, the power of the converter P for one period is expressed by the ratio

$$P = E^2 f C_R = E I_d = U_{OUT} I_{OUT} = const, \tag{19}$$

The first conclusion that can be drawn from the expressions for W and P is that at a constant operating frequency, supply voltage and capacitance of the resonant capacitor, the transmitted power in the load is constant and does not depend on its parameters. Maintaining a constant power means that the output DC voltage U_{OUT} is self-aligning with the load parameters.

To set the power level in accordance with Equation (19), the value of the capacitor C_R is most often changed. For this purpose, the developed electronic keys are used, for which there are author's claims regarding their use for similar purposes. They consist of only one transistor (IGBT or MOSFET) with a reverse diode and a series-connected capacitor C_R, the capacitance of which is in accordance with the desired power (Equation (19)). Figure 4 shows the schematic diagram of four electronic switches, which participate in the circuit of the converter of Figure 1 and are connected to points a and c. The presented circuit contains half as many elements as the traditional electronic switch, while filled with two oppositely connected transistors with reverse diodes and, therefore, has less static losses compared to other embodiments [24,25].

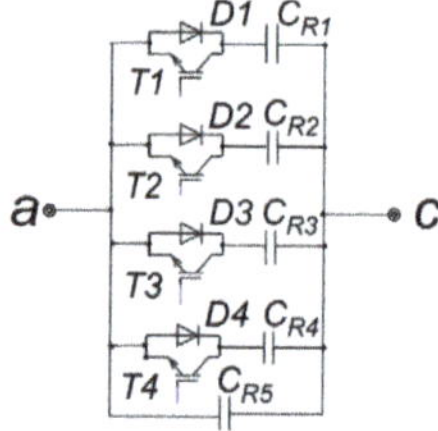

Figure 4. Electronic semiconductor switches for setting the power level.

The combinations between the turned on and turned off capacitors make it possible to set fifteen power values in the range $0 \div P_{NOM}$. No galvanically separated signals are required to control the individual transistors.

It is important to note that the switching of the capacitors C_{R1}–C_{R4} from Figure 4 can be done during the operation of the converter without receiving current and voltage overloads from the transistors.

The second characteristic feature of the converter is obtained by substituting the expression for the load current $I_{OUT} = U_{OUT}/R$ into Equation (19). In this way, a ratio is obtained that shows the relationship between the input and output voltage:

$$U_{OUT} = E\sqrt{C_R.R/T} \tag{20}$$

The conclusion to be drawn from this expression is that by changing the operating frequency, the output voltage can be invariably maintained when the value of the load and/or the input voltage changes. Figure 5 shows the dependence of the output voltage in relative units as a function of frequency at different loads. The information from this characteristic is used in the design of the converter because it takes into account, on the one hand, the relationship between the value of the load and the capacitance of the capacitor C_R setting the power and, on the other, the dependence of output voltage on frequency and input voltage.

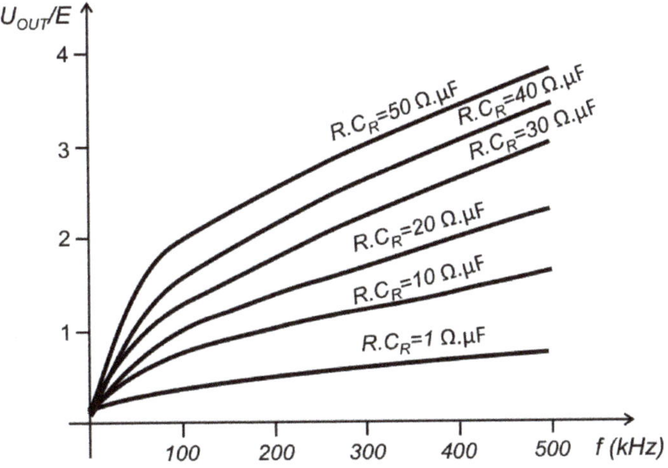

Figure 5. Control characteristics of a converter with energy dosing.

The regulation or stabilization of the output voltage is carried out via feedback, changing the operating frequency of the converter. The analytical dependence of the regulation law can be deduced by differentiating the output power expression with respect to the control frequency f. After some transformations using the expressions:

$$U_{OUT}\, I_{OUT} = C_R E^2 f, \tag{21}$$

$$U_{OUT}\, di_{OUT} + I_{OUT}\, du_{OUT} = C_R E^2 df, \tag{22}$$

$$du_{OUT} = di_{OUT}(R//1/\omega\, C_F) \tag{23}$$

$$I_{OUT} = U_{OUT}/R \tag{24}$$

it is found that
$$\frac{du_{OUT}}{df} = \frac{E}{2}\sqrt{\frac{C_R R}{f}}\,\frac{1}{1+\omega C_F R/2}. \tag{25}$$

This expression represents the transfer function of the control system and represents the law according to which the operating frequency must be changed when the load parameters change in order to stabilize the constant output voltage.

4. Investigation and Comparison of the Characteristics and Converters with and without Energy Dosing Used in the Fast Charging Stations of Electric Vehicles

Based on the analytical expressions obtained in the analysis of the studied scheme (Figure 1), its characteristics were obtained. For greater objectivity of the formulated conclusions, the characteristics of RI with ED without reverse diodes (RD) of power transistors (PT) and two other competing schemes that are often used in charging stations as high-frequency sources [20–22,26–29]: a half-bridge RI with RD and a full-bridge parallel current fed inverter (PCFI). The use of the same input data for the indicated schemes allows for objective comparisons and definitions of their advantages and disadvantages. In Tables 1–3 and the numerical values from the research are presented, as in the column "Type", the conditions under which the results are obtained are given. In these studies, emphasis was placed on the behavior of the circuits when changing the equivalent load parameters, which was achieved by varying the air gap of the contactless transmitter and/or the load resistance. Specifically for each scheme, they were the following:

- For the RI with ED—"0"—nominal mode, the following tests were done when the load parameters change compared to the nominal ones under the following conditions: "1"—without the RD of the PT and the operating frequency equal to the nominal one; "2"—with the RD of the PT and the operating frequency equal to the nominal; "3"—with the phase between the current and the voltage of the parallel load circuit, other than for a pause φ_0 when changing the operating frequency; and "4"—with the phase between the current and the voltage of the parallel load circuit equal to the pause φ_0 when changing the operating frequency.
- For the RI with RD—"0"—nominal mode, the following tests were done when the load parameters changed compared with the nominal ones under the following conditions: "1"—at the operating frequency equal to the nominal one and "2" and "3"—with a change in operating frequency;
- For the PCFI—"0"—nominal mode, the following tests were done when the load parameters changed compared to the nominal ones under the following conditions: "1"—at the operating frequency equal to the nominal one and "2"—with a change in the operating frequency.

The main conclusions that can be drawn were about the reliable operation of the RI with ED, which for these modes, was determined by the conditions for the switching PT. The most favorable were those in which the current and/or voltage were zero (ZCS, ZVS, ZCSZVS). Depending on the equivalent parameters of the AC circuit, the converters operated in one of three modes. If the switching mode at zero current was conditionally accepted as nominal (basic) due to the change of the load or the operating frequency, different changes in the values of the main electrical and phase ratios were observed.

Table 1. Characteristics of the RI with ED.

	Resonant Inverter with Energy Dosing, Characteristics at the Following Element Values: $C_{R1} = C_{R2} = 1.5$ µF, $L_R = 11.1$ µH, $C = 26$ µF																		
R_T (Ω)	0.025			0.0375			0.0425			0.0575			**0.05**	0.0625			0.075		
L_T (µH)	1.15			1.725			1.955			2.645			**2.3**	2.875			3.45		
Type	1	2	3	4	1	2	3	4	1	2	3	4	**0**	1	2	3	4	1	2
P (kW)	15	4	15	20	15	8.5	15	18	15	12	15	17	**15**	15	9.9	14.6	14.5	15	10.6
f (kHz)	20	20	27	28.3	20	20	22	23.5	20	20	21.2	22	**20**	20	20	19.5	19.6	20	20
U_{Cm} (V)	73	73	224	270	152	149	216	256	192	191	231	248	**236**	252	209	224	225	258	217
I_M (A)	280	280	151	148	182	183	154	165	161	162	152	162	**159**	213	175	159	158	251	210
I_{ON} (A)	0	0	0	0	0	0	0	0	0	0	0	0	**0**	0	0	0	0	0	0
I_{OFF} (A)	207	209	120	50	140	146	79	0	97	90	41	0		0	0	8	0	0	0

Table 2. Characteristics of the RI with an RD.

	Resonant Inverter with Reverse Diodes, Characteristics at the Following Element Values: $L_R = 28\ \mu H$, $C_{R1} = C_{R2} = 0.86\ \mu F$, $C = 15\ \mu F$, $C_F = 2000\ \mu F$																		
$R_T\ (\Omega)$	0.045			0.675			0.0765			*0.09*	0.1035			0.1125			0.135		
$L_T\ (\mu H)$	2.05			3.075			3.485			*4.1*	4.715			5.125			6.15		
Type	1	2	3	0	2	3	1	2	3	*0*	1	2	3	1	2	3	1	2	3
P (kW)	20.4	15	15	15	15	16.4	33.1	15	15	*15*	8.4	15	15	6.5	15	15	3.6	10.5	15
f (kHz)	20	19.8	23.7	20	18.2	23.5	20	22.3	17.6	*20*	20	18.2	23.5	20	17	23.5	20	15.5	23.9
U_{Cm} (V)	2510	2251	1293	677	1620	656	1252	656	1486	*919*	670	739	1547	738	827	1711	804	800	1800
U_{OBP} (V)	249	209	249	305	235	321	411	306	246	*548*	242	289	400	222	288	395	178	241	355
I_M (A)	482	420	272	136	270	107	207	109	238	*31*	101	111	327	104	126	370	127	109	393
I_{ON} (A)	450	390	0	15	226	0	0	18	187	*28*	54	14	311	69	30	350	109	33	370
I_{OFF} (A)	0	0	270	0	0	0	45	0	0	*28*	0	0	0	0	0	0	0	0	0

Table 3. Characteristics of the PCFI.

	Full-Bridge Parallel Current Fed Inverter, Characteristics at the Following Element Values: $L_1 = 1.6$ mH, $C = 1.73\ \mu F$												
$R_T\ (\Omega)$	0.432		0.648		0.743		*0.86*	0.994		1.08		1.296	
$L_T\ (\mu H)$	19.9		29.85		33.83		*39.8*	45.87		49.75		59.7	
Type	1	2	1	2	1	2	*0*	1	2	1	2	1	2
P (kW)	127.5	15	25.5	15	13	15	*15*	32.5	15	50.5	15	99	15
f (kHz)	20	26.05	20	23	20	21.65	*20*	20	18.7	20	18	20	16.5
U_{Cm} (V)	1800	876	1100	922	816	923	*919*	1455	924	1919	935	2894	938
U_{OBP} (V)	-	-	-	-	0	500	*548*	1248	559	1800	550	2875	551
I_M (A)	255	31	52	31.2	27	31.2	*31*	68	31	103	31.7	198	31.7
I_{ON} (A)	250	29.5	50	28.9	26	29	*28*	62	28	95	27.5	186	28
I_{OFF} (A)	250	29.5	50	28.9	26	29	*28*	62	28	95	27.5	186	28

The other conclusions were as follows:

(1) As the load parameters (R) decreased, the load circuit and the equivalent AC circuit gained an inductive misalignment. For this reason, the frequency of the RI's own oscillations decreased. The current of the PT, when it was switched off by the control pulses, was different from zero, and this created conditions for switching on the reverse diode of the inoperative transistor. The described phenomenon is the reason for the partial reduction of the RI power, as the second recharging of the resonant capacitors in the considered half-period began. As the operating frequency increased, the power could be set equal to the nominal one, and at the same time, the effect of reducing the switching current of the PT-I_{OFF} was obtained. When the resonance in the load oscillating circuit was reached again, both currents I_{OFF} and I_{ON} became equal to zero. Such a favorable development of the processes was observed when reducing the load parameters to 25–30%. With an even greater reduction in the load, for example, by 50% to achieve $I_{ON} = I_{OFF} = 0$, it was necessary to increase the frequency, which led to an increase in power by 30% above the nominal value (Table 1).

(2) When the inverter was unloaded (R increased), the load oscillating circuit was capacitively disrupted. The frequency of the inverter's natural oscillations increased and the current of the PT naturally became equal to zero before the termination of the control pulse. For this reason, conditions were created for the occurrence of a mode with a natural shutdown of the PT until the power drops, as noted above. By reducing the operating frequency until the resonance in the load oscillating circuit was restored, the ZCS mode could be re-established ($I_{ON} = I_{OFF} = 0$), but the power, although insignificant, was reduced (by 5–10%) in accordance with Equation (19).

From the point of view of the operational characteristics and reliability of RI with ED, it is important that in all operating modes, it is not possible to exceed the set power determined by Equation (19). In practice, energy dosing protects the PT from overload and ensures the natural adaptation and self-alignment of the RI to the load. This, in turn, determines a major positive quality of an RI with ED that is not possessed by a traditional RI.

In the case of similar changes in the load, the circuit of the RI with ED without an RD of the PT always worked with the nominal power without the need to change the operating frequency. The technical applicability of this circuit variant was determined mainly by the voltage on the PT, which was obviously greater than that in the RI with ED and the OD of the PT.

Another important result of the presented characteristics should be noted, which was that switching the PT currents on and off in both cases (when increasing and decreasing R) could be maintained equal to zero if the operating frequency changes in one direction or another until resonance was obtained in the load circle (Table 1). The fact that the most favorable modes in terms of power and currents of switching on and switching off of the PT were obtained by the resonance of the load circuit, facilitates the synthesis of control of the power electronic converter.

The second group of conclusions was about the qualities of the other two schemes and their comparisons with the RI with ED: (1) In the half-bridge RI with RD, with the change of the load parameters, the power changed more strongly (Table 2), and to return to the nominal value, it was necessary to regulate the frequency more deeply, which significantly increased the voltage on the resonant capacitors U_{CRm} and currents on and off (i.e., the PT-I_{ON} and I_{OFF}). Generally speaking, this scheme could not cover a range of variation of the load parameters greater than 15–20% while maintaining a favorable operating mode of the PT. (2) In the full-bridge PCFI, there was an even stronger change in power and the need for deeper regulation of the frequency for its stabilization (Table 3). As R increased, the voltage of the oscillating circuit U_{Cm} increased and, accordingly, the forward and reverse voltages of the semiconductor devices increased. In some of the modes, the circuit time for PT recovery disappeared. Based on the presented results, it was concluded that both alternative RIs had significantly less coordination (regulatory) capabilities than the RI with ED.

A visual comparison is shown in Figure 6, where the characteristics of the output power changing for the load parameter variations for the full-bridge-current-fed inverter (curve 1), half-bridge RI (curve 2), RI with ED and reverse diodes (curve 3) and RI with ED without reverse diodes (curve 4).

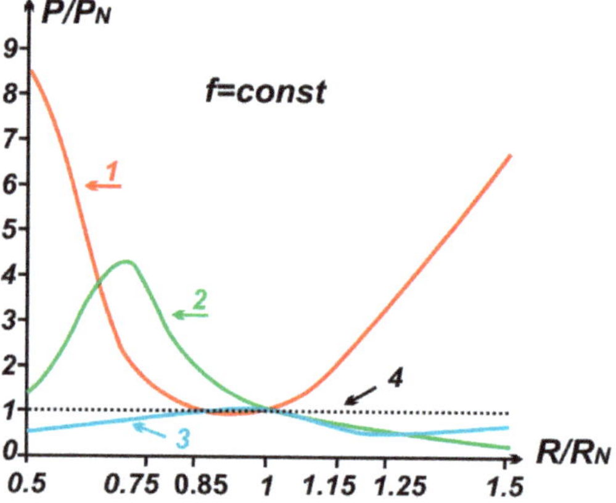

Figure 6. Comparison of the output characteristics P/PN = f(R/RN) of the full-bridge-current-fed inverter (1), half-bridge RI (2), RI with ED and reverse diodes (3) and RI with ED without reverse diodes (4).

An important point here is that the last two characteristics (curves 3 and 4 in Figure 6) guaranteed practical adaptively of the inverter to the load and its changes, due to which, it could also operate without regulation when the load changes within the limits stated above, i.e., from 0.5 to 1.5 times relative to the nominal value. This contributes to a significant improvement in the technical and operational qualities of the high-frequency source and provides the opportunity to work with a wide range of loads, which is typical for contactless charging stations for motorcycles and electric vehicles. Based on these characteristics, it was concluded that energy dosing schemes are advantageous over other types of resonant converters.

On the other hand, according to the characteristics of Figure 6 for the most frequently used topologies based on resonant inverters with reverse diodes (most often LLC), the conclusions made in [30–32] regarding the need to use optimal control in order to realize the charging cycles were confirmed.

5. Experimental Results

With the methodology thus created, a full-bridge RI with ED, used for a contactless charging station of a motoped, was designed and studied. It is shown in Figure 7 and consisted of the following main blocks: an electronic converter (inverter), together with the control system and drivers; a power supply; a measuring system; and a load composed of a transmitter and receiver. The power circuit of the inverter was implemented according to Figure 1.

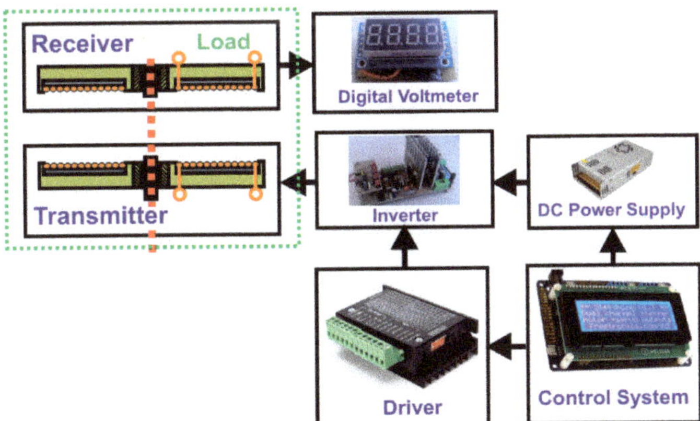

Figure 7. Block diagram of the experimental setup.

During the implementation of the experiment, the following input data were used: output power P = 2.5 kW, operating frequency of the inverter f = 30 kHz, power supply voltage E = 300 V, resonant capacitor C_R = 0.95 µF, resonant inductor L_R = 16.31 µH, W_1/W_2 = 1, ω_0/ω = 1.2–1.4 and equivalent load R = 8.9 Ω.

Figure 8 shows a photo of the stand on which the experimental results were obtained.

Figure 8. A PI with ED that was used for the realization of a contactless charging station of a motoped.

Furthermore, simulation studies of the system were performed with these data. In Table 4, the results obtained from the stand are compared with those from a program for the simulation of power electronic devices. The table shows that RI actually had dosing properties, which was determined by the average value of the consumed current I_0, equal to the input current $I_{in} = P/E = 8.3$ A, which was practically the same value from the analysis and the computer experiment.

Table 4. Results from the design, the computer experiment and the practical research.

Value	ϑ_d, (°)	I_{in} (A)	U_{out} (V)	I_{out} (A)	I_{VTm} (A)	I_{VT} (A)	I_{VD} (A)
Calculated	92.1	8.3	148	16.6	42.9	9.3	1.02
Simulation	93.5	8.1	145	16.2	41.1	9.1	1
Stand	95	8.5	147	16.5	43.6	9.6	1.1

This basic property of the circuit was also confirmed by the load voltage U_{outm}, which also coincided with the calculations and the experiment. Obtaining the set power could be shown by the following result of the table, corresponding to the physical processes in RI with ED, namely, that the difference between the currents of the transistors and diodes, i.e., I_{VT} and I_{VD}, were equal to I_{in} and corresponded to the energy consumption of energy in the interval $\vartheta = 0 \div \vartheta_d$ and a short circuit of the supply DC bus in the interval $\vartheta = \vartheta_d \div \pi - \varphi_0$. In addition, compliance with the boundary conditions for switching and periodicity could be used as an additional criterion for the reliability of the obtained results. In this case, they were expressed as obtaining the set operating mode of the RI in which the PT turned on and off at zero current (ZCS) (see Figure 9a). It should also be noted that in the calculations, the parameters of the current pulse through the key devices were obtained with great reliability, as the difference with the measured values did not exceed 5%.

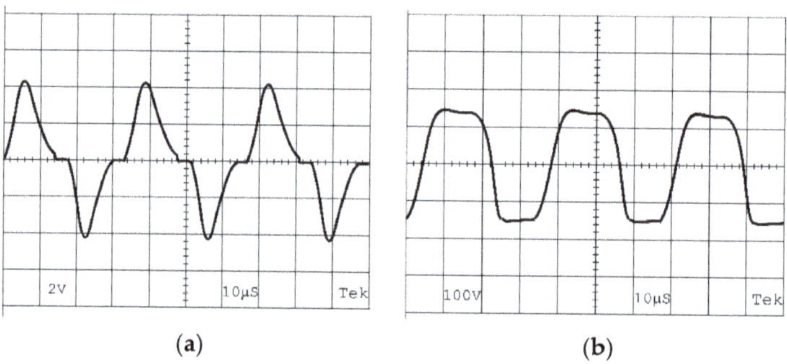

(a)   (b)

Figure 9. Graphical results obtained from the operation of the stand: (**a**) current through the resonant inductor (43.6 A maximum value) and (**b**) voltage on the resonant capacitor (maximum value 150 V).

Figure 9 presents the results of the following measurements: (a) current through the resonant inductor and (b) voltage on the resonant capacitor, performed with the help of the experimental bench under the conditions described above and operating frequency 30 kHz.

The presented experimental results fully supported the conclusions made in the analysis of the power scheme and confirmed the validity of the design methodology.

6. Discussion

The main results obtained from the theoretical and applied scientific problems developed in this article were as follows:

(1) A methodology for designing an RI with ED was developed that provided switching of semiconductor devices at ZCS and ZVS and was characterized by a satisfactory accuracy of not less than 5%, as shown by computer and practical experiments.

(2) A study of the current pulse of the RI was made and its parameters were determined in order to obtain the minimum installed reactive power in the alternating current circuit and electrical loads and losses in the transistors.

(3) The property of the RI with ED was shown to maintain a constant power in the load, regardless of the change of its parameters, when switching the semiconductor devices at ZCS and ZVS. This adaptability and self-matching properties make it very flexible and convenient to use as a wide-range power source for charging stations, including contactless charging.

From the comparisons made with studies of power circuits used for the realization of charging stations presented in [27,28,30–32], it is necessary to conclude that the main alternative of the considered converters with energy dosing was the resonant inverters with reverse diodes and voltage-fed inverters. In order to work with soft commutations, the latter requires the addition of additional resonant circuits (so-called quasi-resonant), and in the case of reverse diode circuits, it is possible to work both in soft commutation mode and to limit the maximum current of the transistors [33]. Unfortunately, in inverters with reverse diodes, power maintenance is achieved through the synthesis of complex control algorithms and controllers. On the other hand, the challenges and limitations associated with the implementation of an RI with ED are related to the fact that no energy is consumed from the power source during the whole period, the lack of limitation of the maximum current through the transistors and the need for sufficient time to dissipate the energy in the resonant inductor when working with high-resistance loads and low power. Regarding this aspect, the achievement of electromagnetic compatibility standards requires the addition of additional modules and ancillary devices.

7. Conclusions

The manuscript presents a study of a charging station for electric vehicles based on a resonant converter with energy dosing. Based on the analysis performed in the established mode of operation of the power circuit, the main relations were derived, through which, the values of the circuit elements were determined. On the basis of the analytical expressions, computer simulations and experiments, the advantages of this class of schemes for the realization of charging stations with different capacities and applications were shown. Regarding a continuation of the current research, the combination of a design with techniques of mathematical modeling and computational mathematics in order to determine optimal values of circuit elements for different objective functions, such as minimum losses, maximum efficiency and minimum dimensions, should be undertaken. In addition, future research could conduct experiments at higher capacities, and in order to achieve greater flexibility, they will be built on a modular principle, as well as the dynamic performance of the system, in order to achieve an aperiodic transition process with minimum duration.

The proposed scheme was also successfully used for contactless dynamic charging of electric vehicles [34]. The main advantage of energy dosing schemes is that the power does not depend on the size of the load. Therefore, during driving, when the equivalent load is constantly changing (due to the coefficient of magnetic coupling), the power transferred to the vehicle is constant.

Author Contributions: N.M. and N.H. were involved in the full process of producing this paper, including conceptualization, methodology, modeling, validation, visualization and preparing the manuscript. All authors have read and agreed to the published version of the manuscript.

Funding: This research was funded by Bulgarian National Scientific Fund, grant number КП-06-H37/25/18.12.2019, and the APC was funded by КП-06-H37/25/18.12.2019.

Institutional Review Board Statement: Not applicable.

Informed Consent Statement: Not applicable.

Data Availability Statement: Not applicable.

Acknowledgments: This research was carried out within the framework of the project "Optimal design and management of electrical energy storage systems", КП-06-Н37/25/18.12.2019, Bulgar-ian National Scientific Fund.

Conflicts of Interest: The authors declare no conflict of interest.

References

1. Pareek, S.; Sujil, A.; Ratra, S.; Kumar, R. Electric Vehicle Charging Station Challenges and Opportunities: A Future Perspective. In Proceedings of the 2020 International Conference on Emerging Trends in Communication, Control and Computing (ICONC3), Lakshmangarh, India, 21–22 February 2020; pp. 1–6.
2. Cai, L.; Pan, J.; Zhao, L.; Shen, X. Networked Electric Vehicles for Green Intelligent Transportation. *IEEE Commun. Stand. Mag.* **2017**, *1*, 77–83. [CrossRef]
3. Kapetanović, M.; Vajihi, M.; Goverde, R.M.P. Analysis of Hybrid and Plug-In Hybrid Alternative Propulsion Systems for Regional Diesel-Electric Multiple Unit Trains. *Energies* **2021**, *14*, 5920. [CrossRef]
4. Boulatov, O.G.; Tsarenko, A.I. *Thyristor-Capacitor Converters*; Energoizdat: Moscow, Russia, 1982. (In Russian)
5. Madzharov, N.D. *Power Electronics*; TU-Gabrovo: Gabrovo, Bulgaria, 2021; ISBN 978-954-683-636-6. (In Bulgarian)
6. Yusop, Y.; Saat, S.; Husin, H.; Nguang, S.K.; Hindustan, I. Analysis of Class-E LC Capacitive Power Transfer System. *Energy Procedia* **2016**, *100*, 287–290. [CrossRef]
7. Triviño, A.; González-González, J.M.; Aguado, J.A. Wireless Power Transfer Technologies Applied to Electric Vehicles: A Review. *Energies* **2021**, *14*, 1547. [CrossRef]
8. Chen, W.; Han, P.; Bazzi, A.M. Evaluation of H-bridge and half-bridge resonant converters in capacitive-coupled wireless charging. In Proceedings of the 2017 IEEE Applied Power Electronics Conference and Exposition (APEC), Tampa, FL, USA, 26–30 March 2017; pp. 3738–3742.
9. Niculae, D.; Iordache, M.; Stanculescu, M.; Bobaru, M.L.; Deleanu, S. A Review of Electric Vehicles Charging Technologies Stationary and Dynamic. In Proceedings of the 2019 11th International Symposium on Advanced Topics in Electrical Engineering (ATEE), Bucharest, Romania, 28–30 March 2019; pp. 1–4.
10. Mohammed, S.A.Q.; Jung, J.-W. A Comprehensive State-of-the-Art Review of Wired/Wireless Charging Technologies for Battery Electric Vehicles: Classification/Common Topologies/Future Research Issues. *IEEE Access* **2021**, *9*, 19572–19585. [CrossRef]
11. Kline, M.; Izyumin, I.; Boser, B.; Sanders, S. Capacitive power transfer for contactless charging. In Proceedings of the 2011 Twenty-Sixth Annual IEEE Applied Power Electronics Conference and Exposition (APEC), Fort Worth, TX, USA, 6–11 March 2011; pp. 1398–1404. [CrossRef]
12. Elekhtiar, A.; Eltagy, L.; Zamzam, T.; Massoud, A. Design of a capacitive power transfer system for charging of electric vehicles. In Proceedings of the 2018 IEEE Symposium on Computer Applications & Industrial Electronics (ISCAIE), Penang, Malaysia, 28–29 April 2018; pp. 150–155.
13. Erickson, R.W.; Maksimovic, D. *Fundamentals of Power Electronics*; Kluwer Academic Publishers: New York, NY, USA, 2000.
14. Hart, D.W. *Power Electronics*; McGraw-Hill: New York, NY, USA, 2011; ISBN 978-0-07-338067-4.
15. Zishan, A.A.; Haji, M.M.; Ardakanian, O. Adaptive Congestion Control for Electric Vehicle Charging in the Smart Grid. *IEEE Trans. Smart Grid* **2021**, *12*, 2439–2449. [CrossRef]
16. Carvalho, R.; Buzna, L.; Gibbens, R.; Kelly, F. Congestion Control in Charging of Electric Vehicles. *arXiv* **2015**, arXiv:1501.06957.
17. Singh, M.; Kumar, P.; Kar, I. A Multi Charging Station for Electric Vehicles and Its Utilization for Load Management and the Grid Support. *IEEE Trans. Smart Grid* **2013**, *4*, 1026–1037. [CrossRef]
18. Sultanbek, A.; Khassenov, A.; Kanapyanov, Y.; Kenzhegaliyeva, M.; Bagheri, M. Intelligent wireless charging station for electric vehicles. In Proceedings of the 2017 International Siberian Conference on Control and Communications (SIBCON), Astana, Kazakhstan, 29–30 June 2017; pp. 1–6.
19. Chen, H.; Su, Z.; Hui, Y.; Hui, H. Dynamic Charging Optimization for Mobile Charging Stations in Internet of Things. *IEEE Access* **2018**, *6*, 53509–53520. [CrossRef]
20. Joos, G.; de Freige, M.; Dubois, M. Design and simulation of a fast charging station for PHEV/EV batteries. In Proceedings of the 2010 IEEE Electrical Power & Energy Conference, Halifax, NS, Canada, 25–27 August 2010; pp. 1–5. [CrossRef]
21. Badawy, M.O.; Arafat, N.; Ahmed, A.; Anwar, S.; Sozer, Y.; Yi, P.; De Abreu-Garcia, J.A. Design and Implementation of a 75-kW Mobile Charging System for Electric Vehicles. *IEEE Trans. Ind. Appl.* **2015**, *52*, 369–377. [CrossRef]
22. Mkahl, R.; Nait-Sidi-Moh, A.; Gaber, J.; Wack, M. A linear optimization approach for electric vehicles charging. In Proceedings of the 2015 3rd International Renewable and Sustainable Energy Conference (IRSEC), Ouarzazate, Morocco, 10–13 December 2015; pp. 1–6.
23. Rozario, D.; Azeez, N.A.; Williamson, S.S. A modified resonant converter for wireless capacitive power transfer systems used in battery charging applications. In Proceedings of the 2016 IEEE Transportation Electrification Conference and Expo (ITEC), Dearborn, MI, USA, 27–29 June 2016; pp. 1–6.
24. Madzharov, N.D.; Tonchev, A.T. Inductive High Power Transfer Technologies for Electric Vehicles. *J. Electr. Eng.* **2014**, *65*, 125–128. [CrossRef]
25. Madzharov, N. High-Frequency Power Source with Constant Output Power. *J. Eng. Sci. Technol. Rev.* **2016**, *9*, 157–162. [CrossRef]

26. Covic, G.A.; Boys, J.T. Modern Trends in Inductive Power Transfer for Transportation Applications. *IEEE J. Emerg. Sel. Top. Power Electron.* **2013**, *1*, 28–41. [CrossRef]
27. Arif, S.M.; Lie, T.T.; Seet, B.C.; Ayyadi, S.; Jensen, K. Review of Electric Vehicle Technologies, Charging Methods, Standards and Optimization Techniques. *Electronics* **2021**, *10*, 1910. [CrossRef]
28. Ahmadi, M.; Mithulananthan, N.; Sharma, R. A review on topologies for fast charging stations for electric vehicles. In Proceedings of the 2016 IEEE International Conference on Power System Technology (POWERCON), Wollongong, NSW, Australia, 28 September–1 October 2016; pp. 1–6.
29. ElGhanam, E.; Hassan, M.; Osman, A. Design of a High Power, LCC-Compensated, Dynamic, Wireless Electric Vehicle Charging System with Improved Misalignment Tolerance. *Energies* **2021**, *14*, 885. [CrossRef]
30. Barsari, V.Z.; Thrimawithana, D.J.; Covic, G.A.; Kim, S. A Switchable Inductively Coupled Connector for IPT Roadway Applications. In Proceedings of the 2020 IEEE PELS Workshop on Emerging Technologies: Wireless Power Transfer (WoW), Seoul, Korea, 15–19 November 2020; pp. 35–39.
31. Leone, C.; Longo, M.; Fernández-Ramírez, L.M. Optimal Size of a Smart Ultra-Fast Charging Station. *Electronics* **2021**, *10*, 2887. [CrossRef]
32. Abraham, D.S.; Verma, R.; Kanagaraj, L.; Raman, S.G.T.; Rajamanickam, N.; Chokkalingam, B.; Sekar, K.M.; Mihet-Popa, L. Electric Vehicles Charging Stations' Architectures, Criteria, Power Converters, and Control Strategies in Microgrids. *Electronics* **2021**, *10*, 1895. [CrossRef]
33. Donati, A.; Madzharov, N.; Melandri, A.; Sighinol, F. Induction Sealing Device for Producoing Pourable Food Packages. U.S. Patent 8,286,406 B2, 16 October 2012.
34. Madzharov, N.; Hinov, N. Flexibility of Wireless Power Transfer Charging Station Using Dynamic Matching and Power Supply with Energy Dosing. *Appl. Sci.* **2019**, *9*, 4767. [CrossRef]

 *energies*

Article

A Bidirectional Grid-Connected DC–AC Converter for Autonomous and Intelligent Electricity Storage in the Residential Sector

Ismail Aouichak, Sébastien Jacques *, Sébastien Bissey, Cédric Reymond, Téo Besson and Jean-Charles Le Bunetel

Research Group on Materials, Microelectronics, Acoustics, and Nanotechnology, GREMAN UMR 7347, University of Tours, CNRS, INSA Centre Val-de-Loire, 37100 Tours, France; ismail.aouichak@univ-tours.fr (I.A.); sebastien.bissey@univ-tours.fr (S.B.); cedric.reymond@st.com (C.R.); teo.besson@univ-tours.fr (T.B.); lebunetel@univ-tours.fr (J.-C.L.B.)
* Correspondence: sebastien.jacques@univ-tours.fr; Tel.: +33-6-667-639-46

Abstract: Controlling the cost of electricity consumption remains a major concern, particularly in the residential sector. Smart home electricity management systems (HEMS) are becoming increasingly popular for providing uninterrupted power and improved power quality, as well as for reducing the cost of electricity consumption. When power transfer is required between a storage system and the AC grid, and vice versa, these HEMS require the use of a bidirectional DC–AC converter. This paper emphasizes the potential value of an almost unexplored topology, the design of which was based on the generation of sinusoidal signals from sinusoidal half waves. A DC–DC stage, which behaved as a configurable voltage source, was in series with a DC–AC stage, i.e., an H-bridge, to achieve an architecture that could operate in both grid and off-grid configurations. Wide bandgap power switches (silicon carbide metal-oxide-semiconductor field-effect transistors [MOSFETs]), combined with appropriate control strategies, were the keys to increasing compactness of the converter while ensuring good performance, especially in terms of efficiency. The converter was configured to automatically change the operating mode, i.e., inverter or rectifier in power factor correction mode, according to an instruction issued by the HEMS; the latter being integrated in the control circuit with automatic duty cycle management. Therefore, the HEMS set the amount of energy to be injected into the grid or to be stored. The experimental results validate the operating modes of the proposed converter and demonstrate the relevance of such a topology when combined with an HEMS, especially in the case of an AC grid connection. The efficiency measurements of the bidirectional DC–AC converter, performed in grid-connected inverter mode, show that we exceeded the efficiency target of 95% over the entire output power range studied, i.e., from 100 W to 1.5 kW.

Keywords: home electricity management systems; bidirectional DC–AC converter; high compactness; high efficiency

Citation: Aouichak, I.; Jacques, S.; Bissey, S.; Reymond, C.; Besson, T.; Le Bunetel, J.-C. A Bidirectional Grid-Connected DC–AC Converter for Autonomous and Intelligent Electricity Storage in the Residential Sector. *Energies* **2022**, *15*, 1194. https://doi.org/10.3390/en15031194

Academic Editors: Massimiliano Luna and Nicu Bizon

Received: 2 November 2021
Accepted: 4 February 2022
Published: 7 February 2022

1. Introduction

The energy policy challenges facing the European Union are greater than ever. One such challenge is the intelligent management of electricity at all levels, from production to final consumption [1,2]. Distribution efficiency and the reliability of service delivery are two indicators that can be used to evaluate the performance of distribution systems [3,4]. The increasing deployment of smart grids has improved their monitoring and controllability [5]. An example of a modern smart grid feature is the ability to assist with load shifting, limit peak demand and automatically identify malfunctions or outages [6]. These issues are all the more controversial as the number of microgrids, i.e., small-scale independent power systems, continues to grow. Two main modes of operation characterize a microgrid: the grid-connected mode and the off-grid mode [7]. In off-grid mode, electricity is either unavailable due to a grid failure or outage, or the power grid is in islanding mode, i.e., completely inaccessible. The performance of standalone microgrids has been regularly analyzed in the literature for several years. These standalone microgrids typically consist of two key components: photovoltaic (PV) arrays and/or wind turbines and energy storage systems, such as flywheels, supercapacitors or batteries, which are used to implement intelligent voltage regulation and load tracking systems [8,9].

Energy management strategies are currently playing an increasingly important role in regulating the power quality of microgrids [10]. One of the challenges is to control energy flows to achieve various operational objectives, such as cost minimization, guaranteed delivery or security. One method to achieve this is to adjust the flow of energy to and from the main grid, the distribution of energy resources and the controllability of loads. In the residential sector, this type of strategy, now known as a home electricity management system (HEMS), is typically implemented to stimulate the integration of renewable energy and to protect the electricity distribution system from potential outages [11].

With the advent of Internet of things (IoT)-based smart buildings, in which all smart appliances are connected, as well as smart meters, we now have not only real-time usage data but also the ability to remotely operate various appliances in a home [12,13]. Much of this data can be used by an HEMS to improve the synchronization of energy transfers between the storage system, power generation and its consumption.

A wide range of techniques is already available in the literature and these techniques can be classified into two main categories [14–18]. Indeed, these approaches are based either on information processing methods or on optimization tactics:

- Information process methods: These methods are designed to model nonlinear systems. For example, HEMS controllers can use artificial intelligence (AI) methods, such as support vector machines (SVM), artificial neural networks (ANN) or recurrent neural networks (RNN), to map the electrical properties of the house using categorized metering data [19]. There are also adaptive neural fuzzy inference systems (ANFIS), which represent a mix of neural networks and fuzzy inference systems, to monitor and predict power consumption [20];
- Optimization process methods: These methods use fixed functions, such as economic feasibility, optimal management or error minimization. Model predictive control (MPC), particle swarm optimization (PSO) and genetic algorithm (GA) are three examples of the most commonly used techniques in HEMS scheduling [21,22].

As shown in Figure 1, Bissey et al. recently proposed an example of HEMS using fuzzy logic control (FLC) [23]. This approach used machine learning and data collected from different houses (i.e., time, day, past power consumption and indoor/outdoor temperature data) to find optimized criteria for the configuration of an HEMS. Although the approach described in this paper is effective, its validity only makes sense if a large experimental database is available to build the fuzzy rule base and membership functions needed to make this HEMS work. Recently emerged deep learning (DL) methods are proving to be effective tools, even surpassing traditional approaches as they greatly reduce the need for human interaction [24,25].

A storage system is required for the installation of such an HEMS. Alternating loads can be powered either directly from the AC grid or through an inverter in case the storage system would relieve the AC grid. To implement these two modes of operation, a necessarily bidirectional DC–AC converter must be operational [26]. Its metrics in terms of compactness, efficiency and output signal quality must be as high as possible. In this context and as illustrated in Figure 1, this paper focuses on the sizing and implementation of such a bidirectional converter.

Multilevel DC–AC converters, introduced in the 1980s, that are widely deployed today and whose performance has been discussed in the literature for many years allow the output voltage to be varied in steps by generating levels [27–29].

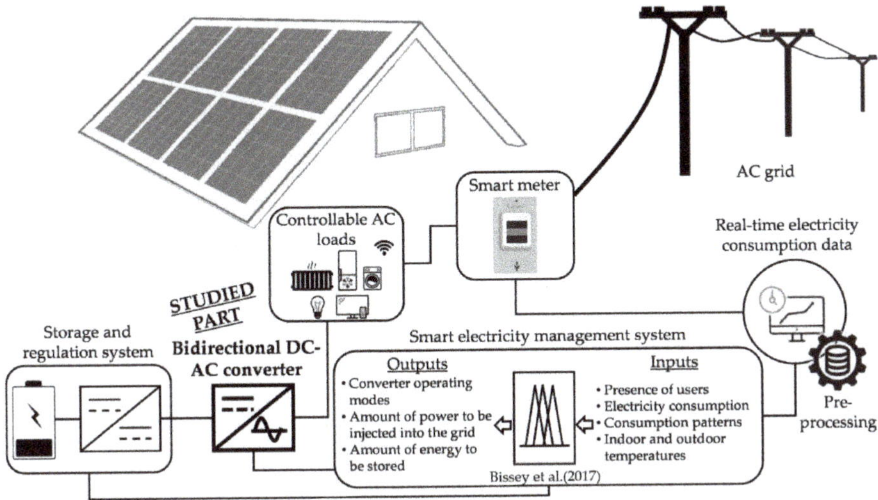

Figure 1. The principle of the smart HEMS implemented in this study [23].

By significantly increasing the number of levels, the output signals become more similar to sinusoids, which drastically reduces their total harmonic distortion (THD) and thus, improves the power quality of the converter and a fortiori of the microgrid. To increase the number of levels, two approaches are classically implemented: increasing the number of voltage sources on the DC bus side or multiplying the number of semiconductor devices to be controlled [30]. Multilevel converter topologies are still widely used for medium and high voltage applications, such as electrical motor drives or grid connected converters, because they generate very low harmonics [31].

Totem pole topologies are also interesting structures dedicated to bidirectional DC–AC conversion. In this type of structure, bidirectional switches based on MOSFETs, thyristors or Triacs are used. Such a topology is generally suitable for high-power applications. Its two main advantages are the reduction in the number of components to be controlled and the simplicity of the use of the converter itself. Recent studies have demonstrated their energy efficiency for particularly high-power densities [32,33].

Despite their high energy efficiency, totem pole topologies cause electromagnetic interference at high frequencies due to the floating points that exist in these architectures. Multilevel DC–AC topologies, widely used today in high voltage, use particularly bulky filters to limit electromagnetic compatibility problems. Thus, to avoid these problems, we sought to implement a topology more suitable for HEMS applications whilst remaining within the specified power range.

This paper aims to prove experimentally the merits of another type of bidirectional DC–AC converter that is more suitable for HEMS applications. The proposed topology, which has not yet been fully explored in the literature [34,35], is formed by the association in series of a DC–DC stage and a DC–AC stage with these two stages being necessarily bidirectional. In inverter mode, the DC–DC stage generates a full-wave rectified sine wave thanks to its pulse width modulation (PWM) control. This signal is then inverted half-period by half-period by the DC–AC stage to create the sine wave signal. In rectifier mode, the DC–AC stage acts as a full bridge while power factor correction (PFC) is provided by the DC–DC stage.

The main contribution of this study is the sizing and experimental validation of the proposed bidirectional DC–AC converter topology coupled with a control strategy for HEMS applications. This converter needed excellent compactness and high efficiency (at least 95%), both at low power (a few hundred watts) and up to 1.5 kW. The HEMS, which is associated with the designed converter, needed to supervise the transition between operating modes, as well as the amount of energy stored or injected into the AC grid.

The remainder of this paper consists of the following sections. Section 2 describes the proposed topology with the above key properties. The sizing strategy of the converter and the main experimental results are presented in Section 3. Finally, a discussion based on the main experimental results that were obtained is proposed in Section 4.

2. Bidirectional DC–AC Converter Topology Proposal and Control Methodology

In order to validate the fuzzy logic-based control technique implemented in our proposed HEMS [23], we had to design a bidirectional DC–AC converter that was capable of functioning as both an inverter and a PFC rectifier. The transition between these two modes of operation needed to be fully automated and without human intervention in order for our HEMS to autonomously store, produce and supply energy for domestic use.

Before validating its operation by appropriate experimental measurements, it is essential to detail the operating modes of the converter, as well as its control strategies.

2.1. Proposed Topology and Details of Its Operating Modes

Figure 2 shows the overall structure of the proposed bidirectional DC–AC converter. The energy transfer between a DC voltage source and an AC voltage source, and vice versa, was the basis of this structure. The association in series of a DC–DC stage and a DC–AC stage ensured this principle of operation with these two stages being necessarily bidirectional.

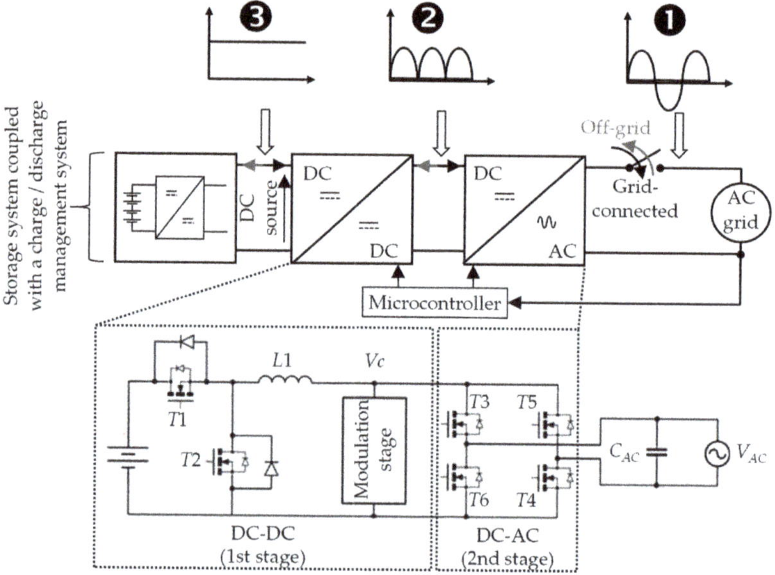

Figure 2. Principle diagram of the proposed bidirectional DC–AC converter.

The DC–DC stage needed to generate a rectified sine wave from the PWM command of the power switches (see ❷ in Figure 2). Since the power devices in this stage switched at a frequency of a few hundred kilohertz to optimize the compactness of the whole converter, the inductance, noted $L1$ in Figure 2, was sized so that the ripple of the current was negligible compared to the sinusoidal component at low frequency (in this case, 50 Hz). Therefore, the DC–DC converter acted as a controllable output voltage source.

By changing the voltage of a modulation stage, the output current could be regulated. This is very interesting, especially when the output current is strongly reduced or in the of variable DC voltage. This strategy allowed the output voltage of the DC–DC stage to be modulated. Specifically, when this modulated voltage was higher than the mains voltage, the output current had a positive value. In the opposite case, the output current was negative.

Thus, two modes of operation were possible:

- Inverter mode: In this case, energy was transferred from the storage system to the AC grid (see the black arrows in Figure 2, as well as Figure 3).
- PFC rectifier mode: In this operating mode, energy was transferred from the AC grid to the storage system (see the grey arrows in Figure 2, as well as Figure 4).

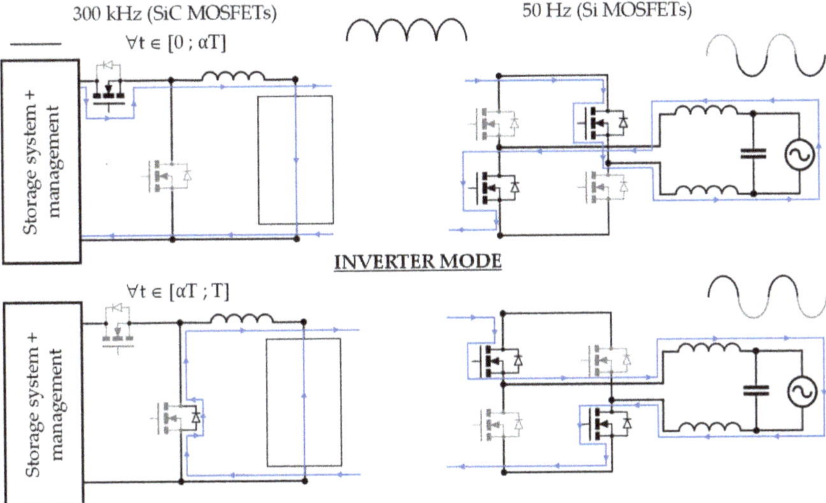

Figure 3. Schematic diagram of the converter in inverter mode.

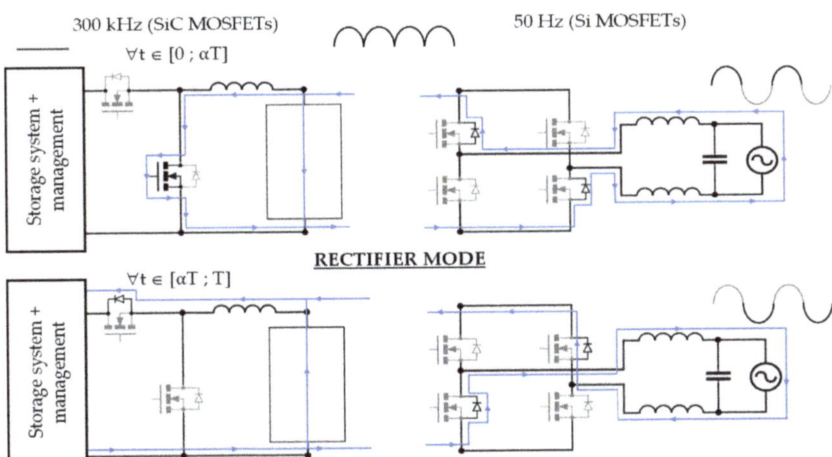

Figure 4. Schematic diagram of the converter in PFC rectifier mode.

The design approach that was chosen allowed the converter to be used in both grid-connected and off-grid modes. Since the microcontroller was synchronized with the AC grid to sensibly drive the DC–DC and DC–AC stages, only the grid-connected mode will be discussed in the remainder of this paper.

The DC–AC stage was responsible for inverting every other sinusoidal half-wave to obtain a full sinusoidal output signal (see ❶ in Figure 2).

Compared to existing voltage source converter topologies, such as multilevel structures, the coupling of a DC–DC stage with an H-bridge has many advantages:

- The standard DC–DC converter and the H-bridge are two very common and mastered topologies;
- Many H-bridge topologies are composed of four or more power components that switch at high frequency. In our proposed architecture, only those in the DC–DC stage (see transistors $T1$ and $T2$ in Figure 2) operated at high frequency, i.e., 150 kHz implemented here. In the DC–AC stage, all components (see components $T3$, $T4$, $T5$ and $T6$ in Figure 2) switched at low frequency, i.e., 50 Hz;
- When switching at high frequency, it is imperative to take into account the delay between the two switching operations in the same branch for safety reasons. Here, the safety delay was easier to regulate since only one stage operated at high frequency;
- In our architecture, the capacitance used at high frequency to modulate the voltage Vc (see Figure 2) was small (about 10 μF). This is not the case in many other topologies in the literature, where the authors consider that the use of an AC capacitor is mandatory.

2.2. Modulation of the Output Voltage of the DC–DC Stage

In a rather classical way, three solutions (see Figure 5) are proposed here to adjust the voltage Vc (see Figure 2) of the DC–DC stage. We assume that the whole DC–AC converter operates in inverter mode (see Figure 3) to briefly explain each solution.

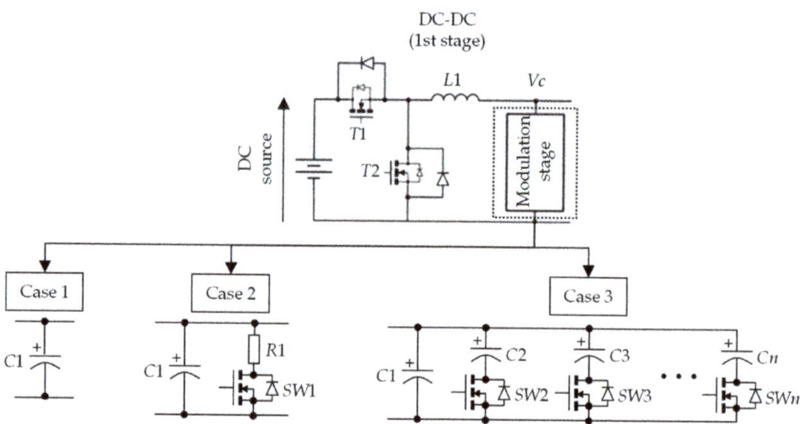

Figure 5. Solutions to modulate the output voltage of the DC–DC stage.

In the three cases of Figure 5, the output capacitor of the DC–DC stage, noted $C1$, was always used. Its value considered here was 10 μF to avoid a strong ripple of the voltage Vc.

The first solution (see Case 1 in Figure 5) was to use the capacitor $C1$ directly. The DC–DC stage then operated as a buck-type step-down converter. Despite the simplicity of this solution since it does not use any power device, it has a major drawback. Indeed, a current could flow inside the transistor, named $T2$ in Figure 2, during the discharge of the capacitor, which could lead to it overheating.

As can be seen in Case 2 in Figure 5, a resistor of a few ohms coupled with a switch could be used to solve the above problem. If the load impedance was too high, then the capacitor used in this case could be discharged through this resistor. This solution was especially interesting when the output power of the inverter fluctuated during operation. However, because of this resistance, this solution strongly penalized the efficiency of the DC–DC stage and, by extension, of the whole DC–AC converter.

The last solution (see Case 3 in Figure 5) was to connect n-quadrupoles in parallel. Each quadrupole consisted of a capacitor and a power switch that were in series. Such a solution aimed to fix the ripple of the output voltage of the DC–DC stage. Even though this solution was slightly more complex than the previous ones, adaptation to many more loads could be achieved. Finally, the efficiency of the DC–DC stage was not penalized so much in this case. In the remainder of this manuscript, this solution is implemented with two quadrupoles in parallel with the output capacitor (see capacitor $C1$ in Figure 5) of the DC–DC stage. In conclusion, the values of the three capacitors, $C1$, $C2$ and $C3$, of the modulation stage were equal to 10 µF, 1 µF and 68 nF, respectively (see Figure 5). The value of each capacitor was determined by the ripple of the voltage Vc (arbitrarily, we considered 1% here) and the estimated current (see (1)).

$$C = \frac{I \times \Delta t}{\Delta V} \tag{1}$$

where:

- I is the estimated current;
- ΔV is the ripple of the voltage Vc, which is constant (i.e., 1%);
- $\Delta t = \alpha T = \frac{\alpha}{F}$ with α as the duty cycle and F as the switching frequency.

2.3. Control Strategies

In this section of the paper, we will describe the control strategies of the proposed bidirectional DC–AC converter. We will only give the principles and thus, we will not detail the control circuit or the AC network connection strategy because several patents are pending.

2.3.1. Inverter Mode

In inverter mode (see Figure 3), energy flowed from the storage system to the AC grid. In this type of operation, the most important objective was to control the current injected into the AC grid by regulating the output voltage of the DC–DC stage. Figure 6 shows the overall architecture of the MOSFET control circuit inside the DC–DC stage (see transistors $T1$ and $T2$ in Figure 2). We recall that this stage supervised the generation of a semi-sinusoidal output signal. The possibility of modifying the injected current was the indispensable aspect of this control strategy. Figure 6 shows that the microcontroller was programmed to adjust the duty cycle from zero to one with great precision.

A minor increase or decrease in this variable caused the current injected into the AC grid to increase or decrease. This allowed us to control the selection of the operating mode and provided us with the ability to switch from inverter to rectifier mode, as we will see later.

The DC–AC stage, on the other hand, was controlled using the frequency measurement of the smart meter with a switching frequency of 50 Hz in order to be synchronized with the grid (see transistors from $T3$ to $T6$ in Figure 2).

2.3.2. PFC Rectifier Mode

In PFC rectifier mode (see Figure 4), energy flowed from the AC grid to the storage system. In this kind of operation, the most important objective was to control the absorbed current, which needed a sinusoidal waveform. This objective could be achieved by adjusting the equivalent capacitance of the modulation stage of the DC–DC stage (see Case 3 in Figure 5). As for the inverter mode, we chose the solution with a capacitance $C2$ equal to 1 µF because it offered the best results. The PFC mode, achieved by the DC–DC stage, was essential to meet the requirements of IEC 61000-3-2.

Figure 7 shows the structure of the control circuit of the MOSFETs inside the DC–DC stage. The regulation of the control circuit was made possible by a current sensor and a voltage sensor to realize the PFC strategy.

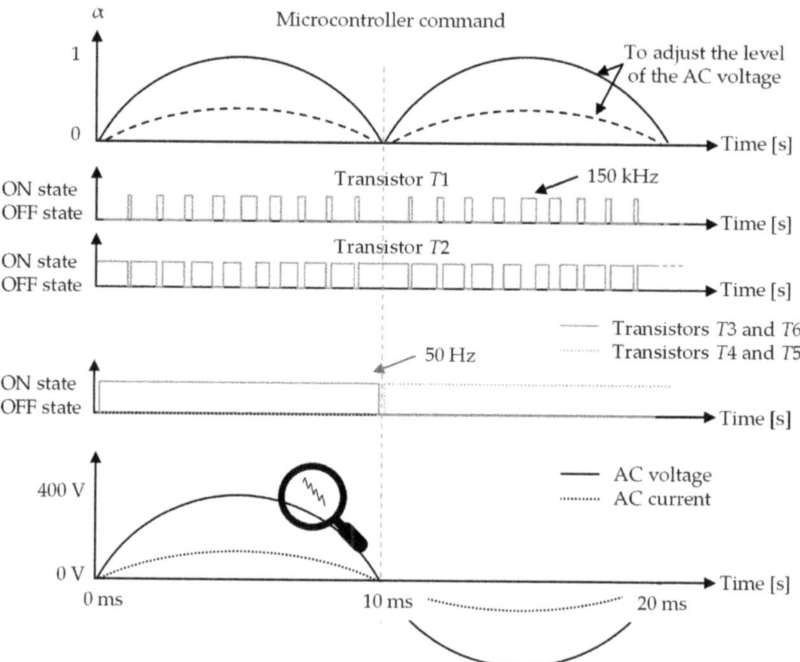

Figure 6. Control strategy of the MOSFETs inside the proposed bidirectional DC–AC converter used in inverter mode.

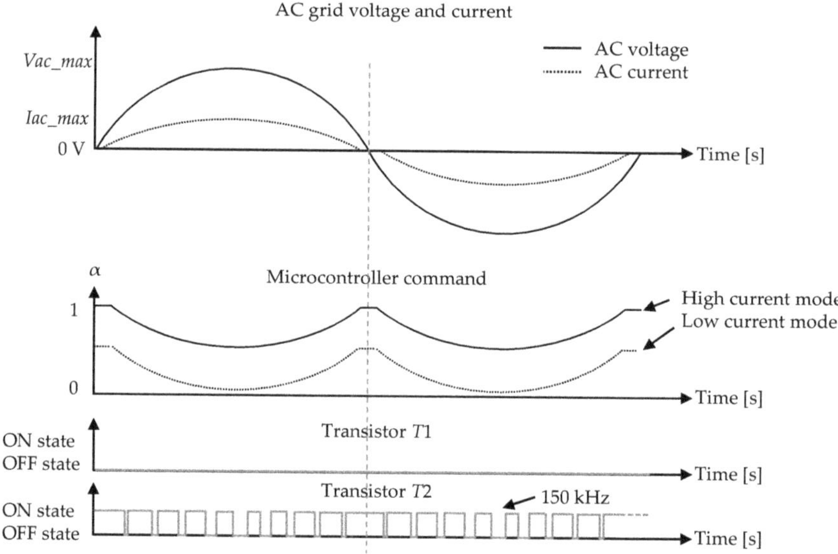

Figure 7. Control strategy of the MOSFETs inside the proposed bidirectional DC–AC converter used in PFC rectifier mode.

The DC–AC stage was controlled by the very same signal as the inverter mode, so it was synchronized with the AC grid at a switching frequency of 50 Hz.

3. Sizing of the Bidirectional Converter and Main Results in Grid-Connected Mode

3.1. Specifications, Key Sizing Steps and Selection of the Main Components

The specifications of the prototype bidirectional DC–AC converter that we realized are listed in Table 1. The electrical waveforms were evaluated based on the topology and control strategies detailed in the previous section.

Table 1. The specifications of the proposed bidirectional DC–AC converter.

	Parameters	Values
AC bus	RMS voltage (V)	110/230
	Frequency (Hz)	50/60
	Grid capacitor (µF) (see C_{AC} in Figure 2)	1
DC bus	DC voltage (V)	400
DC–DC stage	Input current (A)	10 max.
	Output current (A)	5 max.
	Switching frequency (kHz)	150/300
DC–AC stage	Output current (A)	5 max.
	Switching frequency (Hz)	AC grid frequency
Targeted power		From 100 W up to 1.5 kW
Targeted efficiency		About 95% over the entire power range considered

The approach to the sizing of the converter was classical because the two stages taken separately are now well-known and mastered [36–40]. Of course, this approach is adaptable according to the targeted application and design constraints. Figure 8 presents the key steps for the sizing of the passive and active components of each of the two stages. In this approach, the following parameters were essential to take into account:

- The static and dynamic losses in the MOSFETs (in the DC–DC stage, which switched at several hundred kilohertz, it was particularly important to determine the switching losses) and the losses in the passive components (it was particularly important to take into account the losses in the core of the DC–DC stage choke due to hysteresis and eddy currents) to reach the efficiency objectives;
- The thermal management of the components over the targeted power range;
- The choice of the technology and the packaging of the components to optimize the compactness and mass of the converter;
- Other constraints in view of the industrialization of the product, such as electromagnetic compatibility problems and also the development cost.

As shown in Figure 8, the main components to be sized were the power MOSFETs of both stages [36–38], the inductor of the DC–DC stage [39,40] and the output voltage modulation of the DC–DC stage [26].

For the power MOSFETs, given the high switching frequency of the DC–DC stage (a few hundred kilohertz) and to optimize the thermal management of the components, we chose silicon carbide (SiC) devices. For the DC–AC stage, which switched at the AC grid frequency, silicon substrate components were perfectly suited. Then, whatever the two stages, it was essential to take into account: the voltage withstand; the current flowing in the drain; the switching characteristics, especially the rise and fall times for the components of the DC–DC stage; the gate charge characteristics; the characteristics of the body diodes; and the thermal characteristics, in terms of junction temperature and thermal resistances.

The choice of power MOSFETs also took into account future normative tests, which are essential before the industrialization of the product, such as immunity to burst transients, resistance to electrostatic discharges and electromagnetic compatibility. As shown in Figure 9 and Table 2, the DC–DC stage used two (see $T1$ and $T2$ in Figure 9) SiC MOSFETs (reference: SCT3080AL; manufacturer: ROHM semiconductor) with, for these two transistors, a nominal drain current and drain-to-source voltage equal to 30 A and 650 V, respectively. Two switching frequencies were considered: 150 kHz and 300 kHz. Switching the components of this stage at 300 kHz allowed for the drastic reduction in the size of the passive components and a fortiori the optimization of the size of the converter. On the other hand, this penalized the efficiency of the system (the desired efficiency of 95% was difficult to achieve) because the switching losses were very significant (see Section 4.2). Therefore, the switching frequency of 150 kHz was implemented experimentally in order to obtain a compromise between the compactness of the system, its efficiency over the entire power range (i.e., from 100 W to 1.5 kW) and the optimization of its power quality; this last point will be studied in more detail in the near future. As for the DC–AC stage, as shown in Table 2 and Figure 9, it used four MOSFETs (designated *T3*, *T4*, *T5* and *T6*) on a silicon substrate (reference: IRFPS43N50K; manufacturer: Vishay Siliconix), each with a nominal drain current and drain-to-source voltage equal to 47 A and 500 V, respectively. The power devices were turned on and off with zero crossing of the AC grid voltage to minimize losses.

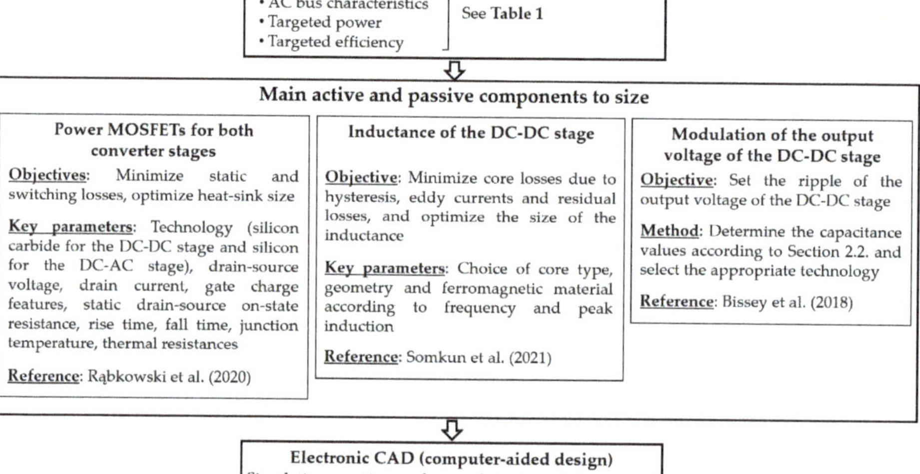

Figure 8. The key sizing steps of the proposed converter [26,38,40].

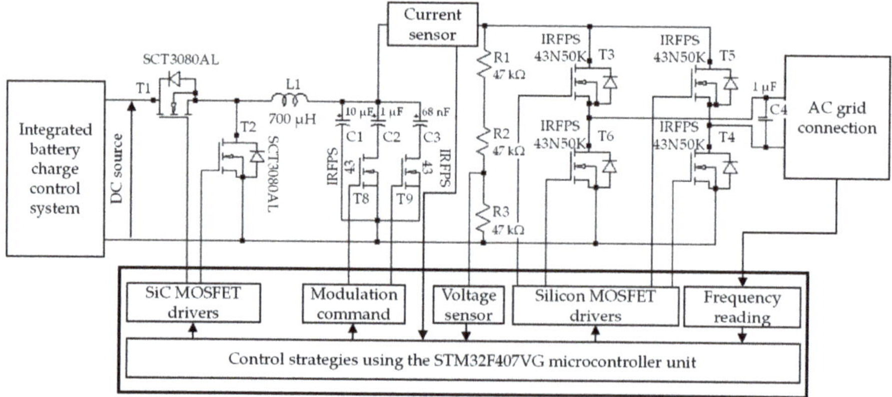

Figure 9. Electrical diagram of the power circuit of the converter.

Table 2. A selection of the main components.

	Parameters	Values
DC–DC stage	Inductance of the DC coil (µH) (see $L1$ in Figure 9)	700
	Power MOSFETs used (see $T1$ and $T2$ in Figure 9)	Two N-channel SiC power MOSFETs of 650 V, 30 A, 80 mΩ (part number: SCT3080AL; manufacturer: ROHM Semiconductor); switching frequency of a few hundred kHz (150 kHz implemented experimentally)
Modulation stage	Capacitances (µF) (see $C1$, $C2$ and $C3$ in Figure 9)	10 ($C1$), 1 ($C2$) and 0.068 ($C3$)
	Power MOSFETs used (see $T8$ and $T9$ in Figure 9)	Two N-channel power MOSFETs of 500 V, 47 A, 78 mΩ (part number: IRFPS43N50K; manufacturer: Vishay Siliconix).
DC–AC stage	Power MOSFETs used (see $T3$, $T4$, $T5$ and $T6$ in Figure 9)	Four N-channel power MOSFETs of 500 V, 47 A, 78 mΩ (part number: IRFPS43N50K; manufacturer: Vishay Siliconix); switching to AC mains frequency (50 Hz here)
Control strategies (SiC and silicon MOSFET drivers, voltage/current sensors, AC grid frequency reading)		Microcontroller based on the Arm® Cortex®-M4 32-bit RISC (reduced instruction set computer) core, operating at a frequency of up to 168 MHz (part number: STM32F407VG; manufacturer: STMicroelectronics)

Concerning the optimization of the DC–DC stage inductance in terms of electromagnetic characteristics, compactness and mass, we took into account the geometry, toroidal or planar and the type of core, as well as the ferromagnetic material used, according to the frequency and the maximum induction [39,40]. Several technologies were studied in order to determine the most suitable inductance for the sized converter. The choice of the 700 µH inductor (see $L1$ in Figure 9) was based on limiting the current ripple in the inductor to less than 20% of the maximum current (i.e., 5 A here) [41]. We also considered its maximum DC resistance (0.12 Ω) and the evolution of its impedance with frequency. Considering all these elements, we chose the WE-FI leaded toroidal line choke from the manufacturer Wurth Electronik.

Regarding the output voltage modulation of the DC–DC stage, the value of the capacitors (see $C1$, $C2$ and $C3$ in Figure 9) was explained in Section 2.2.

Concerning the servo strategy of the proposed converter, a differential measurement was used to determine the AC bus voltage as a function of the phase potential (VL) and the neutral point potential (VN). This differential measurement was performed with a voltage divider bridge of the same resistance value, i.e., 47 kΩ (see $R1$, $R2$ and $R3$ in Figure 9). To measure the current flowing between the DC–DC stage and the DC–AC stage, as shown in Figure 9, a 15 A closed-loop Hall transducer (reference: LKSR 15-NP; manufacturer: LEM) was used in series with the inductor of the DC–DC stage.

Finally, an STM32F407VG microcontroller unit from the manufacturer STMicroelectronics drove the entire converter and provided automatic AC grid connection and disconnection. The control board with the microcontroller, the MOSFET drivers and the control strategies of the power components are currently under patent and cannot be detailed in this manuscript.

3.2. Experimental Test Setup and Standby Mode

A comprehensive experimental process was adopted to validate the two modes of operation of the bidirectional DC–AC converter proposed here when connected to the AC grid and in a power range up to 1.5 kW. The converter demonstrator naturally included the DC–DC and DC–AC stages, but also an adaptive filter, as well as a power supply (i.e., +5 V and +12 V) to power the onboard electronics.

As shown in Figure 10, four 12 V, 7 Ah batteries (reference: NP7-12; manufacturer: YUASA) were used to simulate the storage system. These batteries were associated in series to provide the overall voltage of 48 V (see ❶ in Figure 10). On the AC side, the connection to the power supply was provided by a single-pole variable transformer (reference: SEC2; manufacturer: LANGLOIS) set at 110 V RMS. A line impedance stabilization network (LISN) (reference: PD30; manufacturer: EMC MASTER; main features: monophasic, 220 V, 10 A, 150 kHz to 30 MHz) was used to control the network impedance and ensure measurement repeatability (see ❷ in Figure 10).

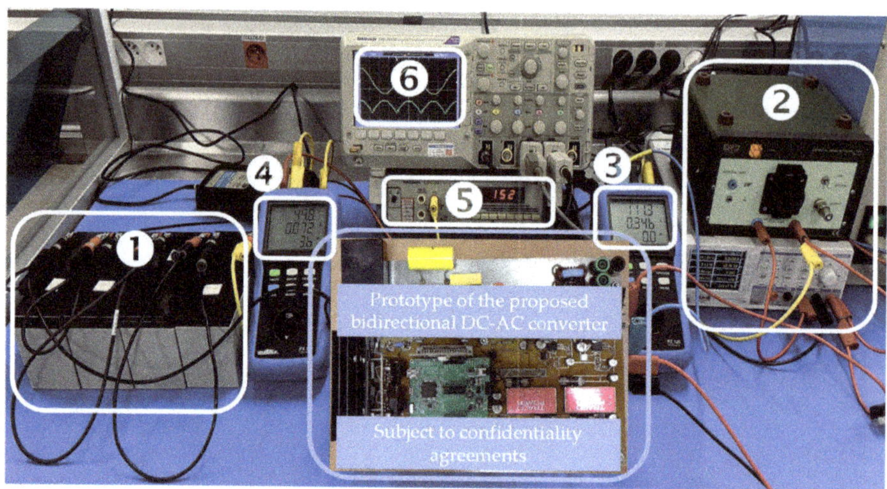

Figure 10. The experimental test setup.

A high-voltage differential probe (reference: P5205; manufacturer: Tektronix) and a 15 A AC–DC current probe (reference: TCP202; manufacturer: Tektronix) were used to measure the output voltage and current, respectively. The output power of the DC–DC stage was measured using a power meter (reference: PX 110; manufacturer: Metrix). For the measurement of the output power of the DC–AC stage, the same type of electronic meter was used.

To connect to the AC grid, the bidirectional DC–AC converter used a patent-pending automated routine. Figure 10 shows the converter in a steady state with the grid at 0 watts (see ❸ in Figure 10) and the batteries powering the system. The power consumption of the converter in sleep mode was 3.6 W (see ❹ in Figure 10). The DC bus was monitored (see ❺ in Figure 10) and displayed the DC voltage (see ❸ in Figure 2), while differential probes were used to monitor the voltage generated between the two stages of the converter, as well as that between the converter and the grid connection (see ❻ in Figure 10, as well as Figure 11).

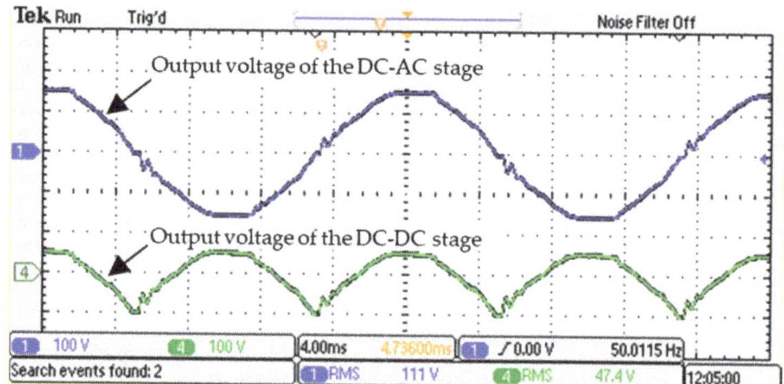

Figure 11. Experimental validation of the standby mode: output signals.

3.3. Experimental Validation of the Operating Modes

3.3.1. Foreword

This section of the paper aims to validate the two operation modes (i.e., inverter and PFC rectifier) of the proposed bidirectional DC–AC converter. To avoid damage to the prototype of the whole converter, the experimental tests were performed at low power. The single-pole variable transformer was used to adjust the grid voltage to 110 V RMS, 50 Hz, while the DC side was connected to the four Yuasa batteries for a total of 48 volts DC. The prototype also included a charge–discharge management system, capable of supplying up to 350 V DC to the whole converter.

The RMS voltage and frequency were monitored by the microcontroller on both sides of the primary switch connecting the converter to the AC grid (see Figure 2). The switch was closed when all criteria for grid connection were met [36], and the voltage on both sides of the converter was balanced using the duty cycle adjustable by the control circuit. Therefore, the bidirectional converter could operate in inverter or PFC rectifier mode, depending on the state of the AC grid. Thus, the HEMS provided a corresponding signal that forced the bidirectional converter to inject electricity into the grid (inverter mode) or to charge the batteries (PFC rectifier mode). Two buttons were added to the microcontroller to substitute the HEMS signal for the following results. The duty cycle could be adjusted, and the power direction and operating mode could be controlled with these two buttons.

Between the batteries and the DC–DC stage, a third stage was introduced for this experiment. The purpose of this stage was to modify the charge and the injection of electricity and to adapt the voltage and the current between the DC–DC stage and the batteries.

The following two subsections examine these two modes of operation in detail.

3.3.2. Inverter Mode

The objective here was to confirm the proper operation of the bidirectional DC–AC converter in inverter mode. Figure 12 shows an example of the experimental results in this mode, where the batteries supplied 485 W to the whole system. In this example, the DC bus voltage was 178 V; this voltage represented the input DC voltage of the bidirectional converter. Figure 12 shows the output voltage of the DC–DC stage, confirming the sine half-wave, here at the RMS voltage of 126 V (measured using ❺ in Figure 10). The output of the DC–AC stage was a 50 Hz sine wave with, here, an RMS voltage of 120 V. This signal was applied to the AC grid and the power was effectively negative in this case (measured using ❸ in Figure 10) since power was being injected (in this example, 410 W). Figure 12 also shows the current injected into the AC grid, here with an RMS value of 2 A. As the current was imposed by the converter, it was necessarily out-of-phase with the voltage. While the inverter was injecting power into the AC grid, the voltage would remain constant even if the current changed.

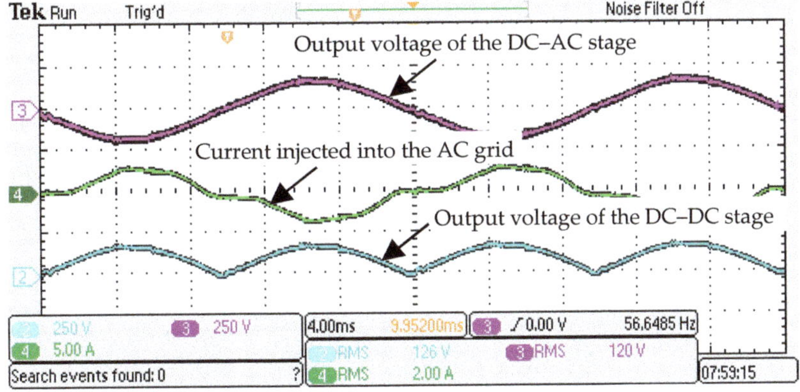

Figure 12. Experimental validation of the inverter mode: output signals.

3.3.3. PFC Mode

The objective here was to confirm the proper operation of the bidirectional DC–AC converter in PFC rectifier mode. First, it was essential to demonstrate that the power factor modification was indeed feasible in this case. The sinusoidal current could be absorbed by appropriately controlling transistor *T2* of the DC–DC stage (see Figure 9).

Figure 13 shows an example of the experimental results in PFC mode. In this configuration, the AC grid delivered a positive power of 80 W to the whole system (measured using ❸ in Figure 10). The power transmitted to the batteries via the bidirectional converter was necessarily negative in this case, which indicates that we were in a charge cycle (indicated by ❹ in Figure 10). Figure 13 shows the sine wave imposed by the AC grid, as well as the half-sine curve between the DC–DC stage and the DC–AC stage, while we monitored the input of the DC–AC stage. Figure 13 shows an RMS voltage of the half-sine wave of 49 V, while the voltage at the DC bus was 166 V (measured using ❺ in Figure 10).

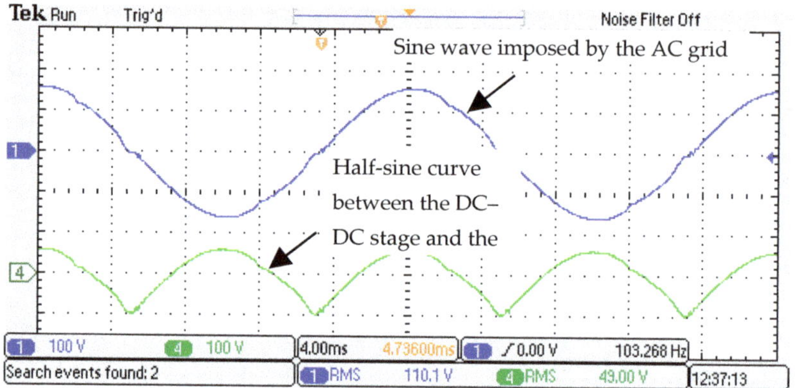

Figure 13. Experimental validation of the PFC mode: output signals.

In this example, the voltage and the absorbed current were imposed in order to follow the charge curve of the batteries.

4. Discussion

4.1. Switching of the HEMS Control

The main objective of this work was to design and implement a bidirectional DC–AC converter that was connected to an HEMS system [23]. The HEMS system supervised the control of several elements of the converter, such as:

- Automatically connecting the converter to the AC grid in standby mode, which ensured a balanced condition between the energy supplied by the batteries and the electricity supplied by the AC grid (see Figures 9 and 10);
- Inverter mode: In this mode, the HEMS determined the amount of energy to be supplied to the grid by changing the duty cycle of the control signals. This allowed us to provide the precise amount of energy required. In the example described in Figure 12, we changed the duty cycle to inject 410 W; depending on the state of the grid, this amount could be as much as 1.5 kW;
- PFC rectifier mode: In this mode, the system operated as a battery management system (BMS). The HEMS could read the state of charge (SoC) of the batteries and adjust the voltage and current sent to the batteries in order to store the precise amount of excess energy available from the AC grid (see Figure 13).

At this stage, the bidirectional DC–AC converter was tested on a 110 V, 50 Hz AC grid. The next experiments will be performed on a 230 V, 50 Hz AC grid. The converter is fully self-adaptive and, therefore, requires no human interaction; all of the above parameters can be modified using real-time data measurements provided by the HEMS.

The experimentally designed and implemented bidirectional DC–AC converter is compatible with the HEMS system that we recently developed in [23]. Therefore, it can be fully used either to store excess energy available on the AC grid in batteries or to inject energy into the AC grid, according to the consumption patterns defined in the house.

4.2. Efficiency of the Whole Converter

This section of the manuscript aims to study in detail the efficiency of the proposed bidirectional DC–AC converter and, more specifically, when it operates in inverter mode. The PFC rectifier mode was regulated by the battery management system and thus, depended on the battery state of charge. The amount of energy stored could be varied but was comparable to the voltage and current measurements in the batteries, as well as the charge curves of the storage system. Combined with the charge control system, the overall efficiency of the system shown in the example in Figure 13 was approximately 85%.

In grid-connected inverter mode, we performed the efficiency measurements at low power (a few tens of watts) and up to 1.5 kW. The measurements were tricky to perform because we needed to ensure that the power injected by the batteries was flowing through the charge–discharge management system and that the power was detected at the output using the DC voltage and current measurements shown in Figure 12.

Figure 14 shows the evolution of the efficiency of the bidirectional DC–AC converter as a function of the rated output power. Figure 14 shows that the average efficiency of the whole converter was about 96.5%, from 100 W to 1.5 kW. It should be noted that it would be difficult to achieve any better given the general architecture chosen, i.e., the series connection of a DC–DC stage and a DC–AC stage, even if the DC–DC stage presented excellent performances (average efficiency of 98.5% over the tested power range). At very low power (e.g., 50 W), although the 95% target was not reached, the values obtained (here about 91.5% at 50 W) were more than satisfactory compared to other topologies discussed in the literature.

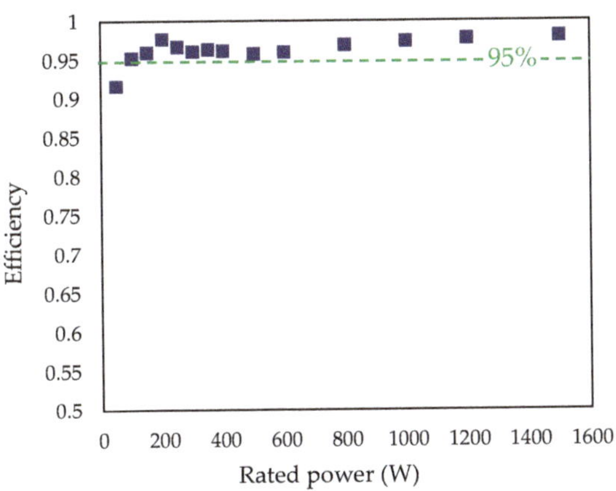

Figure 14. The efficiency of the bidirectional DC–AC converter in inverter mode (experimental results).

The losses were distributed mainly by the MOSFETs and the passive devices, especially the inductance of the DC–DC stage. The classical ferrite core was banned from this prototype because it caused too many losses in the iron. Therefore, we chose a high flux core that was ideal for high frequencies. To be more precise, the losses in the MOSFETs represented 1.1% of the target power, while the losses in the DC–DC stage inductor represented 2.4% of the target power. Conduction losses and switching losses could be estimated from the key characteristics (including drain-to-source on-state resistance, rise time and fall time) of each MOSFET. These losses represented 0.1% and 1% of the target power, respectively. By adjusting the reference of the SiC devices used in the DC–DC stage, for example with 36 A, 900 V SiC MOSFETs from the manufacturer Cree, it was possible to reduce the losses of the MOSFETs and gain nearly 0.5 percentage points in efficiency.

Compared to the first converter presented in [26] (this converter was experimentally tested in islanding mode), we intentionally decreased the switching frequency of the MOSFETs in the DC–DC stage from 300 kHz to 150 kHz. At 300 kHz, the converter efficiency lost, on average, 0.3 efficiency points over the output power range studied here. By switching to 150 kHz, we optimized the efficiency and, in addition, anticipated future electromagnetic compatibility issues, which we will detail in the near future.

5. Conclusions

In this paper, the main objective was to present and experimentally validate a bidirectional DC–AC converter, connected to the AC grid and suitable for HEMS applications when an energy storage system is required. The proposed topology was based on two necessarily bidirectional stages associated in series: the first being a DC–DC stage composed of two silicon carbide power MOSFETs, which were controlled at high frequency (the frequency of 150 kHz was implemented experimentally); the second being an H-bridge composed of four MOSFETs on silicon substrate, which were controlled at the frequency of the AC grid (i.e., 50 Hz here). With this converter architecture, the DC–DC stage regulated the DC voltage and established the positive parts of an AC waveform, while the DC–AC stage reversed it to obtain the sinusoidal voltage. Thus, the proposed structure provided an excellent AC waveform, but the latter depended mainly on the DC–DC stage.

A complete experimental procedure was defined and implemented to validate the operating modes of the bidirectional DC–AC converter, i.e., the inverter mode and the PFC rectifier mode, especially in the case of a grid connection. The energy efficiency of the whole DC–AC converter operating in inverter mode was evaluated and exceeded the target of 95% over the entire power range studied, i.e., from 100 W to 1.5 kW.

The three main outcomes of this study are summarized below:

1. The complexity of the topology is reasonable, so it can be recommended as an alternative solution for HEMS applications;
2. In the case of stand-alone inverter operation and unlike traditional H-bridges, the proposed converter does not require the use of a bulky filter, which optimizes its compactness. The optimization of the compactness of the whole system is made possible by the use of silicon carbide MOSFETs in the DC–DC stage, switching at several hundred kilohertz (the frequencies of 150 kHz and 300 kHz were investigated here and the 150 kHz frequency was implemented experimentally);
3. The proposed bidirectional DC–AC converter can operate in a grid-connected configuration, with the ability to charge batteries during off-peak hours and use the energy from those batteries during peak loads. The operating mode of the entire converter is controlled by the previously presented HEMS system [23].

The experimental results demonstrate the expected performance of such a system, both in terms of its operating modes and its high energy efficiency. The bidirectional switching of the components of this architecture is autonomous in adjusting the control commands of the microcontroller, as well as controlling the amount of energy to store or inject into the grid. The algorithm of the control strategy, the communication between the developed system and the smart meter are being patented.

The short-term perspectives of this work are as follows. The bidirectional DC–AC converter was developed to operate in both grid and off-grid mode. The islanding mode will be presented in another paper. In addition, the electromagnetic compatibility aspects will have to be realized in order to conform with the standards of connection to the AC grid that are in application. In the longer term, the whole converter and the HEMS will be installed in real smart homes and the profitability of our approach will be evaluated.

Author Contributions: Conceptualization, J.-C.L.B. and S.B.; methodology, J.-C.L.B., S.J., S.B. and I.A.; software, S.B., C.R., I.A. and T.B.; validation, I.A., T.B. and J.-C.L.B.; formal analysis, J.-C.L.B. and S.B.; investigation, S.B., C.R., I.A., T.B. and J.-C.L.B.; resources, J.-C.L.B.; data curation, , I.A., S.J. and J.-C.L.B.; writing—original draft preparation, S.B. and S.J.; writing—review and editing, I.A., S.J. and

J.-C.L.B.; visualization, I.A., S.J. and J.-C.L.B.; supervision, J.-C.L.B.; project administration, S.J. and J.-C.L.B.; funding acquisition, S.J. and J.-C.L.B. All authors have read and agreed to the published version of the manuscript.

Funding: These research activities were supported by "Région Centre Val-de-Loire" (research project number: 2015-00099656). The authors of this paper thank our colleagues from this institution who provided insight and expertise that greatly assisted the project.

Institutional Review Board Statement: Not applicable.

Informed Consent Statement: Not applicable.

Conflicts of Interest: The authors declare to respect the confidentiality clauses currently defined in the collaboration contract defined above.

Abbreviations

The following abbreviations are used in this paper:

AC	Alternating current
AI	Artificial intelligence
ANFIS	Adaptive neural fuzzy inference systems
ANN	Artificial neural networks
BMS	Battery management system
CAD	Computer-aided design
DC	Direct current
DL	Deep learning
FLC	Fuzzy logic control
GA	Genetic algorithms
HEMS	Home electricity management systems
IoT	Internet of things
MOSFET	Metal oxide semiconductor field effect transistor
MPC	Model predictive control
PSO	Particle swarm optimization
PFC	Power factor correction
PV	Photovoltaics
PWM	Pulse width modulation
RISC	Reduced instruction set computer
RMS	Root mean square
RNN	Recurrent neural networks
SiC	Silicon carbide
SoC	State of charge
THD	Total harmonic distortion
Triac	Triode for alternating current
VSC	Voltage–source converter

References

1. Javaid, N.; Javaid, S.; Abdul, W.; Ahmed, I.; Almogren, A.; Alamri, A.; Niaz, I.A. A Hybrid Genetic Wind Driven Heuristic Optimization Algorithm for Demand Side Management in Smart Grid. *Energies* **2017**, *10*, 319. [CrossRef]
2. Torriti, J. The Risk of Residential Peak Electricity Demand: A Comparison of Five European Countries. *Energies* **2017**, *10*, 385. [CrossRef]
3. Kornatka, M.; Gawlak, A. An Analysis of the Operation of Distribution Networks Using Kernel Density Estimators. *Energies* **2021**, *14*, 6984. [CrossRef]
4. Elfeki, I.; Jacques, S.; Aouichak, I.; Doligez, T.; Raingeaud, Y.; Le Bunetel, J.-C. Characterization of Narrowband Noise and Channel Capacity for Powerline Communication in France. *Energies* **2018**, *11*, 3022. [CrossRef]
5. Markkula, J.; Haapola, J. Shared LTE Network Performance on Smart Grid and Typical Traffic Schemes. *IEEE Access* **2020**, *8*, 39793–39808. [CrossRef]
6. Lu, H.H.-C.; Chu, C.-C.; Nwankpa, C.O.; Wu, C.W. Guest Editorial Complex Network for Modern Smart Grid Application—Part 2: Stability, Reliability and Resilience Issues. *IEEE J. Emerg. Sel. Top. Circuits Syst.* **2017**, *7*, 345–348.
7. Iqbal, Z.; Javaid, N.; Iqbal, S.; Aslam, S.; Ali Khan, Z.; Abdul, W.; Almogren, A.; Alamri, A. A Domestic Microgrid with Optimized Home Energy Management System. *Energies* **2018**, *11*, 1002. [CrossRef]

8. Kim, Y.-S.; Hwang, C.-S.; Kim, E.-S.; Cho, C. State of Charge-Based Active Power Sharing Method in a Standalone Microgrid with High Penetration Level of Renewable Energy Sources. *Energies* **2016**, *9*, 480. [CrossRef]

9. Khan, S.; Khan, R. Elgamal Elliptic Curve Based Secure Communication Architecture for Microgrids. *Energies* **2018**, *11*, 759. [CrossRef]

10. Jelić, M.; Batić, M.; Tomašević, N. Demand-Side Flexibility Impact on Prosumer Energy System Planning. *Energies* **2021**, *14*, 7076. [CrossRef]

11. Dinh, H.T.; Yun, J.; Kim, D.M.; Lee, K.; Kim, D. A Home Energy Management System with Renewable Energy and Energy Storage Utilizing Main Grid and Electricity Selling. *IEEE Access* **2020**, *8*, 49436–49450. [CrossRef]

12. Floris, A.; Porcu, S.; Girau, R.; Atzori, L. An IoT-Based Smart Building Solution for Indoor Environment Management and Occupants Prediction. *Energies* **2021**, *14*, 2959. [CrossRef]

13. Pavón, R.M.; Alberti, M.G.; Álvarez, A.A.A.; del Rosario Chiyón Carrasco, I. Use of BIM-FM to Transform Large Conventional Public Buildings into Efficient and Smart Sustainable Buildings. *Energies* **2021**, *14*, 3127. [CrossRef]

14. Leitão, J.; Gil, P.; Ribeiro, B.; Cardoso, A. A Survey on Home Energy Management. *IEEE Access* **2020**, *8*, 5699–5722. [CrossRef]

15. Mariano-Hernández, D.; Hernández-Callejo, L.; Zorita-Lamadrid, A.; Duque-Pérez, O.; Santos García, F. A review of strategies for building energy management system: Model predictive control, demand side management, optimization, and fault detect & diagnosis. *J. Build. Eng.* **2021**, *33*, 101692.

16. Atef, S.; Ismail, N.; Eltawil, A.B. A New Fuzzy Logic Based Approach for Optimal Household Appliance Scheduling Based on Electricity Price and Load Consumption Prediction. In *Advances in Building Energy Research*; Taylor & Francis, 2021; pp. 1–19. Available online: https://cogentoa.tandfonline.com/doi/abs/10.1080/17512549.2021.1873183?journalCode=taer20 (accessed on 21 December 2021).

17. Bot, K.; Santos, S.; Laouali, I.; Ruano, A.; Ruano, M.d.G. Design of Ensemble Forecasting Models for Home Energy Management Systems. *Energies* **2021**, *14*, 7664. [CrossRef]

18. Andriopoulos, N.; Magklaras, A.; Birbas, A.; Papalexopoulos, A.; Valouxis, C.; Daskalaki, S.; Birbas, M.; Housos, E.; Papaioannou, G.P. Short Term Electric Load Forecasting Based on Data Transformation and Statistical Machine Learning. *Appl. Sci.* **2021**, *11*, 158. [CrossRef]

19. Franco, P.; Martínez, J.M.; Kim, Y.-C.; Ahmed, M.A. IoT Based Approach for Load Monitoring and Activity Recognition in Smart Homes. *IEEE Access* **2021**, *9*, 45325–45339. [CrossRef]

20. Hosseinnezhad, V.; Shafie-Khah, M.; Siano, P.; Catalão, J.P.S. An Optimal Home Energy Management Paradigm with an Adaptive Neuro-Fuzzy Regulation. *IEEE Access* **2020**, *8*, 19614–19628. [CrossRef]

21. Zupančič, J.; Filipič, B.; Gams, M. Genetic-programming-based multi-objective optimization of strategies for home energy-management systems. *Energy* **2020**, *203*, 117769. [CrossRef]

22. Yousefi, M.; Hajizadeh, A.; Soltani, M.N.; Hredzak, B. Predictive Home Energy Management System with Photovoltaic Array, Heat Pump, and Plug-In Electric Vehicle. *IEEE Trans. Ind. Inform.* **2021**, *17*, 430–440. [CrossRef]

23. Bissey, S.; Jacques, S.; Le Bunetel, J.-C. The Fuzzy Logic Method to Efficiently Optimize Electricity Consumption in Individual Housing. *Energies* **2017**, *10*, 1701. [CrossRef]

24. Khan, M.; Seo, J.; Kim, D. Towards Energy Efficient Home Automation: A Deep Learning Approach. *Sensors* **2020**, *20*, 7187. [CrossRef] [PubMed]

25. Bhatt, D.; Hariharasudan, A.; Lis, M.; Grabowska, M. Forecasting of Energy Demands for Smart Home Applications. *Energies* **2021**, *14*, 1045. [CrossRef]

26. Bissey, S.; Jacques, S.; Reymond, C.; Le Bunetel, J. An Innovative Bidirectional DC-AC Converter to Improve Power Quality in a Grid-Connected Microgrid. *Preprints* **2018**, 2018070252. [CrossRef]

27. Ali Khan, M.Y.; Liu, H.; Yang, Z.; Yuan, X. A Comprehensive Review on Grid Connected Photovoltaic Inverters, Their Modulation Techniques, and Control Strategies. *Energies* **2020**, *13*, 4185. [CrossRef]

28. Bughneda, A.; Salem, M.; Richelli, A.; Ishak, D.; Alatai, S. Review of Multilevel Inverters for PV Energy System Applications. *Energies* **2021**, *14*, 1585. [CrossRef]

29. Alotaibi, S.; Darwish, A. Modular Multilevel Converters for Large-Scale Grid-Connected Photovoltaic Systems: A Review. *Energies* **2021**, *14*, 6213. [CrossRef]

30. Martinez-Rodrigo, F.; Ramirez, D.; Rey-Boue, A.B.; De Pablo, S.; Herrero-de Lucas, L.C. Modular Multilevel Converters: Control and Applications. *Energies* **2017**, *10*, 1709. [CrossRef]

31. Lourenço, L.F.N.; Perez, F.; Iovine, A.; Damm, G.; Monaro, R.M.; Salles, M.B.C. Stability Analysis of Grid-Forming MMC-HVDC Transmission Connected to Legacy Power Systems. *Energies* **2021**, *14*, 8017. [CrossRef]

32. Jacques, S.; Reymond, C.; Le Bunetel, J.-C.; Benabdelaziz, G. Comparison of the power balance in a Totem-Pole Bridgeless PFC topology with several inrush current limiting strategies. *J. Electr. Eng.* **2021**, *72*, 12–19. [CrossRef]

33. Amiri, P.; Eberle, W.; Gautam, D.; Botting, C. An Adaptive Method for DC Current Reduction in Totem Pole Power Factor Correction Converters. *IEEE Trans. Power Electron.* **2021**, *36*, 11900–11909. [CrossRef]

34. Yang, Z. Bidirectional DC-to-AC inverter with improved performance. *IEEE Trans. Aerosp. Electron. Syst.* **1999**, *35*, 533–542. [CrossRef]

35. Koutroulis, E.; Chatzakis, J.; Kalaitzakis, K.; Voulgaris, N.C. A bidirectional, sinusoidal, high-frequency inverter design. *IEE Proc. Electr. Power Appl.* **2001**, *148*, 315. [CrossRef]

36. Zhang, J.; Shao, J.; Xu, P.; Lee, F.C.; Jovanovic, M.M. Evaluation of input current in the critical mode boost PFC converter for distributed power systems. In Proceedings of the 16th Annual IEEE Applied Power Electronics Conference and Exposition (Cat. No. 01CH37181), Anaheim, CA, USA, 4–8 March 2001.

37. Ivanovic, Z.; Blanusa, B.; Knezic, M. Power loss model for efficiency improvement of boost converter. In Proceedings of the 2011 XXIII International Symposium on Information, Communication and Automation Technologies, Sarajevo, Bosnia and Herzegovina, 27–29 October 2011.

38. Rąbkowski, J.; Skoneczny, H.; Kopacz, R.; Trochimiuk, P.; Wrona, G. A Simple Method to Validate Power Loss in Medium Voltage SiC MOSFETs and Schottky Diodes Operating in a Three-Phase Inverter. *Energies* **2020**, *13*, 4773. [CrossRef]

39. Van den Bossche, A.; Valchev, V. Modeling Ferrite Core Losses in Power Electronics. *Int. Rev. Electr. Eng.* **2006**, 14–22. Available online: https://www.praiseworthyprize.org/jsm/index.php?journal=iree&page=issue&op=archive&issuesPage=1#issues (accessed on 21 December 2021).

40. Somkun, S.; Sato, T.; Chunkag, V.; Pannawan, A.; Nunocha, P.; Suriwong, T. Performance Comparison of Ferrite and Nanocrystalline Cores for Medium-Frequency Transformer of Dual Active Bridge DC-DC Converter. *Energies* **2021**, *14*, 2407. [CrossRef]

41. *CLC/FprTS 50549: Part 1-Requirements for Generating Plants to be Connected in Parallel with Distribution Networks—Part 1: Connection to a LV Distribution Network—Generating Plants up to and Including Type B*; European Committee for Electrotechnical Standardization: Brussels, Belgium, 2012. Available online: http://tinyurl.com/TS-50549-1 (accessed on 21 December 2021).

MDPI

St. Alban-Anlage 66

4052 Basel

Switzerland

Tel. +41 61 683 77 34

Fax +41 61 302 89 18

www.mdpi.com

Energies Editorial Office

E-mail: energies@mdpi.com

www.mdpi.com/journal/energies